Trails in Academic and Administrative Leadership in Kenya

This book is a product of the CODESRIA Higher Education Leadership Programme.

Trails in Academic and Administrative Leadership in Kenya

A Memoir

Ratemo Waya Michieka

Council for the Development of Social Science Research in Africa
DAKAR

© CODESRIA 2016
Council for the Development of Social Science Research in Africa
Avenue Cheikh Anta Diop, Angle Canal IV
P.O. Box: 3304 Dakar, 18524, Senegal
Website: www.codesria.org

ISBN: 978-2-86978-642-4

All rights reserved. No part of this publication may be reproduced or transmitted in any form or by any means, electronic or mechanical, including photocopy, recording or any information storage or retrieval system without prior permission from CODESRIA.

Typesetting: Alpha Ousmane Dia
Cover Design: Ibrahima Fofana

Distributed in Africa by CODESRIA
Distributed elsewhere by African Books Collective, Oxford, UK
Website: www.africanbookscollective.com

The Council for the Development of Social Science Research in Africa (CODESRIA) is an independent organisation whose principal objectives are to facilitate research, promote research-based publishing and create multiple forums geared towards the exchange of views and information among African researchers. All these are aimed at reducing the fragmentation of research in the continent through the creation of thematic research networks that cut across linguistic and regional boundaries.

CODESRIA publishes *Africa Development*, the longest standing Africa based social science journal; *Afrika Zamani*, a journal of history; the *African Sociological Review*; the *African Journal of International Affairs*; *Africa Review of Books* and the *Journal of Higher Education in Africa*. The Council also co-publishes the *Africa Media Review*; *Identity, Culture and Politics: An Afro-Asian Dialogue*; *The African Anthropologist, Journal of African Tranformation, Method(e)s: African Review of Social Sciences Methodology*, and the *Afro-Arab Selections for Social Sciences*. The results of its research and other activities are also disseminated through its Working Paper Series, Green Book Series, Monograph Series, Book Series, Policy Briefs and the CODESRIA Bulletin. Select CODESRIA publications are also accessible online at www.codesria.org.

CODESRIA would like to express its gratitude to the Swedish International Development Cooperation Agency (SIDA), the International Development Research Centre (IDRC), the Ford Foundation, the Carnegie Corporation of New York (CCNY), the Norwegian Agency for Development Cooperation (NORAD), the Danish Agency for International Development (DANIDA), the Netherlands Ministry of Foreign Affairs, the Rockefeller Foundation, the Open Society Foundations (OSFs), TrustAfrica, UNESCO, UN Women, the African Capacity Building Foundation (ACBF) and the Government of Senegal for supporting its research, training and publication programmes.

Dedication

This book is dedicated to my beloved wife, Esther Nyabonyi Michieka, my children and my grandchildren, who encouraged me as I worked through the various drafts.

Contents

Acknowledgements ix
About the Book x
About the Author xi
Abbreviations xiii
Preface xv

1. Immersion into University Leadership 1
2. Cultural and Religious Foundations 5
3. Foundations for Future Academic and Leadership Pursuits 25
4. First Job Placement 51
5. University Education and the Foundations of Academic Leadership 63
6. Growth as an Academic Leader and Researcher 83
7. The Triple Balancing Act: Academics, Family and University Administration 111
8. Engagements with Academic Leadership at the Grassroots 129
9. A Surprise Appointment to the Office of Deputy College Principal 139
10. Elevation of the University College into a University 145
11. Tenure as Vice-Chancellor 161
12. Vice-Chancellorship and Networking for University Development 183
13. University Leadership and the Donor Community 201
14. University Leadership and Quality of Academic Programmes 211
15. Scholarship and Academic Community Service 227
16. Exiting University Administration 237
17. Leadership in the Corporate World 247
18. Back to the Ivory Tower 287
19. Some Reflections on Leadership and Governance of Higher Education Institutions 293

Acknowledgements

The writing of this book was a pleasant undertaking which I thoroughly enjoyed. I had to conduct research on topics outside my academic discipline but which were dear to my heart. I would like to acknowledge the contribution of several individuals who assisted me at the drafting stage of this book.

My sincere thanks go to the many students whom I interviewed and the elders in my rural home who shared with me the traditions and culture which bound us together. I thank them. I am grateful to Catherine Muhandiki who typed the manuscript.

I am highly indebted to my colleague, Amateshe Kisa, who read the manuscript and made excellent suggestions which enriched the contents of the book.

Finally, I wish to thank Prof Ibrahim Oanda, the coordinator of the CODESRIA/Carnegie Higher Education Leadership Project/Programme for overseeing the progress of this work and making time to go over drafts of the manuscript. It would not have been easy to complete the book without their generous facilitation.

About the Book

Trails in Academic and Administrative Leadership is a book with an insight through which the author narrates his experiences. The autobiography gives an account of a young school boy who was dedicated to factual knowledge as he strove to perform his duties with integrity. The account details the events which led the author to become a scholar, a professor, a vice-chancellor and a chairman of Council. The many organizations which he assisted to build and/or revitalize are clearly identified. His early experiences, right from the formative stages to his time at the university, created a formidable character of an administrator who nurtured young institutions into world-class centres. His training in the USA exposed the author to international scholars and university linkages and he ended up working in Nigeria. This is the maturity stage of the author.

The book covers experiences which positively impacted on staff and students later in their lives. The quality of academic programmes and faculty are paramount in the industrial take-off of developing countries. A whole chapter is dedicated to this important sector of educational development. Environmental pollution and degradation is given prominence as many countries are hurting in this respect. Some of the socio-political experiences with far-reaching implications are discussed, and procedures of academic promotions are highlighted in the book. A comparison between management and leadership is given prominence in the context of political expediency.

This book is not only meant for scholars, but also decision-makers in various managerial and academic leadership positions.

About the Author

Ratemo W. Michieka is a Professor at the University of Nairobi and the Chairman of Kenyatta University Council. His teaching career spans over 35 years, punctuated with other national appointments. Michieka holds a PhD from the prestigious Rutgers University, USA, where he was recognized as a Distinguished Alumnus in 2003 for the exemplary work he has done in Kenya's educational development. He did his postdoctoral research at the International Institute of Tropical Agriculture (IITA), Ibadan, Nigeria, prior to his departure to his home country, Kenya.

Michieka was the founding Vice-Chancellor of the Jomo Kenyatta University of Agriculture and Technology. He also served as the Chairman of the Inter-University Council of East Africa (IUCEA). He was also the Director-General of the National Environment Management Authority (NEMA) in charge of the country's environmental protection among many other responsibilities.

Michieka has published many papers locally and internationally, and has written a number of books. Michieka was a member of the United Nations Framework Convention on Climate Change (UNFCCC); and is a Fellow and Honorary Secretary of the Kenya National Academy of Sciences (KNAS).

Abbreviations

AAU	Association of African Universities
AICAD	African Institute for Capacity Development
ACU	Association of Commonwealth Universities
CAES	College of Agriculture and Environmental Sciences
CATS	Continuous Assessment Tests
CAVS	College of Agriculture and Veterinary Sciences
CBD	Centre for Biological Diversity
CBPS	College of Biological and Physical Sciences
CCK	Communication Commission of Kenya
CEES	College of Education and External Studies
CHE	Commission for Higher Education
CHS	College of Health Sciences
CHSS	College of Humanities and Social Sciences
CIMMYT	International Maize and Wheat Improvement Centre
CITES	Convention on International Trade in Endangered Species
D-G	Director-General
DFID	Department of International Trade
DP	Deputy Principal
DVC (AA)	Deputy Vice Chancellor, Academic Affairs
EAC	East African Community
FAO	Food and Agricultural Organization
FKE	Federation of Kenya–Employees
GEF	Global Environmental Facility
HELB	Higher Education Loans Board
ICARDA	International Centre for Agricultural Research in the Dry Areas
IDRC	International Development Research Centre
IITA	International Institute of Tropical Agriculture
IUCEA	Inter University Council for East Africa

JAB	Joint Admissions Board
JEPAK	Japan Ex-participants of Kenya
JICA	Japan International Cooperation Agency
JKIA	Jomo Kenyatta International Airport
JKUSO	Jomo Kenyatta University Students Organization
JKUAT	Jomo Kenyatta University of Agriculture and Technology
KBC	Kenya Broadcasting Cooperation
KADU	Kenya African Democratic Union
KARI	Kenya Agricultural Research Institute
KATA	Kenya Agricultural Teachers Association
KBS	Kenya Bureau of Standards
PCPE	Kenya Certificate of Primary Education
KEMRI	Kenya Medical Research Institute
KENET	Kenya Education Network
KCSE	Kenya Secondary Certificate Education
MBWA	Management by Walking Around
NACOSTI	National Council for Science, Technology and Innovation
NEC	Nyamagesa Education Club
NYC	New York City
OCS	Officer Commanding Station
SAA	South African Airways
SSA	Sub-Saharan Africa
UASU	University Association Staff Union
UIP	Universities Investment Project
UNEP	United Nations Environment Programme
UNFCCC	United Nations Framework Convention on Climate Change
UoN	University of Nairobi
USAB	University Students' Accommodation Board
USHEPIA	University Science, Humanities and Engineering Partnerships in Africa
UWV	University of West Virginia
VC	Vice Chancellor
WSSEA	Weed Science Society of Eastern Africa

Preface

This book presents my detailed profile as a young man who grew up in rural Kenya, rose through the academic ranks to become an authority in administrative and academic management. The book begins with my enrolment at the primary school where I started to play leadership roles at a tender age. I met challenges as I tried to show justice in my role as class representative. As I proceeded to high school, I joined one of the prestigious secondary schools where I met other students from the whole country and had no option than to cope with the cultural richness and complexities of my country, Kenya. I also met British teachers who were on their way out as Kenya had attained its independence.

The experience I gained at high school hardened my stand in life. Over the years, I learnt to be fair and mindful of my student fraternity. The continuous administrative experience gained from youth assisted me to manage bigger institutions later in life. The university training in the USA impacted strongly upon me up till this day. Discipline in research and cultural interaction was a life-long experience. The Nigerian post-doctoral fellowship squarely put me on an African soil with a difference.

I offer an introduction to early human culture and cultural value as I progressed through my academic career. Similarly, I offer rare insights into the human nature during my management undertaking. I do not intend to cover all the work that I performed as a vice-chancellor during a very tumultuous political period in Kenya, but wish to show that academic life can be challenging, yet profitable, in nurturing the young faculty and students. I have demonstrated that African universities can be run with the least political interference, yet this was not the case in the 1980s and 1990s in Kenya. Both internal and external influences made the roles of university administrators difficult and unpleasant.

I also demonstrate in the book the role of international linkages. Development partners had their agenda and vice-chancellors had to balance between these agenda and what was good or beneficial for their institutions. They were driven to hard walls at times.

As an administrator in the environmental authority, I went through very difficult moments trying to do justice to the Kenyan environment and conservation of natural resources. Despite the many hurdles I encountered as the Director General, I was able to articulate environmental matters across Kenya and beyond. My role as a chairman of Council and management gave me a total insight of university intricacies and management. These are complex systems.

I have compared the roles of university Councils and other organizations that constitute a complete university. This is also well articulated in the book. The role of University Council members is underestimated, yet they are the counting officers of the universities.

I have discussed the importance of maintaining quality education, standards and international competition. The role of ICT and modern communication techniques must be embraced in the present-day technology.

From my own experiences, I cannot overemphasize the importance of integrity, humility, meritocracy, time management, and fair judgement. Our universities have reasonably well set out administrative structures, but individual leadership has a lot to do with the success or failure of mission and vision. There is more in our universities than the apparently meaningless routine cycles.

I trust that our university administrators and scholars will find this book useful in their careers.

1

Immersion into University Leadership

It happened on a day like any other in 1989. There were precise things I looked forward to that day, but the heralds I was about to receive were not part of them. Presently, the phone rang and, from the other end, the Vice-Chancellor of the University of Nairobi, requested me to come to his office for some unspecified consultation. It was after I got there that he informed me of the decision by the Chancellor and President of the Republic of Kenya to appoint me the Deputy Principal (Academic) for Jomo Kenyatta College of Agriculture and Technology (JKUCAT), then a constituent college of Kenyatta University. The Vice-Chancellor followed this with a piece of advice on why I needed to accept the appointment. Nevertheless, he still gave me some time to reflect on it, as was the practice then regarding appointments to senior university administrative positions.

I left his office with a heavier load on my mind than when I came in. But, in the end, I did accept the appointment and with that singular decision, I set myself on a career path that I had neither envisioned when I joined the academic community nor been prepared for. For the next thirteen years, I had plenty of experience to learn on the job in order to strengthen the foundations of what later became Kenya's fifth public university. I had no choice because I was determined to make a difference as the founding Vice-Chancellor of this budding institution.

This was, however, during one of the most turbulent periods in Kenya's higher education history. When Kenya attained independence in 1963, it inherited a university education system that consisted of one University College (Nairobi), then a constituent college of the University of East Africa with an enrolment of 602 students. The pressure to expand the university system to produce the workforce required for the socio-economic development of the country started to be felt immediately thereafter. The government established the University of Nairobi as a public university in 1970, and in the course of the third decade of independence, three other public universities were established, namely, Moi University in 1984, Kenyatta University in 1985, and Egerton University in 1987. Towards the end of the 1980s, the rapid expansion in the tertiary education subsector had begun to tell seriously on the nation's finances owing to government's financial limitations. The trend had set in since 1974 when government introduced a university student loan scheme designed to form a revolving fund from which those who qualified and were admitted to university would benefit. However over the years, recovery of the loans from past beneficiaries was very inefficient and the original purpose was defeated. Towards the end of the 1980s, therefore, the government found itself facing immense pressure to expand university education in the midst of economic austerity.

With pressure from donors and resistance from students and parents, the government, through Sessional Paper No. 6 of 1988 on '*Education and Manpower Training for the Next Decade and Beyond*' (Kamunge Report) proposed a reduction of student allowances. The paper also recommended that the loan scheme be continued but managed by a commercial bank in order to improve recovery. From 1988 to 1991, the government tried to implement the various proposals with so much resistance from students, leading to institutional closures and sometimes violent confrontations between the police and students that resulted in loss of lives. The burden of implementing the various government decisions regarding funding university education fell on university leaders. Given the nature of appointment to university leadership then, which was highly politicized, and with the president still the chancellor, a university leadership that failed to contain student riots and successfully implement the reforms would be interpreted as sabotaging government policy and being lukewarm in its support for the political establishment. This is the context within which I was being immersed into university leadership. On the one hand, and as deputy principal, I had the task of developing the academic programmes of the college and successfully navigating it to attain full university status. But, on the other hand, the broad issues of funding and the subsequent student riots tested our mettle as university administrators.

The above scenario did not help the political mood in the country then. The 1989-1993 period witnessed heightened political activities in the country that eventually forced the political leadership to give in to the demands for political pluralism. Any resistance or demands from students and academics for better funding of the universities was usually dismissed casually by politicians as the work of external forces or political saboteurs. This was the case with the student riots, and the Kenyan media then carried headlines that castigated the students as unpatriotic and being influenced by foreigners. For example, the *Kenya Times*, a newspaper owned by the then ruling party, in its issue of Monday, 8 July 1991 had a headline, '*Foreigners instigated varsity rioting – Ndoto*'. Similar headlines were in the Standard newspaper such as, 'House hits at students' (Standard, 4 July 1991); '*Students should apologize*' (*Standard*, Sunday, 7 July 1991); 'Don't incite students' (*Standard*, 11 July 1991). I had not anticipated such scenarios nor did I envisage I was going to be dealing with student riots this early in my career.

After completing my doctoral studies, my first engagement was with the International Institute of Tropical Agriculture (IITA) in Ibadan, Nigeria, as a researcher. The persuasion of colleagues from the University of Nairobi to join the Faculty of Agriculture as a lecturer in the Department of Crop Science promised me a career in university teaching and research, but not administration. Hence, to succeed as an administrator, I had to rely on the basics of administration I had learned in my earlier life, my sheer willingness to learn on the job and my determination to succeed.

My second major assignment as the Director-General, National Environment Management Authority (NEMA) was not an easy task either. The transition from academia to a heavily-controlled government parastatal was not an easy task. Transforming the mind-set of Kenyans to appreciate the importance of environmental conservation was a herculean undertaking. This is when I recognized the finite difference between a manager and a leader. I had to portray both traits in different endeavours. I had no choice but to cross paths with powerful individuals and make decisions in favour of saving Kenya's natural resources, notwithstanding the consequences. As I took over this assignment, there was no prior training just like that of the university leadership. On-the-job training and the use of common sense were the driving forces behind my success in both careers.

2

Cultural and Religious Foundations

Throughout my life as an academic and university administrator, I have had to draw a lot of inspiration from the teaching I received in my early upbringing. On reflection, I have realized that the earliest socialization within my father's household and my experiences during my primary and secondary school days contributed immensely to shaping my academic and administrative career. I was born in Nyamagesa Village, Nyaribari Masaba, Kisii County. I am the ninth child of the late Mzee Patroba and Beldina Nyatero Michieka. We were a family of twelve children, but some have already passed away.[1]

My ancestry is from a very large clan, the Kerindo family that was named after my great-grandfather. The Kerindo family is one of the largest and most prominent in the larger Kamba Nane (Eight Brothers) clan. The clan is composed of very controversial, short-tempered, intelligent and arrogant offspring who can react against the slightest form of provocation. This is perhaps indicative of my great-great-grandfather's reputed short-temper. The larger Nyaribari is inhabited by several clans that allow inter-marriages between people from unrelated clans. Mzee Kerindo had four wives whose names I cannot remember. I am told he lived in Kitutu Chache, the original home of Omogusiias, a people that is also referred to as *'Enda y'Enchogu'* which literary means *'The Elephant's Womb'*. Mzee Kerindo, the son of Obare (my great-great-grandfather) sired my grandfather Okioga Kerindo, who was his eldest son.

I was told that Kerindo loved Okioga because of his courage and obedience. He was a reserved son. Okioga married three wives; one of them, Kemunto, gave birth to my late father, Patroba Michieka in 1907. There were two other brothers and two sisters, namely: Mouti, Ongeri, Nyamaera and Nyakerario. In lineage, therefore, my grandfather, Okioga, commanded respect from all his brothers and step-brothers since he was the eldest son.

I never saw my grandfather alive but I learnt he was a very intelligent man and he helped sort out village disputes. My mother told me that he used to wear a blanket and a hat and owned several heads of cattle. He had some brothers including

Ombati Nyarangi Kerindo, Mayore, Achuti, Israel Onywere, Paul Mogaka Siro, Arika, Miranyi; and several sisters.

There was large exodus of people to the east, Masaba, in the early 1900s in search of new frontiers for agriculture and livestock rearing. Those who moved from their ancestral land, Nyaura village, in the outskirts of Kisii town, did not sever relationship with their relatives. My parents were peasant farmers who also migrated from Nyaura village in the environs of Kisii town and moved to Masaba in search of bigger land to cultivate. In 1915, Mzee Michieka and his cousins of the larger Kerindo ancestry travelled 30 kilometres to Nyamagesa Village and occupied virgin land that was then considered unsafe for fear of wild animals and skirmishes with the neighbouring Maasai and Kipsigis tribes.

The group which moved to the new frontiers settled there and embarked on farming and cattle-keeping. Knowing that my father had occupied land in Nyamagesa against all odds made me realize and appreciate the power of bravery in any of life's undertakings. As young boys, we were told several stories which taught us leadership traits, guidance, humility and bravery. The stories would later serve some instructive ends in my career.

Specifically, I come from one of the clans referred to as *Kamba Nane*, which means *eight brothers*.[2]

The Abagusii are a patrilineal people. My grandmother Kemunto was Okioga Kerindo's second wife, and gave birth to Mouti, Ongeri and three sisters. I was able to see only one of the sisters, Nyamaera, who passed away in 2012 at the age of 115. She was the fourth sibling in the family, followed by my father. Some of the information I have used was provided by her before she died. She was a great aunt who used to tease me a lot by referring to great things she and my father did during their youth. She nicknamed my father *Onchana*, the youthful name he used to be called when he was hunting game. He was a very accurate shot. During those days, my father was considered a great hunter and could spear wild animals from a long distance.

If it were today, his talent of being a good shot would be considered and tapped. Some of my cousins today serve in the armed forces and acquired their abilities from watching my father hunting. Others are among the best artisans in Kenya. Some of the grandparents whom I interacted with included: Ombati Nyarangi, Mayore, Onywere and Maturi.

My grandfather's brother, Achuti, who lived before I was born, was a great and unique man. He was one of the Gusii warriors who resisted the British invasion of Gusiiland in 1908. He fought alongside Otenyo Nyamaterere who led the resistance against the British. The British, led by a fierce soldier named Geoffrey Alexander Nothcorte (also referred to as Nyarigoti – being the Gusii pronunciation of his name) faced intense resistance from the unarmed Kisii warriors. Otenyo was killed in a fierce battle against the British soldiers and it is believed that his head was taken to London. Just like Dedan Kimathi is respected for the role he played in the Mau Mau rebellion, Otenyo and by extension Achuti, are credited with the expulsion of the British from Gusiiland.

Achuti missed death by a whisker when he was shot twice; one bullet glazed his forehead and formed a lasting scar, and the other one was lodged in his right thigh but was later operated on and removed by a local Kisii 'surgeon'. He limped for the rest of his life! My father and his elder brothers who lived with Achuti until the latter's death in 1950 have great memories and respect for his bravery. He advised us to be brave like Achuti in the defence of our rights. Many of my family members do not condone mediocrity and cannot accept any injustice lying down. They speak out for or against any issues which affect them.

The Abagusii people were among the first communities in Kenya to fiercely defend their land and livestock using crude weapons against very sophisticated British firearms. That is why the Gusii highlands, which have a perfect climate for all types of good farming activities, were spared. A few thousand hectares were, however, taken along the eastern part of the slopes. The Mau Mau rebellion affected the Abagusii people just like it did other tribes in Kenya. The Mau Mau historians often downplay the roles that this community played. I was always inspired by my step-grandfather, Achuti, even during my early school days in Kisii where I first came into contact with British teachers.

Nyamagesa Village where my late father, Mzee Patroba settled with my uncles and their cousins is located on a hill overlooking the magnificent Maasai Mara plains to the east. The place is hilly, punctuated with beautiful valleys that are endowed with several rivers, streams and swamps. The terrains are challenging for any form of ground transport. If one flies over the hills, one sees continuous greenery punctuated by iron-roofed houses. As in many parts of the larger Gusiiland, Nyamagesa has deep loam soils suitable for agriculture, and cultivation of various crops is practiced, as well as cattle keeping. Gusiiland receives an annual rainfall of about 1500 mm which is evenly spread over the year. When it rains in Gussiiland, it pours. The tropical weather, which is characterized by plenty of sunshine, induces high crop productivity in the fertile soils. It is rare to hear of famine or food shortage in Gusiiland unless a catastrophe occurs. In many cases, poor seed quality or unprecedented floods may occasion crop failure, and hence hunger and subsequent famine.

My father had two wives – my own mother and my stepmother, Rachel Nyarangi, who also had four boys and four girls. During our upbringing, my step-brothers and sisters did not distinguish between their biological mother and my mother. According to our culture, we were all brothers and sisters who lived and ate together in one large homestead.

We grew up knowing that we were one large family under one caring father. We were always together in whichever activities we carried out. The only clear distinction amongst us was the age difference. Five of my elder brothers were naturally advanced in age compared to me. My eldest brother William Nyamwange was closer to my father in age and I respected him because of that, especially during my youth. He was my disciplinarian and mentor. I never challenged him. His word to me and my other brothers was final. In fact, he used his position as an elder brother to always remind and intimidate us that he was supreme in decision-making.

Three of my elder brothers were close in age and they equally respected him. They always worked together in the farm when they were out of school. The brother before me, the late Hezron Tirimba, was so close to me that I regarded him as my age-mate despite having my late sister, Mary Bonareri, between us. My younger brother, Amenya, was equally close to me and very supportive. He often came to my rescue whenever I was in trouble. I, however, regarded him as a junior sibling whom I would bully at times. We were nevertheless very close and I could not do anything or go anywhere without making him aware. We were each other's keeper.

Religious Life

We grew up in a strict Christian family. My father and mother were staunch Seventh Day Adventist (SDA) followers who believed and followed the teachings of the Bible. My mother was a deaconess and a church building promoter. The early SDA missionaries instilled religious doctrines into the Nyamagesa village community which changed traditional beliefs.

One great church elder that made a major difference in our community deserves unparalleled recognition: my uncle, the late Augustine Ogero. He made the Nyamagesa community and its environs what it is today. Elder Augustin Ogero was a teacher, a preacher and a strict disciplinarian during his 'reign'. He was a proclaimed law implementer, law executor for the community, especially for the young boys and girls. He shaped the characters of both the youth and elders alike. I grew up knowing that Christian and education doctrines went hand-in-hand.

I went to church at an early age of five years on condition that I was not to go and tend our cows with my elder brother, Tirimba. Going to church then was, and even today still is, a silent mandatory obligation of the SDA community. The rules and regulations of the church, and especially the Ten Commandments, are always in force and must be adhered to at all times. The community elders quoted and reminded us of the same messages in every single encounter.

I used to be asked to read the Bible in special youth congregations and recite several biblical verses in front of the entire church. I developed confidence in standing and talking to crowds of grown-ups. I was considered a brave, young boy who could face the audience without stammering or getting nervous or confused. I became brave and bold at a very early age. Public speaking, as I realized later was not just an automatic art, but a skill that needed to be acquired through practice early in life. This was best developed at a tender age and built on as one advanced in maturity. Reading the verses required some degree of confidence and bravery, lest one trembled and skipped lines. I was always eager to lead the children's Sabbath school in my church. I must, however, admit one thing: one must go through some fear and nervousness initially before one gains full confidence to address a congregation.

My Early Schooling

I first went to school in 1957 at the age of seven. Elders in our community were always reminding us of the benefits of acquiring a good, sound education. We were told to go to school and learn as this was the only sure way to good living. We were further exhorted to be obedient to all who were older than us. In addition to these basic requirements, we were taught to obey, remember and recite the Ten Commandments. Taking alcoholic drinks was condemned, and those who were found doing so were considered rebels who ought not to mingle with others. Incidents of deaths occasioned by consuming illicitly-brewed alcoholic drinks in Kenya threaten the well-being of its citizens. However, up to the time of writing this book, alcohol abuse in Nyamagesa Village was not as pronounced as in other surrounding villages. Instead, the area has for long been, and still is, more famous for the academic excellence of its youths.

Any elder who came across anyone's child misbehaving had an automatic right to mete out instant punishment necessary to correct the child. The community respected elders' judgement at that level. I, honestly, did not know how this right was accorded to these elders. The youths feared and respected anyone senior to them. If a youth absconded from school, the whole community would want to discipline the concerned youth. I did not want to fall victim. Indeed, the elders and the community at large made us what we are today. I pay tribute to their early wisdom that education was the answer to a better life in future.

My father told me one afternoon in January 1957 that I would go to Ibacho Primary School the following week. This was exciting news as I knew that I would get a new uniform but no shoes, and meet new pupils and teachers. My elder brothers had already been through this process and I saw no big issue about it. I was sure that I was no longer going to tend our cattle in the mornings, and I was going to be able to read and write! I looked forward to being a class one pupil. There were no nursery classes then. I also knew that Saturdays were my free days and I would be reading my verses to the congregation in church. I would also get exercise books and a slate to write on. My most exciting expectation was the opportunity of graduating to dipping a nib into an inkpot and then writing in an exercise book! I had seen what my elder brothers were doing, how well they were dressed and the neat writing they displayed on the blackboard we had at home. I had also witnessed how well they were respected and treated when they came home from either day or boarding schools. The uniform they wore and shoes were so neat that they contrasted sharply with my tattered shirts and shorts.

My first day at Ibacho Primary School arrived. I got up very early, wore my uniform and walked there barefoot. It was some three kilometres (1.5 miles) away up and down some hills. Some four boys from my home area had also been admitted to the same school. This was the closest school to Nyamagesa Village. Gesusu and Nyanturago primary schools were farther away than Ibacho and our parents opted for it since it was also opening its doors for the first batch of Standard One pupils that year. During those days, there were no baby classes or nursery schools. One therefore went straight into class one.

My friends, Ratemo Monyenge, Tirimba Okongo, Paul Mageka, Nicodemous Mose, Aska Monyenye and I all met at the grounds of Ibacho Primary School on the first day of opening. We all followed the tyre track of the bicycle used by my uncle and teacher, David Onyiego Onywere (now late), who had been posted there to teach. I recall very vividly that it had rained heavily the previous night and after wading through the morning dew and stepping in the mud, one left temporary marks which those coming later could track. I knew this because cattle stolen by the Maasai and Kipsigis was tracked by the hoof marks they left behind until they could be recovered from a hideout. My father told me that cattle foot marks and warm cow dung droppings assisted them to trace stolen cows. Later on, I was occasionally involved in tracking stolen cattle from either Maasailand or Kipsigis area. Only brave men could pursue cattle raiders.

We reached Ibacho Primary School dressed in blue uniforms, which I thought were smart. I recall many of us in Standard One considering ourselves the greatest as there was no other class ahead of us. The classroom was grass-thatched, with an earth-smeared floor and a makeshift blackboard. There were twenty of us who had come from all corners of the surrounding villages. I recall that we were sixteen boys and four girls in the class. Soon after, two girls quit and went to get married.

My most astonishing observation was the age difference between a group of us from Nyamagesa and those we met from other places. We were so young that they called us lads, heckled and subjected us to doing all manner of clean-ups in the mornings. Bullying of pupils existed during our primary school days. We could not report the boys as we feared they would attack us on our way home. This behaviour hardened us at an early age. We had good teachers, some of whom had come from teacher training colleges, and they protected us particularly during break and physical education periods.

My luck came when I was appointed the class head – a monitor. My role was to pick chalk for the teacher, wipe the blackboard, keep order in class and carry out the roll-call. I was also charged with maintaining silence in class and reporting any noise makers to the class teacher. I had some authority and recognition amongst my Standard One peers. I do not, however, recall reporting cases of indiscipline to any class teacher, except when I was in Standard Four, handling disks. This is where any pupil speaking in his or her mother-tongue was given a disk-like object which in turn he or she would give to the next pupil who made the same mistake. All the teacher needed to do was to ask for the disk from the pupil who got it first. That pupil would mention who he or she had given it to, and it would be followed all the way to the last pupil who had it that day. This was meant to discourage the usage of mother-tongue in school, and promote the use of English.

We used to arrive early in school, go to parade for inspection, do some marching or mark-time, sing the National Anthem then entitled '*God Save the Queen*' and enter our respective classes marching in style. Our lessons lasted for half a day then, which meant they had to stop at noon. As a monitor, my role again was to return the remaining chalk, a checked roll-call book, a duster and the teacher's ruler to the office. There were no other privileges accorded to me except being told that I was a good class monitor. I appreciated the compliments. Today, I do acknowledge the good deeds of my workers and often thank them more handsomely than just making verbal compliments.

These were my routine duties every morning. There was no incentive for my dedicated services as is done nowadays, but I enjoyed my work. The only advantage we had as class monitors was that we had a special place on parade and would recommend disciplinary measures for misbehaving pupils. We were also respected by the immediate communities for the role we played. The rural people knew our roles in school. Prefects had more powers than monitors.

Our half-day schooling was fun. I remember my mother, Mama Beldina Nyatero, waking up every day at 5 am to fend for us before we went to school. Considering the distance from my village to Ibacho Primary School, walking all the way to be in class at 7.30 am was not easy for young boys and girls. My mother was very special when it came to feeding her children. She could cook '*ugali*' (maize meal) for us to eat that early before we left for school. The half-day school was only applicable to Standards One to Four. Due to lack of classroom space, we had to leave the classrooms for the upper primary classes (Standard Five to Standard Eight) to use in the afternoon.

We covered all periods as prescribed in the syllabus during the morning hours. Classes ended at 12.00 noon and we had to trek home over the hills, valleys and rivers to Nyamagesa Village. There were many times when it would rain on us in the mornings and afternoons. This, however, was not a big deal as the rain would drench our clothes and they would still get dry on our bodies. Most days in Kisii are punctuated by sunny and rainy periods.

Back home in the afternoon, we had our chores to contend with. I used to get home at about 1 pm, have lunch if it was there and then head straight to tend our cattle. I would relieve my father who would then go and attend to other more demanding chores. I was reliable and timely in relieving him. My father knew that come 1 pm, I would be there without fail. This allowed him to plan his afternoon accordingly. My other brothers were either in full-day schooling or boarding and were therefore not available. My elder sister Ruth was in another school and if she came early, she assisted mother in the kitchen or went to fetch firewood, water or vegetables. School days were so programmed that we could not be available for any other duties. I recall that at the end of the day, late in the afternoon, my father would call me and teach me how to write numbers and letters.

My father had a very good handwriting but was slow. He had learnt the basics and could keep records. My mother was illiterate, but had very sharp memory and did cram parts of the new and old testaments in the Bible. My other afternoons, when cattle were kept in their shed, were spent in practising in the church choir, weeding farms and cleaning up our homestead.

The young boys from my village who I walked with to school formed a defensive clique to protect ourselves. We walked to primary school together, returned home in a group and played makeshift soccer balls as a group. We were taught to be social by our parents who encouraged us to study as a team. I was always ready to volunteer on any task either in school or at home. We were taught to look out for the girls who went to school with us in case they would be molested by other boys. It was interesting to note that other boys from neighbouring hills viewed us as potential enemies who could harm them because we were always together.

The grouping was an advantage to us since no one dared to bother us. The other boys' houses were in close proximity to the school and we envied them because they were able to dash home for tea break and leave us hungry in the school compound. Some pupils who were lucky to have relatives around the school or were related to the teachers could be invited for tea or porridge. Occasionally, a sympathetic teacher could call me to his house and serve me with tea which I appreciated. The early school education I have narrated was typical of many primary schools in Gusiiland. My experience from Standards One to Four, was exploratory.

I believed in excelling and keeping my school uniform clean. I would remove it as soon as I arrived home, and do homework: reading, writing and arithmetic. I set aside reading time after the cattle came back home. My brothers and sisters who went to other schools encouraged me to study in the evenings. I used to peruse through their work and learn from them. We had a makeshift type of table where I placed a small kerosene lantern. It produced reasonable light for reading and writing. Our parents encouraged us to study hard and bought us reading materials. I had to strive to be among the first three at the end of each term. As a monitor, it would be most embarrassing to perform poorly because other pupils would heckle me, and call me all sorts of names. In fact, they would attribute bad results to over-indulgence for being a monitor. I was, therefore, under continuous pressure to excel lest I would be called names even by the community.

My formative schooling was eventful. I met other boys and girls from the neighbouring hills, compared notes, talked about our parents and bonded. My circle began to grow through networking and making friends. The youngest boys in our class performed very well during the end-of-term examinations. I was one of them. One thing that was very clear in our interactions was not to leave anyone behind as we traversed the rivers and hills to our homes. We were together at all times except on weekends. The teachers knew us as a group of lads who were polite and always punctual.

The mentality of teachers then was to cane pupils for any mistakes they committed. Caning was a typical way of disciplining the youth, especially late-comers. I detested being caned and did everything possible within my powers to be on the right side of the school rules. I was, however, given several strokes for not marching properly. My father never wanted his children caned in school. He believed that teachers could hurt them, and this often happened to other pupils. I later saw the sense in his belief and followed his philosophy; I never wanted my children caned either.

We feared all teachers like hell. They always carried canes provided to them by monitors or prefects. There were times that I would bring a small piece of stick to be used for caning pupils and the teacher would turn on me with the same stick and whip me for bringing a tiny cane. The sizes of caning sticks varied with the teachers. By the way, boys were caned on the bottoms and girls on the palms. Latecoming, noisemaking, failure to deliver firewood to school, failure to sweep the dusty floors, failure to bring water to teachers' houses, failure to respond either 'Yes, Sir' or 'Yes, Madam' and other minor offences could land one in trouble. Punishment was meted out in broad daylight at the parade by either the headmaster or the duty master.

During my primary education, I encountered various influential teachers who had a strong bearing on my early education and character.[3] I have named the teachers not in any order of seniority but on account of certain attributes which changed my life. Some of the teachers stayed in our school for a long period while others were transferred to other schools. There were numerous incidences which I will highlight in later sections of this book. Most of the teachers were helpful to me and liked to assign me duties which they believed I could accomplish.

As a class prefect, having been promoted from the cadre of a monitor to that of a prefect, I was assigned to ensure that the small school 'library' was well maintained. Some junior reading books like *Gulliver's Travels* by Jonathan Swift were stocked here. I kept all records of the reading books, borrowers, date taken and date returned. I was trusted and no books got lost. The only advantage I had was to borrow several short storybooks to read at home because I had access to them. Many titles were of English short stories. I could take two or three books for the weekend and return them on Monday after reading them. I actually became fluent in the language and was able to write and read well.

As a library prefect, I would also carry books to class, distribute them to students during a particular reading period and collect all of them at the end for safe keeping.

I was strict while distributing and collecting them as some students could easily take them away. I was the accounting officer of the books for my classes. Honestly, there was nothing like a library. Books were stacked in piles in carton boxes arranged according to the different subjects. The rooms were dusty and crowded. Leaking roofs would destroy some. Those were the typical primary school libraries in those days.

I was responsible for reporting lost or torn books to avoid being caned or punished. During the discharge of my duties, there was one critical observation I made: never to be late in carrying out my chores. The last thing I wanted was to be declared late in accomplishing any assignment. I do not recall any single day when I reported to school late. In fact I would be there very early in the morning, open the classroom door, which was partially broken, and then start cleaning the blackboard. I developed the habit of being punctual to class.

I became known for keeping time, and my additional role was to ring the school bell for parades and lesson period change-over. If I knew I could be late for any reason, I quickly abandoned what I was doing and would be on time for the next duty. Part of the failure of many of my friends was the non-compliance with time demands and not observing strict deadlines.

Other than caning, there were other ways of meting out punishments to latecomers. For instance, one had to dig up and produce a live mole. This was a mammoth task which caused several pupils to quit school. Some students could opt for ten or twelve canes on a chilly morning, rather than unearth a live mole from the deep underground. Tirimba Okongo and Makori Nyagwencha had to go through this punishment and they dug up the whole field without any success. I did not want to be found on the wrong side of the rules and go through this harrowing punishment.

School drop-outs became rampant, because of excessive punishments, but many of us persevered. In fact, we assumed that severe punishments were the norm. We heard of more serious punishments of pupils in other schools where parents went to fight teachers using pangas or machetes. Other punishments demanded the presence of one or both of your parents or a guardian. This was the ultimate embarrassment to anyone in school. At this point, one was threatened with expulsion or severe punishment if one was to be allowed to continue with school. The demand was so painful to the pupils that nobody wanted to commit an offence to necessitate the summoning of a parent or guardian. It was degrading, debasing and disrespectful.

Luckily for me, I did not have to summon my parents to Ibacho Primary School. I had enough advice from home and I consequently behaved. It would have been absurd for a class leader to be requested to bring a parent to school on disciplinary matters. All these problems were manifested in all classes from Standards One to Four and later continued to upper primary.

My entry into Standard Four meant that we stayed in school all day from 7 am to 4 pm. Those of us who proceeded to Standard Five were about twenty-five and

I recall some of them. At this level, other pupils joined us from the neighbouring schools due to various reasons. We considered ourselves great to be in the upper primary classes.

I made several friends with whom I did many things in common: sharing books, the occasional meal and home visits. This was where I met mature entrants who had repeated classes several times. Some of them were married men. I thought I had seen senior class mates at lower primary. I now had to contend with older boys as a prefect. Some of these were uncontrollable and no one dared them, not even teachers! I still recall several names of my Standards Four to Eight classmates.[4]

Standards Five to Eight classrooms were small and always packed to capacity. The names I have given were those I could remember from Standard 5. To be honest, as we progressed to Standards Six, Seven and Eight, I could only recall a handful of us sitting for the primary school leaving examination which enabled me to join the secondary school. Many of them dropped out due to several reasons, which included academic pressure, indiscipline, poor performance in exams and transfer to other schools. The girls were either married off or just quit school. Others repeated classes but we still kept in touch.

I was close to my fellow students and they liked me as a good young boy who would not report them when they made mistakes. They therefore felt protected. This assisted me in remembering their names and the areas they came from. Some of them travelled much longer distances than us. I still served as the bell-ringer and class prefect but avoided any confrontations whatsoever. Again, my role as a class leader did not affect my academic performance.

These earlier experiences later came to enrich the perspectives I developed in university administration. For example, during my tenure, sometimes tribal politics emerged among the student community, who would sometimes be on opposing camps depending on which political party was more acceptable to senior political figures from their community. I remember one case when a group of students, one from the Kikuyu and the other from the Kalenjin community fought the whole night over differing political ideologies. The incident involved about 80 students.

The campus security alerted me early in the morning of the incident. I drove to the campus at about 4 am and summoned all the students to my office. I gave the students a long talk and advised them against what they had been involved in, emphasizing the importance of nationhood and the negative effects of parochial tribal political affiliations. Though the matter was serious, the students appreciated my counsel and apologized. The matter rested there and I gave the students another chance. Over time, they atoned and never got involved in similar skirmishes again. My verbal talk served as their punishment.

The manner in which punishments were meted out to mischievous pupils in upper classes was brutal to say the least. The training of teachers during those days was to literally apply the saying that *'spare the rod and spoil the child'* to its extreme. During my primary school days, I underwent and also witnessed fellow pupils undergo forms of extreme corporal punishment which had an everlasting imprint in my later life. Some of the punishments were so unfair that, as children, we could tell the injustices and biases in each mistake. Three of these incidents are illustrative.

Incident One: I had an encounter with one teacher while serving as a class prefect in Standard Eight. As a prefect, I did not condone any latecomers since this was the school rule dictated to us by the head teacher. My duty was to comply and execute. I would write the names of latecomers and hand them over to the duty master for necessary action.

One afternoon, a teacher's wife who was in the same class with us came late. The mature lady strolled into class way past 2 pm after lunch. I think it was a Mathematics lesson which was almost coming to an end. In my young innocence, and being a prefect of a class and a law abiding kid, I loudly told the husband who was on duty that the lady had come late and deserved a punishment just like any other latecomer. Several late comers had been caned a few minutes before she arrived. She too, I reasoned, must be subjected to all the school rules which governed us. This almost put an end to my school career and life! I had stepped on a hot and glowing wire!

The husband/teacher, who was known for his hideous anger and tyranny landed on me with countless strokes, slaps, kicks, squeezing and curses. This was being done in front of my classmates who had a lot of respect for me. He thundered and cursed on top of his voice as he hit me. I recall him asking if I knew whether the lady was his wife or not. All the pupils in the adjacent classrooms were interrupted by his shouts.

The deputy headmaster came over from the office to find out what was happening! Everybody was alerted about the commotion in our class. One teacher; Mr. Hezekiah Mobisa Ombworo, walked into the classroom and he looked stunned. He saw me sobbing and bleeding from the upper lip where I had been given an astounding jab. He looked surprised and angry. I walked out of the class, followed him to the office and explained to the headmaster who appeared equally surprised at the merciless beating I had gotten.

I told them that I was being beaten because I had reported a lady who arrived late in class. I did not know that laws were applied discriminately. I was observing the rules and practice of the law with the hope that justice could be effected uniformly.

I was made to know that laws and justice did not apply to all. Some people were above the law and the weaker ones suffered the consequences. Much later in my high school, I came across this truth that all animals are equal but some are more equal than others. This was in George Orwell's *Animal Farm*.

I felt demoralized and still wondered what wrong I had done to deserve such a brutal beating. I later learnt that there were more teachers and pupils who sympathized with me than the duty master and his arrogant wife. But did the lady classmate, the wife of the teacher, feel any mercy for me as I was being hammered savagely by her husband for reporting her lateness? I went home, a sad and bruised boy, and told my father about the incident. He was disturbed about it, but I told him that I was okay and would continue with my school work and avenge in other academic ways. Perhaps this was a booster to me later in life.

I still vouched to keep the law and continued to perform my duties unperturbed. I convinced myself to let the bygones be just that, as I pushed ahead in my learning habits with zeal. That beating was perhaps an impetus to make me work hard and pass my examinations. In fact, I passed my Standard Eight examinations well and the lady failed miserably.

I concluded that for every negative and/or destructive activity, there was a stronger and more powerful positive thrust/success. This became my guiding principle which I have always shared with my family. Let no failure deter you from moving forward. For every single drawback, there are numerous successes awaiting you.

Incident Two: My very close friend called Martin was a brother-in-law to a teacher who taught us Mathematics. Martin made a mistake of answering a question rudely to this brother-in-law while in class. Incidentally, the same teacher was the one who had whipped me earlier in the term for reporting his wife. The teacher asked Martin to go in front of the class.

The teacher carried a cane at all times. He commanded him to bend and receive some of the strokes cane. Martin tried to resist, but the teacher pulled him by the pants, the way the Kenyan law enforcers lift criminals by the pants. The teacher whipped Martin fiercely. He floored him and stepped on his well-pressed shorts. The smart shirt and shorts gathered dust from the floor.

I sympathized with him, having gone through a similar experience myself. Again the teacher thundered all sorts of abuses and curses in front of a stunned class. He later remarked that he had taught him a lesson for being rude and disrespectful.

This kind of punishment made me wonder whether there was any other humane mode of discipline. We were not mischievous in any way to deserve harsh corporal punishment. During our school days, pupils grew up in fear wondering when their turns for beatings would come. Many, therefore, quit schools altogether and went off to graze cattle or simply wandered about the countryside. I kept asking myself a basic fundamental question: why could a teacher apply such a serious corporal punishment to a young pupil without first explaining the reason? One can spare the rod and still have a disciplined youth. That is why my father never wanted his children caned.

Incident Three: Pupils in upper primary classes were banned from speaking their mother tongue, *Ekegusii*. Any offender was punishable by receiving several strokes or in the form of carrying a disc. To deter the habit, a disc was given to students found speaking *Ekegusii*. The discs were administered by class prefects and the school head boy. Each morning, prefects, me included, were called into the headmaster's office and handed a disc to give to the first student caught speaking *Ekegusii*.

The first to receive the disc hid it and was always on the lookout for another vernacular speaker to whom the disc was passed. This process was repeated until the end of the day where all disc-holders' names were called in reverse order. They were then paraded in front of the school for caning. This practice was implemented for a few years before being discontinued. It was a colonial mentality in the guise of promoting the English language in African schools. Its primary purpose was to promote speaking in English at the expense of *Ekegusii*.

The practice did not work as well as expected because many pupils performed poorly in English. Since I had the responsibility of handing out the discs, I opted to speak English or keep quiet altogether like some of my schoolmates. The school appeared to have zombies walking around the compound without uttering a word. This behaviour made our life miserable as we had to wait until we were out of the school compound to talk.

My earnest assessment of the situation was that one could not push an idea down others' throats. Instead, people should have the freedom to choose to agree and adopt a new idea. There have been several situations where good ideas have not been readily embraced due to the mode of execution. This knowledge came in handy during my tenure as the University Vice-Chancellor and the Director-General of the National Environment Management Authority (NEMA). I let my staff implement new programmes which they thought would enable them to work efficiently. My duty was to accord support and general guidance.

Incident Four: When I was in Standard Five, I had the opportunity to stay in a teacher's quarters instead of trekking home every day. I stayed with my cousin, Mr Nicodemous Mose, who was physically challenged. He had had a fire accident at infancy and both his legs were damaged just below the knee. He had to use special tailor-made round shoes for walking, while supporting himself with two sticks. He would walk to school very slowly and had to leave his house around 5 am.

The opportunity for us to stay in an Ibacho Primary School teacher's compound was a relief for him. I offered to live with him in one small kitchen where we shared a bed. We were pretty close.

Unfortunately, next to our kitchen lived a naughty lad, Peter Onyonta, who did not like my cousin Nicodemus. He always abused him, beat him up and ran away because Nicodemus could not run after him because of his condition. The young boy took advantage of Mose's disability and did all sorts of silly things to him. He was a year older than me and would not listen to my pleas to stop his habitual harassment. I sympathized with my cousin and considered taking action against Peter. He always made me feel very bad whenever he harassed and ridiculed Mose.

One afternoon, he joined us where we were seated and started abusing Mose. This was his way of provoking Mose whenever he saw him. I thought to myself: enough is enough. As soon he called him names, I interjected and told him to stop the abuses. My interjection aggravated the situation and he slapped Mose in defiance.

There was a pile of poles next to where we were sitting. I stood up, pulled out one of them and gave Peter a thorough beating. He cried out in agony. Then, a teacher who was passing by saw the commotion and intervened. I narrated what had transpired and what informed my violence.

The teacher listened attentively and further caned Peter for his naughty behaviour. He never harassed or assaulted my cousin again and kept to himself for the whole duration he was a student at Ibacho Primary School. He respected both Mose and me, and learnt the lesson of being polite and considerate to the disabled. I have always been very considerate of the physically challenged. That was one of my father's teachings.

My quick and prompt reaction over issues that I did not agree with would later arise during my tenure in university administration. There were occasions that I acted swiftly to save a bad situation, though sometimes I had to rein in my quick action and consult widely before making a final decision. On one occasion, the teaching staff went on strike at the national level over poor remuneration. This was in 1994, when I had just been elevated from Principal to the founding Vice-Chancellor of the University.

This was during the then one-party state when the President was still the Chancellor of all public universities and issues of university autonomy were really constrained by the political system. In those days, a strike at the university would be interpreted by political operatives as a weakness of the VC who had been overwhelmed by the situation. Sometimes this would lead to the sudden removal of one as VC. The consequence of this culture was that some VCs would go overboard and work against the interests of their staff and students just to appease the political system. During this strike, several VCs dismissed a number of high-calibre academic staff. In fact, 1994 is known as a time when public universities in Kenya lost qualified academic staff to other universities, especially in Southern Africa, while a growing number left to join the then emerging consultancy career.

The exodus was sparked by what was seen by the academics as the unwillingness of the university administration to lobby for better working conditions and remuneration from the political establishment. The exodus has had its impact till date, with many qualified academics unwilling to go back to the institutions in the face of severe shortages. On my part, I did not rush to make this decision. In fact, not a single academic was dismissed. Rather, I convinced the University Council, the employer, that there was no need to dismiss any member of staff over grievances that were genuine, as terms of service were then poor. I also had extensive discussions with officials of the chapter staff union and assured them that the university would not penalize them, but would continue to explore ways of improving their terms of service. Later on, my initiatives such as the staff housing project that I initiated and which is discussed later in this work, was my way of keeping this promise to the staff, and it materialized.

But, sometimes, such wide consultation does not always work and, on some occasions, I had to take immediate action without consultation. An illustration is how I handled cases of staff that were cheating on their medical claims. I remember a case where one lecturer lied that his six-month-old baby had undergone extensive surgery at their rural home, some 500 km from Nairobi. Because of the urgency, I immediately approved the claim without any background check and verification from the University finance officer and the medical officer.

Later on, the university medical officer, a Doctor Were, requested to examine the baby who had undergone surgery. To our dismay, the university doctor established that no such surgery had been performed on the baby; rather the said member of staff had cheated and claimed payment from the university. The university incurred an expense because of my quick action, based on my humane reaction to a situation I thought demanded a quick action. Although later on the university recovered the money and sacked the member of staff involved, I learnt a lesson that, in administration, it is not good to make hasty decisions based on emotions. The purported urgency of the situation does not matter.

Academic Competition

During our primary and secondary education days, teachers always reminded us that education was the only weapon we could arm ourselves with. The future would be demanding. Time and again they reminded us that a good foundation was built on education. Education was a ticket to a better life. We believed so and worked hard.

I strived to remain in position one, two or three at the end of each term. Three and above were very competitive positions. Besides competing for positions, we also aimed for high marks. I tended to compete for these positions with my friends and the top five positions were always shared amongst us. The teachers knew this and encouraged that competition to the fullest. There were other bright repeaters who were a challenge. Keara Marando, Caleb Ong'era, Omato Mokaya Kariuki, Makori Nyagwencha, Matthew Aburi, to name but a few.

What motivated us to work hard? We were brought up in strict Christian families. We all attended the same church, Nyamagesa SDA, and had very supportive parents who knew the importance of early child development. They also extolled hard work. The community as a whole stressed the value of education. Our elder brothers and sisters had already attained the highest possible academic levels at that time. We were aware of the national schools and aimed at joining them.

One day an idea occurred to me to form a study group since we used to spend many hours together. I shared the idea with my friends, arguing that instead of each one of us studying independently, we could assemble in one hut and study together as a team. We could share scarce books in turns, save on kerosene, encourage one another through discussions in certain subjects, encourage the weak ones to work hard and finally do well in the final examinations, and hence be role models to the other youth. The idea was embraced by all of us and adopted! It was a simple plan of pooling our reading resources together. First, we used our homes as meeting venues on a rotational basis and later chose one central venue. The requirements were to include all those interested in the plan and who were Standard Eight candidates from the area. Wherever we went, one of our mothers would provide supper which we ate communally.

Our reading group started in January to prepare for the final primary examinations. It was strictly evening preps, six evenings a week except Fridays, as this was our Sabbath eve. Our aim was to meet from about 6 pm to dawn and proceed to Ibacho Primary for day classes. After visiting and inspecting all homes to ascertain suitability, we opted for Mzee Stephen Monyenye's which was centrally located for our studying plans. My home was farther down the valley and was not considered central enough for overnight studies. Darina, Monyenye's wife, was a nice lady and welcomed all of us to use her son's house for a period of one year. This was 1965.

Our study group comprised five candidates: Ratemo Monyenye, Tirimba Okongo, Samuel Ogoti, Paul Marsh Onyambu and I. We called ourselves The Nyamagesa Education Club (NEC); coincidentally, the small transistor radio we used to listen to the news was also called NEC. Each one of us had to eat from his house before gathering for the sessions. However, one could bring food for all to share. The purchasing of all other amenities was shared equally. These amenities included: kerosene, match boxes, a lantern and at times tea leaves and sugar for late-night hot beverages to keep us warm and awake. We did not have any elaborate sleeping arrangements but there was one bed which was shared by all the five of us in turns. The sleeping period could not exceed three hours per person and one had to declare his time for sleeping.

The hut where we studied was divided into two rooms, a study room with a large table and five chairs. The other space was occupied by a small bed. The bed literally measured about one metre wide by two metres in length. It was connected by a gunny bag and a tiny mattress fitted on it. For the whole year, we were determined to perform well and excel in the examinations, hence the perseverance.

We borrowed several books from older boys, acquired past examination papers and their model answers from many teachers who resided nearby. We were lucky to have an English and Mathematics teacher who resided nearby, Naftal Onyambu Onkwani (now deceased), who used to give us impromptu mock tests, mark and return them to us. I still regard the late Onyambu as one of our greatest mentors in our early years.

These earlier experiences, studying in hardship came in handy during my tenure in University administration when we had to implement cost-sharing policies. The experiences enabled me to implement cost-sharing in a manner that did not affect students so much, especially regarding issues related to cafeteria prices and establishment of bursary schemes for needy students at the university level. My elevation to the office of VC was at a time that Kenya was implementing the World Bank-supported cost-sharing programme in all public universities, a situation that was met by resistance from students and parents as well.

Mr Onyambu was one of the most polite educators that we ever had in the community and was always available for consultations after his normal teaching assignments. We were sure of one thing: examination would take place in November and each day counted. Time was of essence. Despite all the nice arrangement we had for preps, there were several challenges. Some of us even fell asleep during the day classes. I remember Tirimba Okongo snoring at 3 pm during a class lesson and the teacher wondered whether he had indeed slept the previous night. He had certainly not.

I knew how to balance my study and sleep periods. I am a light sleeper to date and can go for hours without being exhausted. At night, I would request to study until 1 am or 2 am and sleep for the rest of the night, get up by 5 am and proceed to school. I cannot remember how we got breakfast. But somehow we ate ripe bananas or chewed sugarcane for lunch.

These arrangements made us influence intra and inter school competition. The community knew of our organized study group and we got a few boys wanting to join us. We allowed them, although late, but on one condition: that they bring their beddings and follow our rules and regulations. We did not want to be accused of discriminating against other boys from the community. They could not, however, cope with the times and conditions of our work so they gave up.

The five of us persevered through the cold nights as the weeks rolled by until we sat for the examinations. The results for the Kenya Primary Examination came out and all of us passed well to join various provincial secondary schools. Two of us joined Kisii High School and three joined Sameta Secondary School. My grades in the Standard Eight final examinations were as follows: An A in Mathematics, A in English, and B+ in General Knowledge for the three subjects that we sat for. I still keep my precious clean certificate.

Future Implications

What lessons did we learn at primary education level? We formed an organized study group voluntarily under my general guidance. We knew the importance of good and quality marks to enable us join good secondary schools. We were disciplined, organized and responsible at a very early age. We practiced cost- sharing, shouldered responsibilities, kept time and fulfilled our promises.

We also allowed and accommodated other boys to join our study group; but they could not cope with our conditions and eventually gave up voluntarily. We were also generous to all. We were exemplary in the area. All of us persevered through the cold, rainy nights typical of Nyamagesa hills.

We ate together as brothers, emulating our mothers and fathers. We were a young community comparable to any large society. We respected one another and agreed to do chores in turns without complaining.

We shared our thoughts and promoted unity, understanding and tolerance. We needed one individual to influence the process. We were each other's keeper. We all proceeded to higher education as we dispersed into various secondary schools. We later became independent and proceeded on individually.

The NEC succeeded in 1966 and was voluntarily disbanded. One surprising thing was that none of us got sick despite the cold nights we used to sit through. I consider the natural foods we ate to have been the explanation for our good health during our youth.

Growing up in rural Gusiiland had its challenges besides going to school. My parents were peasant farmers who relied on a meagre income from farming. Tea and pyrethrum were the two cash crops we relied on. We also cultivated subsistence crops like maize, beans, vegetables and potatoes. I used to volunteer for house or farm-related chores whenever I had an opportunity. We still had to balance between going to school, weeding, ploughing, picking tea and pyrethrum, and ensuring that home work was done. My late elder brother, William Nyamwange, was very particular about time and wanted us to excel in class work.

My personal belief was that I had to balance all these chores by not wasting any time on unnecessary engagements. Besides being focused on schoolwork, I was also involved in extra-curricular activities like sports and athletics. My sporting talents were, however, not that spectacular. I did, however, have good vocals and I used to be in the church choir.

The late Uncle Augustino Ogero was a self-proclaimed disciplinarian of Nyamagesa. Nobody dared him and his word was judgement. As a church elder, Mzee Ogero commanded adherence to the SDA doctrines and ensured that the Sabbath Day was kept holy as prescribed in the Bible.

He used to ring the bell every Friday at 6 pm to signify the commencement of Sabbath. All systems stopped, no manual work was done, but prayers started. Both the youth and elders feared and respected him. He was a teacher, an elder and a disciplinarian. He was responsible for the elitism of Nyamagesa Village.

I was told that while he was a teacher, my late brother, William Nyamwange, David Onyiego and Justice James Onyiego Nyarangi strayed from his class one day and went to watch a football match in Kisii Stadium, some 20 kilometres away. He gave them a record beating. The three boys were beaten so hard that they got permanent scars on their legs, buttocks and hands. But they were all grateful for his role as a teacher. I grew up knowing my boundaries and avoided crossing paths with the late Uncle Augustino Ogero. Children belonged to the community in Nyamagesa Village and punishments were the order of the day. I performed my duties on time and reported back the results to whoever had assigned me the work.

Early childhood discipline makes a world of difference for future generations.

Notes

1. They included William Nyamwange,(late) Tabitha Mocheche, Joel Onami, Andrew Okioga, Samuel Clement, (late) David Ombogo, Ruth Moraa,(late)), Hezron Tirimba (late), Mary Bonareri (late) myself and Amenya, Grace Kemuma. My step mother had 8 children: Helllen Nyakerario, Thomas Ogutu, Askah, Gladys (late), Stanley, Annah, Mogaka(late) and Samuel.
2. The eight brothers include descendants from: Bomobea, Bonyamoyio, Bonyakoni, Mwamonda, Mwamoriango, Mwaboto, Bogeka and Bonyamasicho
3. Machoka Singombe, David Onyiego Onywere, Zablon Anyieni, Johnstone Nyonga, Reuben Oanya, Hezekiah Mobisa, William Nyamwanye (brother), James Kabuna, Charles Magati, Barnabas Tureti, Johnstone Rayori (later Senator), Bathsheba Matonda (the only lady), Samuel Rogoncho, Samuel Kenanda Meraba, Gilbert Nyangweso, Hezekiah Michoma. I recall them because I interacted closely with all of them at my very early age.
4. They included: Omato Mokaya Kariuki, Zablon Chanai, James Chanai, Caleb Omboto Ongera, Matthew Aburi Nyatundo, Samoita Botange, Basweti Mariga, Ogutu Marindi, Ogutu Kombo, Ogutu Onsinyo, Ogutu Kiyiete, Omare Onsinyo, Sunusunu Matunda, Tabitha Nyangate, Dinah Onyambu, Prisca Mokua, Wilikister Joel, Samuel Oanda, Anyona Martin, Matabuta Marangeti, Ogutu Ondimu, Kebwaro Ondimu, Miriam Ogero, Monicah Monyenye, Nyanduko Ogutu, Omare Maaga, Omare Mangongo, Isomba Nyatundo, Miencha Orora, Robert Nyagaka, Melchzedik Anyona, Nyasikera Motari, Makori Nyagwencha, Samuel Mirieri, Samuel Nyabera, William Onsare Mabeya, Tirimba Maturi, Nicodemus Mose Ogutu, Peter Marando, Keara Marando, Onchari Nyakundi, Nemwel Nami Matunda, Tabitha Nyangate, Dinah Onyambu, Prisca Mokua, Wilikister Joel, Samuel Oanda, Anyona Martin, Matabuta Marangeti, Ogutu Ondimu, Kebwaro Ondimu, Miriam Ogero, Monicah Monyenye, NyandukoOgutu, Omare Maaga, Omare Mangongo, Isomba Nyatundo, Miencha Orora, Robert Nyagaka, Melchzedik Anyona, Nyasikera Motari, Makori Nyagwencha, Samuel Mirieri, Samuel Nyabera, William Onsare Mabeya, Tirimba Maturi, Nicodemus Mose Ogutu, Peter Marando, Keara Marando, Onchari Nyakundi, Nemwel Nami, Mecha Nyakundi, Peter Onyambu, Batsheba Ontiri, Rael Ontiri, Matundura Ontiri, Miencha Orora, Onkoba Nyambane, Ratemo Onchera,Ratemo Ondara, Anyona Marangeti, John Arumba, Kebwaro Masese,Marucha Okero, Marucha Ombworo, Tariani Ombworo, Dinah Peter. Plus the Nyamagesa boys.

3

Foundations for Future Academic and Leadership Pursuits

The Role of Parents

Our parents brought us up well. Despite their very meagre earnings from farm produce, my father made every effort to pay for basics such as school fees, limited clothing, reading and writing materials. My elder brother, the late William Nyamwange, completed his teacher training course in the early 1950s. He assisted father in paying our school fees and other expenses. His siblings are forever grateful for the role he played. Nyamwange never finished any talk without stressing the importance of education. For those of us who performed well and joined government-supported schools, the fees were minimal.

I joined Kisii High School which was a boarding school where residential expenses were subsidized. Gusii customs dictate that young boys become men at some stage by going through an initiation ritual ceremony. Elaborate procedures of circumcision were performed on ten of us. This was an all-men affair to mark the passing of age from boyhood to manhood.

Twelve of us, all of the same age group, gathered in one hut where we were coached, told the importance of the ritual and taken across hills and valleys for circumcision. The coaching and vows were so critical at this early stage of between 12 and 16 years that no boy would miss this rite of passage. The process is important and necessary for the development of a boy to a man. I do not know of any boy who did not go through this important cultural passing ritual.

When the boys were paraded, the more experienced and elderly men would look for one brave boy to lead the rest. They would have known us earlier and determined who would lead the group. This was a serious procedure which had to succeed. The inductees had to be brave and the lead boy had to be brave too in order to encourage the rest. After several routine requirements, and having been in cold waters in the very early hours of the morning, we were paraded, lying on the ground with our faces firmly stuck on the ground. I had been informed in advance that I would lead the group for the cut! I obliged.

The person who performed the cut used a special sharp knife and no anaesthesia was used. It was normally done at about 4 am to avoid excessive bleeding and fatigue. The story is that I led a group of twelve young boys to be initiated; anyone who attempts to run away or cry could be 'speared' to death or condemned culturally. The ceremony was a success and it brought joy to all initiates despite it being a very painful experience. Women, as earlier stated, were excluded from this affair but they brought porridge made from a cereal, *wimbi*, to be given to the young men who had bravely gone through initiation. Upon circumcision, we were declared men who could protect the nation and could sit at the periphery during men's meetings or gatherings. We were proud and confident.

My father taught us humility and respect. I remember on many occasions when we went for cerebrations where food was be provided. He would advise me not to queue, not to rush or stampede to the serving point. In case food ran out, we would go home and eat there.

He also told us to be cautious of great talkers. They could be liars. One never rushed to conclusions on important matters. One had to always do one's research.

He used to send me on his errands despite the many sons he had. The reason was that I never questioned him. I remember one time when he asked my elder brother, Tirimba, to go and guard/sleep at our community shop in Masimba. Tirimba refused and got a thorough beating. I had to go there even though it was not my turn that week.

I always believed in timely completion of a task. Assignments must meet deadlines at all costs. My father liked achievers and those that he could trust.

My father was an opinion leader, soft-spoken, and was involved in several village meetings. He was consulted by other elders on delicate issues like family and land disputes. I do not recall him being angry or picking a quarrel with the neighbours. All his brothers, my uncles, remained in Nyaura, but would occasionally come to visit us at Nyamagesa and spend some days with us. I recall one of my uncles, Mouti Okioga, who resembled my father and spoke softly like him.

One of my other uncles from my mother's side, Birari Achimba Nyamwange, deserves special mention. He was considered a healer, a herbalist and a fortune teller. Whenever he came home to visit, my father accorded him the highest and most respectful reception. I held him in the highest esteem for his healing prowess. Father regarded him so.

Whenever he came to Nyamagesa from his Taracha village, we knew we could have the best meals during his stay. A goat or sheep had to be slaughtered in his honour. We benefited from his visits. The rest of my mother's siblings were sisters. Visitors were a blessing to our house and were always welcome.

My father had one trait which was liked by many of his peers. He was a meticulous grass-thatcher. Many houses in Gusiiland in the 1950s and 1960s were grass-thatched and this was done by a gifted or trained thatcher. *Mzee* believed in quality and style of grass-thatching with the finest finishes. His competitor in the area was *Mzee* Thomas Ogutu. The rest of the thatchers were considered amateurs and whichever roofs they put up leaked.

Quality of thatching was measured by non-leaking roofs. Any drop of rain into the living or sitting rooms was considered a shoddy finish and the thatcher had to redo the leaking section. As a young boy, I used to hang around when roofing was being done. Special types of grass referred to in Ekegusii as *ekenyoru* (fine fescue) *or ekebabe (imperata cylindrica)*, both of which are slowly disappearing, were used for thatching. The buildings by the *wazee* (old men) taught me the quality of finished products and the value of social interactions with the elders.

My father was always occupied, putting up roofs for neighbours. By the way, they did the work for no pay. They rotated from one homestead to another as need arose.

An interesting observation is that, even today, only men and not women in our community are involved in building houses. Culturally, a goat or cock was slaughtered on completion of one house and the builders enjoyed a sumptuous meal to celebrate the achievement. Gusii culture dictates that a cock should be slaughtered solely for the main person who has thatched the roof. He is like the lead architect.

My father used to enjoy this privilege, but always shared his traits with his peers. The final celebration is done after a special sharpened stick (*egechuria*) is strategically mounted at the tip of the house which signifies completion. This tradition has several cultural implications which I cannot explain here. The feasting after house completion compares favourably with the present-day house-warming, or opening a highway, a school, a church or a hospital. People celebrate these achievements.

Whenever I put up a new building, I do the same for the contractors, masons and carpenters. We slaughter a goat to celebrate the house's completion. Some people do spread the remains of goats' guts around the house and compound as a blessing. These traditions and beliefs still exist and allow for closer family ties as the people gather for the event. This is part of the blessings for a long life.

My father passed away on 12 October 1982 after a long illness. He had been treated in Kisii General Hospital before he was transferred to Kenyatta National Hospital in Nairobi. I remember making arrangements for his admission to the hospital. My younger brothers, Amenya and Thomas, drove him to our house in Hurlingham, Nairobi, at about 3 am.

My wife Esther and I received him briefly and he found our first-born, Nyakundi, awake and suckling. He held him in his hands, spat saliva on his face and blessed him. This is the traditional way of wishing newly-born baby luck. Nyakundi was born on 11 March 1982.

This was the last time that he saw the young boy despite promising us that he would come back to our house and visit! We were devastated when he passed away. I recall that we (the sons) visited him in Kenyatta National Hospital. He told us that we must work as one family entity. Our family was very large and *Tata* (father) had many children and grandchildren. There was no need to cause differences amongst ourselves. He said he had brought us up well; we were then settled with our families and he asked us to respect each other. Let there be nobody who could cause any rift.

His burial ceremony was attended by people from all walks of life. We lost a father and a great advisor who loved all his children equally.

My mother possessed great knowledge on the identification of plants, and taught me especially those which had healing properties. She knew virtually every plant by its local name as well as its uses. Her brother, the mentioned uncle, was versed in plant identification for medicinal properties. Through my interaction with them, I came to know plants for colds, blood clots, eye treatment, stomach ailments, worms, malaria and diarrhoea. Whether they healed or not is still questionable when we compare them with western medicines.

However, current research in herbal medicine seems to affirm that many of these plants possess medicinal properties. What I came to know for sure was that some of the herbs mother prescribed worked. I occasionally administer them to my family, especially when we are in the rural areas, and they work. Currently, I teach courses in weed science and plant taxonomy and have found the early lessons from my mother very useful. I now know which plants may have medicinal properties despite classifying them as weeds.

My mother used to amaze me at times. She could get out and call passers-by for lunch or dinner in our house. She knew some people who were needy and needed a meal. Mother was regarded as a pillar in social and family matters in Nyamagesa Village. The young ladies who were married into our community consulted her for any social and domestic issues. She was so liked that the whole community paid great tributes during her funeral in February 1992.

When Mama Beldina Nyatero passed away, we were so devastated that we did not know whom to go to for advice. I recall my wife, Esther, and I driving all night to Kendu Mission Hospital to join the rest of the family where she passed away. The loss of a mother is devastating; we could already feel the vacuum in the family. We lost a village pillar whose name later spread by many families naming their siblings after her name as is customary to *Gusii* traditions. The children who were named after her are so special to me and I always remember her through them.

My mother taught us several survival tactics. She believed in hard work and getting up early to accomplish plans for the day. She believed in feeding us well to avoid being sick or being tempted to eat anywhere else and contracting germs. She entertained all sorts of people, both family members and distant relatives. Many great scholars from Nyamagesa community who excelled in their careers passed through our home. They include Lawrence George Nyairo (a great teacher), the late Justice James Onyiego Nyarangi (mother's cousin) and several ladies who were married off from our home. They always remind me of my mother's generosity and loving care.

Among the many things that my parents taught me was the origin of Omogusii, names and their meanings, different types of animals and birds. All these were relayed to me in my mother tongue. My mother would name the twelve calendar months in Ekegusii. I cannot recall all of them now. However, I can name and specify the world compass in my local dialect. North is known as *Irianyi*, South is called *Sugusu*, East is *Mocha* and West *Bosongo*. The best way I remembered these was to coin Christian songs with directions or watch where the sun rose from the East and set in the West.

My mother was very good in short-term weather forecasting, especially when to receive rains. She relied on the noises made by insects and frogs to predict when rains would occur. The appearance of fire ants or army ants was a definite and sure indication that rain was eminent and would fall within a few days. She could also detect it through her body reactions. She was fairly accurate in her predictions and would be on target many times. I used to follow and sometimes record her predictions on rainfall patterns in our area. Indigenous knowledge is a useful tool in many instances, but little attention is paid to it.

I learnt how to collect and conserve rainwater from her. She used to collect water by any utensils available. She had big earthen pots which could store cold water for weeks without it getting spoilt. We therefore did not have to go to Riagayi River to fetch water that frequently. This must have been my early training in water conservation. I am an ardent fanatic in rainwater harvesting and can use any means possible to collect all water falling from a roof.

When I was in NEMA, I appealed to the Ministry of Water to advance the notion of water harvesting from all house roofs for irrigation or general washing. It has not been possible to preserve Abagusii culture. This includes the naming culture, among others. It has been eroded over time due to less emphasis on the importance of culture conservation by the people. Kenya has an educational policy on the promotion of local languages and dialects. The urban influence has had a negative implication on the same. Resistance from some teachers has not helped either. This is an absurd trend of events and one day, the future generations will have no cultural identity due to its erosion.

The worst thing that has happened during my adult life is witnessing a systematic, diminishing of African traditions, cultures and some languages. According to UNESCO, a number of languages in Kenya have either become extinct (such as *Omotic, Ong'amo and Sogoo*) or they are in serious danger of becoming extinct (*Bong'om, Suba, El Molo and Terik*). The westerners are preoccupied with rediscovering themselves and tracing their ancestry, whereas we in the developing world are busy dismantling our own cultures and values. The consequences are grave if we do not put mechanisms in place to preserve and promote what is left of our cultural identity and pride.

Consequently, during my tenure as VC, I encouraged members of staff to embrace African indigenous knowledge and approaches to enrich teaching of the sciences. For example, we designed a whole course in the horticulture department that focused on traditional methods of preserving indigenous vegetables. This culminated into creating a renowned scientist in this area, Prof. Mary Abukosa Onyango, who won several science awards in the area of indigenous crops and vegetables. Some of these vegetables were becoming extinct and the idea we had then was to popularize them and promote their use for their nutritional and health benefits. I also initiated a botanical garden in the university which still thrives and promotes research, learning and conservation of Kenyan herbs which are medicinal but are threatened with extinction.

These efforts were later acknowledged and strengthened by the UNESCO, Director General, H. Matsuura, who created a Chair in the Horticulture Department. I will discuss the benefits of this chair to the university later in this work. I also initiated the teaching of Kiswahili, as an African language though the university was largely offering science-based disciplines. I thought this was my good contribution to harnessing the benefits of African languages in academic discourse.

Kiswahili language is now being promoted all over East Africa, particularly Uganda, Rwanda, Burundi and the Democratic Republic of Congo (DRC). The advocacy for promoting all Kenyan languages and dialects has gained momentum. There is a move to ensure that schools in rural areas should teach in their mother tongues in the early classes. The move is commendable.

My mother's actions enlightened us on how to share the little we had with others. She always reminded us that one would never be fully satisfied to share the little one had, but one had to be generous with the little that God had provided one. One statement she used to implore all of us was that "energy is never exhausted no matter how much work you do, it regenerates; you may die without using it if you are lazy". It simply meant that working hard will not kill you. The philosophy compares well with the current multitasking notion, the more one does, the more one accomplishes.

My late brother-in-law, Senior Chief Timothy Omwenga Rogito, was a great teacher. Whenever I visited him he used to sing for me and my nephews, Samuel, Kemunto and Onchuru the song which was widely known by many educators: *"Early to bed and early to rise, makes a man healthy and wealthy and wise."* I was in Standard Six when I heard the song and memorized this stanza. I still cite it to my children. It may not make sense to many but I use it as a planning philosophy and guide.

Mother was an early riser and all of us followed suit. She used to remark that if by 12 noon you have not met your day's target, forget it, the whole day is gone. You cannot recall, hold, chase or borrow time. It goes and lapses for good. Since then time keeping to me has never been compromised. This is an attribute that I acquired from my primary school days. I have kept the habit and believe that the few successes I have achieved are mainly due to my strict observance of time and deadlines.

The Seventh Day Adventist Church usually holds its annual camp meetings in August. The meetings normally last for a week. My father's expansive land used to be a meeting venue in the early 1960s. This was because we had a large plantation of black wattle and eucalyptus trees to provide day-long shade for the congregation. This was the time when I knew the importance of trees in microclimate amelioration. The notion of tree planting in the late 1980s and 1990s, therefore, was not a surprise to me as my father had thousands of them planted in our land as early as from the 1960s.

The annual camp meeting exposed me to hundreds of people who came from several areas of Nyanza Province. They included church choirs, elders, pastors, secondary and primary schools students. I was a young boy then, tending cattle all day long. But during this time, I would let the cattle loose, hang around the venue and listen to the Bible citations and excellent choirs. The different choirs from churches were the most attractive during this fete. I could imagine angels descending from heaven and witness visions.

As a young boy then, I was so moved by the events and decided to join the Nyamagesa SDA Church Choir. The choir master was James Obare Nyarunda. I do not remember how I was able to cope with singing, but I recall that too much time was spent on practising singing. Every evening of the working days was spent in the church singing on top of our voices. I was there punctually and sang for one year; but after joining Kisii School, I discontinued.

I then became involved only as a leader in counting the camp attendants on Saturdays. I thought that the time spent in singing was too much for other chores like doing homework and washing my clothes. I reduced the attendance periods to concentrate on my school work, but continued to perform other church functions. It was not easy either since there was too much memorizing of hymns and verses. I slowly pulled out but attended church services as usual. Time management became crucial and I had to choose between singing and recitations or school work and church service. Something had to go.

The many experiences I went through as a young person hardened me a lot. I could give guidance in decision-making at an early age. I took risks in implementing my duties and occasionally stepped on other people's toes inadvertently. The punishments we went through made me realize that hard work pays. Being obedient, time-observing, and keeping on the right side of the law made me excel in my work since I did not have to look back for fear of some error committed.

The stress we go through now is a manifest of upbringing. I suffered uncalled for humiliation early, but bore it like a warrior and moved on. My integrity and inner strength were displayed early and are visible traits to date. I believed in friendship even as a child and the many friends I made during my school days became an asset later in my adult life. Fair play in decision-making overrides many human virtues.

The immeasurable lessons from my parents made me the person I am now. My parents and siblings have contributed to my public image, career and character. As I compiled this book, one of the greatest writers, Chinua Achebe, passed away in the USA in late March 2013. I read one of his books for my KSCE literature examination. The book, *Things Fall Apart* was a classic in its own category. I literally memorized the book and the characters mentioned in it. Little did I know that I would go and work in Nigeria and visit his ancestral land. It was amazing for me to cross the River Niger at Onitsha City to visit Chinua Achebe's birthplace.

A lot has been written about Achebe's works, and suffice it to say that those of us who had the opportunity to read Chinua Achebe's *Things Fall Apart* (1958) discovered the gaps and inadequacies created by the colonial powers. The very strong characters depicted like Okonkwo show the richness of culture in Nigeria. The strong belief that the white man came to Africa and did little to develop it is evidenced in the book. The eroding of African cultures is sweeping across the continent as this can be seen in the disappearance of local languages and adoption of western culture. My stay in Nigeria will form a separate chapter in this book. The way many Nigerians revere their traditions and culture is of great cultural and symbolic significance.

I scored impressive results in my primary school leaving examinations at Ibacho Primary School. I was selected to join Form One at the then Government African School (GAS) Kisii in January 1966. The names were displayed on the students' noticeboard and a relative delivered the good message to me. My parents, brothers and sisters were excited over the good results and I had to visualize my life at a boys' boarding school.

The fact that I was going away from my parents' comfort to join a group of students drawn from all over Kenya scared me. Luckily for me, there were several senior boys from my village who were already there. I had been told about bullying by Form Two students and wondered what that entailed. James Nyarunda, Ratemo Obare, Nyaora Mirieri and many more were already there and I knew they would protect me. My elder brother, Nyamwange, was very happy that I joined Kisii High School which was government-supported and we paid very little fees. He was relieved of this burden. My father planned for my trip, bought a small wooden box with a tiny padlock, my first pair of black shoes, stockings and off I went. He gave me some money to cover my fare and no pocket allowance.

I took a bus from Keroka bus station to Kisii School. Kisii town was about 20 miles (30 km) from my home and the road was murram-covered. The inviting letter clearly indicated that we were fully boarded; meals and beddings were provided by the school. We were given uniforms, toiletries, all writing materials, a bed, mattress and beddings. We actually ate better at school than at home. We had breakfast comprising porridge, ten o'clock tea, assorted lunch, four o'clock tea, dinner and occasionally cocoa, tea or milo at night. In addition to the said menu, some privileged students had facilities for brewing their own coffee and had the means to buy bread to go it with late at night. I will allude to this selected group of students later.

I checked into school and found several other Form One students from other parts of Kenya. I met the best of the best from other schools, particularly South Nyanza, Rift Valley and Western Kenya. I had been told by my two elder brothers (Nyamwange and Onami) who had studied there about the strict rules imposed on students by the British teachers and the stiff competition by students from other schools. I was also briefed about weekend movies in the compound, dancing, entertainment and students' mischief. I was reminded time and again that I was going out to join other students from various communities and must take care of myself and be a responsible man. I was no longer a young boy running around doing some home errands.

The media of communication in Kisii High School was English and Kiswahili. That was going to be the nature of my life over the next four years. I had to live with a Kenyan community and not a Kisii one. I arrived in the compound and went through the registration formalities, after which I was given a bed, mattress, two sheets and two blankets and was booked in Ruri House.

There were several other dormitories in the compound which included Manga, Sameta, Wire, Kionganyo, and Homa where other students were distributed. The newcomers, monos as they called us, were taken round the school facilities for orientation. We visited the library, the school clinic, agriculture farm, chemistry and physics laboratories.

In the meantime, other senior students were in classes for prep which we were informed was mandatory up to 10 pm. The senior boys who took us round the campus told us of the importance of upholding the good name of Kisii High in terms of performance and reputation. I was satisfied with the briefings as we got to interact with students from other tribes. All of us Form One students looked new and were impressed at the facilities here compared to those found in primary schools. We later had supper before we retired to bed around 10 pm. This was an excellent orientation.

High School Experience

Kisii High was the turning point in my life and career development. I developed my life-long traits there. I recall that my school registration number was 1878 and I took my PhD degree in 1978 from Rutgers University, USA. I will always remember these two magic numbers.

As a fresh entrant from the rural area of Nyamagesa, I had to plan my life and career interests from the onset. I had to work hard and interact with the other boys on a daily basis. I made many friends with whom we shared common interest, like attending Sabbath regularly and forming new study groups.

Kenya had gained independence when I joined Kisii High School and a good number of teachers were still from the United Kingdom. The headmaster, Mr Bowles, who addressed us was so brief that I did not follow all that he said. He knew that house masters and senior boys would expound on his speech as we were being oriented during the week. Mr Bowles however, a chain cigarette smoker, was always available for consultations and used to receive parents and visitors in his office regularly.

His deputy, Mr Brown, was a proud young man who taught English with a typical Queen's accent. His interest was to train us to speak and write like them. Many school mates strived to imitate the Queen's English at the expense of Kiswahili. In any case, the subject was compulsory and if one failed it, one's grading was affected; one was automatically awarded a lower grade. There was, hence, justification to read and write English and pass it well. This was my second encounter with brainwashing the African mentality. The primary school disc incident was still fresh in my mind. In the secondary school it was even worse, one had to pass the English language or flunk altogether.

I settled down quickly during my first year in Form One and embarked on my class work. Form Two students bullied us. We had to clean halls, wash utensils, mop floors and do all sorts of work because we were 'monos'. My dormitory, Ruri, had outside bathrooms which were used by about 60 students and were not the cleanest at times.

The dormitory was shared by all students from Form One to Form Six. This was an excellent arrangement because we truly interacted freely with senior boys who were very mature and always ready to protect, advise and even mentor us. They actually assisted in controlling the Form Two students who were mistreating the Form Ones. Students from the Luo community were the majority in Kisii High School, followed by Kisiis and Luhyas. There were other tribes from various regions of Kenya. Again, this was the best combination of students from all over the country.

We lived and interacted as one large family; a fact that the then headmaster, Mr Bowles, enjoyed referring to during parade times. He was fond of encouraging us to see one another as brothers and not Luo, Kikuyu, Kalenjin, Luhya or Kisii. We were all one large Kenyan community. He was aware of the tribal and cultural differences amongst us, but played them down.

This approach and encouragement had implications on the stability of the school and discipline. I later appreciated the importance of a multi-cultural, multi-tribal and ethnic cohesion in Kenya. The universities being set up these days are dismantling this grand original cohesion by being tribe-oriented. I made several friends with whom I tagged during my first year in school, and with whom I later came to serve in various leadership roles in the public service.[1]

It is important to demonstrate a typical week of a Kisii High School student in an attempt to understand the behavioural pattern which instilled discipline in us. We had classes from Mondays to Fridays and half-day on Saturdays. Seventh Day Adventist students were exempted from the morning studies on Saturdays.

During the week, we normally got up at 5 am after hearing a wake-up bell and embarked on morning cleaning. I used to have cold showers which were not a bother to me then. Temperatures in the region could sometimes drop to 9° Celsius and showers could be very cold. We had breakfast at 6 am and went to parade on Mondays at 7 am. We were usually addressed by the duty master or headmaster, and classes started promptly at 7.30 am. There were tea and lunch breaks with classes ending at 4 pm. The prefects on duty ensured strict adherence to the times. I had to be alert at all times as I did not want to be late for any function.

The head boy, prefects and monitors were so powerful that they could freely mete out punishments to the boys. We were about one thousand boys, each of whom had his own character. Every dormitory had a house prefect in charge of order and discipline. At the same time, we had other prefects in charge of the library, games, dining halls/kitchen and dispensary. They were all equal in the discharge of their duties. They were usually selected from the senior forms – Three, Four, Five and Six. There were exceptions. For example, the dining hall and kitchen had to have two prefects and two assistants because of the load of work involved.

I was lucky to be appointed as a dining hall assistant during my second year of study. We were nominated by the headmaster in consultation with the house master. I had therefore some more privileges than the rest of my class mates. Later in the third form, I was elevated to prefect status. I started my management responsibilities in earnest at a very early stage. My duties included all issues to do with food, hygiene and water. As a kitchen prefect, I had to attend to all these issues including management of old cooks.

But my primary objective in Kisii High School was to do well in studies and join a university. The responsibilities did not therefore deter my performance; if anything they gave me extra impetus. Since there were four of us responsible for dining hall matters, we agreed to co-operate under the general supervision of he senior prefect, Samuel Langat. We agreed not to disrupt our lessons for any reasons.

I was in charge of food rations to the students, receiving raw and dry foods. They included meat, flour, rice, beans, maize, bananas, beverages, cooking fat, salts, curry powders, and greens. We agreed that the suppliers must come at a specified time on certain days of the week or during weekends to avoid interruption of classes. Occasionally, the supplier could deliver foodstuffs during class hours. I made it clear that I would not leave my class to receive any food supply. Either the supplier had to come at lunch hours or between 4 pm and 5 pm on specific days. We alternated receipt of merchandise with my colleagues. The school bursar was to oversee that what we received was properly recorded for payment. There was no compromise on quantity or quality. The ledger book was under lock and key and I used to carry the key at all times when on duty.

The bursar was not an honest person. He would tempt us to receive the rations during class hours so as to possibly inflate the quantities. The suppliers also wished to deliver their merchandise at odd hours to influence us and inflate records. We detested this habit and threatened to quit our roles. I was brought up in a strict Christian family and my parents provided for the basics. We were not ready to jeopardise our school tenure.

While at school, we were provided monthly with a bar of soap, toilet papers, and uniforms and as prefects, a lantern and kerosene for night use. In fact there was no need for pocket money. All meals were provided and they had a balanced ration. The little money I was provided with was used to buy peanuts at break time or a loaf of bread for night snack. As a young boy, I would be induced into making a wrong entry into the ledger. That was tantamount to committing a crime and contrary to the norms of the society and Christian beliefs. This has been my driving principle unto this day.

I was respected as a prefect but the role could be overwhelming at times. I was in charge of the cooks and it was my responsibility to give them the daily meal portions from the store for the 1,000 students. I used to weigh out the day's raw food early in the morning for the whole day's preparation. There were special diet cases which were a bother at times. A few students had health issues, so doctors had to recommend specific diets for them. It was my duty to ensure that cooks prepared the menu for these students.

I recall one, Okitete Orwa, who was always complaining of his rice and beans not having been cooked on time. He liked me so much that even after high school, he kept in touch and would call me to discuss and reflect on the past. He became a lecturer at one of the Kenyan universities.

I was very popular among the students because I was always on their side, especially when the bursar wanted to reduce food quantities. I recall a case when they planned a riot because the headmaster had discontinued the 10 o'clock tea. In protest, they boycotted lunch and demanded for reinstatement of the tea as well as an accompaniment of either bread or sweet banana.

The headmaster called all the prefects to the boardroom to sort out the problem. Mr Gabriel Walobwa, the headmaster then, had just been transferred to Kisii High School. I gauged him as a polite man and kind of a pushover. With this in mind, I calmly told him and his deputy that the radical change in the menu was not appropriate in the middle of the term since the budget of supplies had already been done and formalized. My colleagues looked at me with surprise. I asserted that the status quo ought to remain until some other time as this would disrupt classes. I even told him that as a kitchen prefect, I knew that meal arrangements had been completed for the term. He obliged.

I delivered the message to the students who resumed classes and appreciated my intervention. I had total respect for my teachers but never cowed from telling them the justifiable truth. They all liked me, and duty masters were particularly happy with the level of judgement I made regarding students' demands.

All prefects were my friends simply because I was generous in the provision of additional meals whenever requested. I was strict and did not allow anybody in the kitchen, as I was already aware of food contamination and kitchen hygiene. This was our top priority as kitchen managers. I do not recall an epidemic due to food or water contamination. I was aware of the consequences of poor cooking environments. This early experience later gave me a good starting point as a Vice-Chancellor in the management of a university kitchen and dining hall. The management experience was further translated into my becoming a Director General, National Environment Management Authority (NEMA). I also applied the same leadership qualities in JKUCAT (then a college before being chartered in the year 1994) when I was a Principal of the University College from 1992 to 1994 and later a Vice-Chancellor from 1994 to 2003.

My involvement in student welfare did not deter my academic performance. In fact, I planned my programme so well that I had extra time to read. All prefects were privileged to have a lantern and kerosene provided free by the school for night duties.

The campus generator was normally put off at 11 pm by a one Mr Obondi. Prefects also had special cubicles in each hall which served as 'offices'. In my case, kitchen prefects had a neat office in the kitchen which was only accessible to us. However, other prefects and their friends used their cubicles for unauthorised night reading after Obondi shut down the generator. I am sure teachers knew about this habit but turned a blind eye. Kisii High School and indeed Kisii town was not connected to the national power grid in the 1960s. We were given free kerosene weekly for dormitory lighting in case of a problem with students or in case of some emergency.

Knowing the embarrassment I would encounter if I performed poorly made me read very hard in the morning hours. My close class mate and friend whom I shared a double-decker with, James Ongwae, wondered where I went every morning. But he suspected it was the kitchen. I indeed went to the kitchen at 4 am and read for one hour uninterrupted before the rest of the boys got up. The reality of the matter was that I knew the amount of time I spent doing chores as a kitchen prefect and I had to compensate my lost time. I had to read at odd times in order to catch up. I knew this was not permitted but my future depended on it.

I already had an experience of reading hard in Standard Eight. I therefore gained seven odd hours a week for revision, by putting in an extra hour daily for seven days. A good number of prefects used their time well and could light their lanterns in the cubicles and study. Our store office was, however, private with no bed nearby to tempt me to sleep. I did my homework and revision without any interruption before the official wake-up time of 5 am.

Students from Ruri House were known to be bright, but rough and mischievous. It had some of the most outstanding academic records for several decades. The late Professor Frederick Angawo Onyango, who later became my Deputy Vice-Vhancellor at JKUAT, and consequently the Founding VC of Maseno University, set an academic record in 1958 which was never broken. He was my senior by several years but each time I go to the records, I still find his marks on top. He had the highest respect for academic performers.

Nurturing

The ages of 14 to 18 are no doubt the formative years. My growth and development in high school literally shaped my future life. My mentors and role models were many; starting with my brothers, uncles, cousins and eventually both primary and secondary school teachers. They all contributed to my academic and wholesome development. My parents and uncles had tremendous impact on my management of both time and social relationships.

Ibacho Primary School was managed by the likes of Mr Machoka Singombe who would assign us 30 mental arithmetic sums to be done in half an hour. If one did not finish and score over 20 of them right, one got canes equivalent to the wrong answers. I learnt to be fast and accurate, but I got whipped too. Later on in upper primary, one re-known disciplinarian, Mr. Stephen Michoma, took us through the final examinations. He was a no-nonsense head teacher and we all saw him as an extraordinary human being. His statements were brief, specific and instructive. Failure to comply meant instant punishment. I owe a lot to these two teachers.

Not many pupils could tolerate both teachers, Singombe and Michoma. I, however, adored them. They had influenced on what I am today and their teaching methods were equally instrumental. So when one talks about mentors and role models, one should look at the wide picture of those persons who touched their souls. Every good performance on my part made me learn something. I would sit down and pray that I could be like so and so. I obviously wanted to teach but not beat my students!

That was the beginning of my philosophy of non-penalty traits. For example, Mr. Charles Magati cheerfully taught us English and handwriting. I sincerely admired his teaching method and sense of humour and even emulated him later in my career. Mr Gilbert Nyangweso taught us handwriting using a blackboard and chalk. We used nibs and ink and produced good handwriting. We were also taught vernacular as a subject.

In Kisii High School, several Kenyan, British, Asian and Americans teachers taught us. My experience with the teachers there was enriching. A British teacher who, for example, taught English Language and Literature used to be so emphatic in the use of correct grammar that we had no choice but to comply. Many of us therefore scored highly in the language. We were thoroughly brainwashed and used to be told that if one failed in English, one automatically got a mere certificate despite excellent performance in other subjects. (You recall the primary school discs?!) We believed in the English language both oral and written. We had been so brainwashed that we forgot to promote our mother tongues.

Of course all other subjects were equally important but English was compulsory in all British colonies. In such circumstances, we had no choice but to emulate these teachers who also made a difference in my life. The teacher of English was sure of our passing and regarded us better in English than typical British youths. We obviously took these remarks positively and excelled. Any compliments from any of our teachers made us work harder. We further became more entrenched in the use of the language.

Kenyan teachers were very good in Mathematics, Geography, Biology, Chemistry and Physics. I took a religious study subject which was taught by Mrs Green and most of us who took the subject scored a grade one. We studied the New Testament and I still recall the style by which she taught the course – it was practical, real and complemented my church lessons.

What was very significant here was the relationship between our behaviour and Jesus. Humility, love and service to the community were emphasized. We argued and debated on how we should behave in society as adults. All those trips that Jesus made in Israel and his encounter with myriad problems made me a daring young man. My visit to Israel with my wife Esther in 2010 reminded me of the lessons that Mrs Green had taught me in 1966–1969.

Humility was the overriding virtue that I learnt. I still read the New Testament with critical reference to the practical nature of Jesus' trip to Israel. Incidentally, I am writing this section of the book during the Easter Holiday (2013) having read many newspaper stories about the death and resurrection of Jesus Christ. I recall the fourteen steps at the Golgotha where he walked through as he was being heckled by the natives. The question I ask myself now is whether we emulate what Jesus did for mankind. What can we learn and construe individually or collectively from His humility?

One of the teachers who taught us Chemistry George Eshiwani, later became a Kenyatta University Vice-Chancellor, whom I worked with as his Deputy Principal in charge of Academic Affairs at JKUAT. He came to Kisii High School for teaching practice and left after about six months. He used to play hockey and coach students after classes. As his Deputy Principal, I had a lot of respect for him and enjoyed learning from him on how to handle staff and student problems. He had a lot of respect for me because I delivered on time any assignments he gave me.

It is him who first told me the famous phrase that managers use: Management by walking around (MBWA). I later perfected on it and, indeed, managed by walking around in my later administrative assignments. I took agriculture as one of my courses in Kisii High School. We were about twenty who opted for it as an examinable subject. The course was taught by the American Peace Corps in selected schools of Kenya. I was very good at the course as we covered practical areas. In addition to theory, we literally went to the farm and handled livestock, chicken, crops, fertilizers, machinery, pesticides, and indeed saw the effect of various experimental treatments. We were also privileged to get driving licenses for tractor and later motor vehicles. My love for agriculture and crop protection started from here.

I wish to narrate two incidences that happened during our learning interactions. An American Peace Corp member called Mr Bean came to class and found us making noise since we were not occupied. This was in 1969. He came to the front of the Machinery Building and shouted at us, "You stupid boys, stop making noise or I shoot you like Rev. Martin Luther King."

The remark shocked us because we were old enough to understand its racial undertones. We took him head on and marched off to the headmaster's office to report him. This man, Bean, was a young BSc holder from Virginia Polytechnic University who took us for granted. We caused havoc and other senior students in Form Five and Six joined us and demanded an apology or we could march to the Kisii District Commissioner's office. The headmaster summoned the teacher and his senior colleague, Mr Price, to explain the matter.

Martin Luther King had been shot dead in the USA and we were aware of racial uprising in that vast country. We read the papers in the library, followed the news and any slightest provocation related to racial undertones was taken head on. Kenya had also gained independence and we were therefore enlightened on the rights of black people here and in the USA. The headmaster persuaded us to calm down as the head of the Peace Corps apologized on behalf of Mr Bean.

Several other Americans present apologized profusely and explained to us the consequences of such utterances. Mrs Meredith, who was one of our History teachers was very concerned about the remark and during her lesson, she could cover the injustices suffered by black people in the USA. Later on in a related incident, I had to quell students who wanted to beat up a Japanese technician for manhandling a Mechanical Engineering student in the laboratory at JKUAT.

The young Japanese demonstrator asked the student to move away from where he was doing an experiment. The student hesitated a bit; and angered by the student's slow response, the Japanese fellow took a pail of used oil and poured it on the student's legs and shoes. The student's shoes, legs, trousers and socks were soaked with dark engine oil!

He came to my office dripping oil. I was angry and demanded to see the offender with his team leader. In my mind I was fully convinced that this silly act was racial in nature and the young man despised Africans. I could not condone that act. I felt sorry for the poor student and calmed him down. I will reflect on these academic injustices when I discuss the issues revolving on my experiences with donors and support to our universities during a time of crisis.

I recalled what the headmaster did and applied similar but a harsher decision on the offender than my headmaster at Kisii High School. I could not accept an apology but have the fellow go back to Japan. I told his team leader in no uncertain terms that the young man had no place in my institution. He had to go. My action did not affect the JICA support programmes; it was a lesson to other nationals who were involved in our projects. The matter ended as soon as he left the country.

All the other experts might have learnt from my action. I compensated the student and gave him money to buy the damaged shoes, socks and pants. I also counselled him and told him the many racial problems that exist around the world, citing South African Apartheid and the USA racial tensions, and even some countries in South America. My act hardened me as I faced more challenging decisions later in life.

School Detentions

Kisii High School had laws and regulations which governed students. The prefects executed them. The headmaster and his staff monitored their effectiveness. Weekends were rather free of the weekdays' rigour. Every Saturday, the headmaster, Mr Bowles, would inspect houses for cleanliness and award a trophy and a certificate at the end of the term to the best house. He would inspect students during assembly, check on uniforms, house floors, the veranda and cleanliness of beddings. I actually thought that was an excellent way to keep us smart. Government African School Kisii, as it was called then, was a renowned national school for good conduct. Teachers wanted to keep the good name. After several inspections and award of marks, the cleanest house and well-behaved boys got an impressive trophy. Ruri House received several such recognitions for most of the period I was there. Sameta used to be a runner-up despite its rowdy boys. Ruri, however, had more cunning and mischievous boys than any other dormitories

The role of prefects was to execute and monitor compliance with the rules. After the Saturday morning parade, boys would have lunch and leave at 2 pm for town, a distance of one mile down the road. The whole town would be dotted with Kisii High School boys, all in the white shirts and grey shorts uniform. It was a punishable crime to wear any other attire than a white shirt, a red striped tie and grey shorts or trousers at all times.

One could not go to town before 2 pm unless one had an exit sheet from the headmaster. Prefects were always on the alert to get offenders, and that is why they became unpopular. There were more than twelve prefects in any one given term, mainly those in Form Four and Form Six. They had special privileges both in the dining halls and halls of residence. They also had red-bonded blazers with a badge inscribed 'PREFECT' for easy identification.

They were not paid any money for their services but were highly respected by the teaching staff and students. These were the advisors (the kitchen cabinet) of the headmaster as they represented various classes and departments. Occasionally, however, some of the prefects would be unreasonable in detaining students or punishing them for mistakes which were trivial. When a boy was detained, he could not go out on Saturdays and would be assigned some manual work to do. There were all types of laborious work he could be assigned to do. I witnessed a few cases where a prefect was beaten up by a student because of unfair punishment. Such cases would be handled by the duty master or headmaster.

Expulsion of students was rare and summoning of parents to Kisii Hign School was not a common demand. Punishment, however, was always meted out to naughty boys. Students' behaviour in general was commendable. We never heard of or witnessed the current vices like drugs or alcohol consumption which are rampant in many secondary schools. We were actually well–disciplined young boys although there might have been isolated cases.

Entertainment

An all-boys' school had to find ways and means of keeping teachers and their students happy. Saturday afternoons were left free, either to go to town or remain in the compound to engage in extra-curricular activities. This included debating with other schools, games, public lectures, athletics, club meetings, church gatherings, choirs, and agricultural clubs and indoor games.

I recall one afternoon when my late uncle, Justice Nyarangi, who was himself an old boy of Kisii High School, came to see me at school. He was a resident magistrate at Kisii Law Courts and was to deliver a speech on the role of courts in Kenya. He called me after the talk, put me in his grey Volkswagen car and drove me around town. I liked the gesture and hoped to own a car like him. He used to talk a lot and told me to work hard and buy a car like the one he had. We had tea in town before he dropped me back to Ruri House where he had resided as a student several years before.

Occasionally, specific girls' schools could be invited for any of the social events like speech days. I recall Ng'iya and Kereri Girls schools visiting us on Saturday afternoons. The interaction period was so short (from 2 pm to 4 pm) that it was practically not feasible to develop any relationship. In any case, their matrons were so watchful that every movement out of the meeting venue was monitored. Their teachers used to shield off the girls from close proximity dancing. One could not be seen approaching the girls.

The most longed for occasion was Saturday night. Students looked forward to a good meal of rice and beef or beans, a hot cup of chocolate; less number of boys eats that night because some had been in town and had met their relatives who had fed them well. We allowed for a generous serving portion for the unlucky boys who may not have had relatives in town. Most important of all was the fact that as a kitchen and dining hall prefect, I allowed for extra serving as long as the food lasted. We apportioned the same food for cooking at all meals. I was popular for this generosity.

Saturday night was also a night of music and movies! Every boy, except certain sects of Christians, attended these events. We had an entertainment prefect, the DJ and selected records. The DJ would spin the then latest hits of the 1950s and 1960s. Many of the records were from British musicians like the Beetles. The bull dance, as it was known then, was so popular that teachers would join the dancing once in a while.

The best dancers then were the Luo, who exhibited the latest dancing swings learnt from Congolese popular musicians. I do not recall having many live bands in Kisii High School. The entertainment usually lasted from 8 pm to 11 pm and all systems stopped. There was a very low-key school band which was led by the head boy, Charles Odera, and assisted by Francis Ongegu Okengo, called 'The Thrashes'. It was in fact not that good and was never liked by the boys. It, however, filled in the free weekend time.

Movies were shown on alternate Saturdays or when they arrived from the UK. Mrs Green was in charge. She used to introduce the films and characters in such eloquent detail that by the time we started viewing them, we already knew every character and role in the movie. Many of these films were drawn from the literature we studied or the thrilling James Bond series. The James Bond – 007 – movies were so popular that many young boys, like John Abbot Nyanchoka, nick-named themselves James Bond and behaved like him in ordinary life. Movies like *Live and Let Die* were featured, and attracted unprecedented interest from the boys.

In addition to weekly movies, plays of Shakespeare were featured on Wednesday nights. All the literature books we used were adapted and acted by selected students. The plays, dances, films, extra-curricular activities made our days as boarders extremely interesting. Sports like rugby, athletics, basketball and volleyball were a thrill whenever we encountered schools like Bishop Cardinal Otunga Secondary School.

I recall that the world record 4 x 400 metre luminary, Robert Ouko, was a class ahead of me. In Rugby, the Kisii High School Sevens were celebrated all over East Africa. I was not good in any of the games, but a good cheering boy. All these activities made us behave well and they reduced the monotony. We studied hard and matured sensibly. We were a community. We were a well-disciplined lot for all the time I was in Kisii High School. There was no corporal punishment as witnessed in the primary schools. Maturity was setting in as we aged. The future plans were drawing near.

High School and Political Turmoil in the Country

One clear Saturday afternoon on 5 July 1969 we received some unbelievable news. It was a school games day and we had visitors who attended the meeting. I was in the dormitory after cheering our house participants. A fellow student walked to my bed where I was having a siesta and asked me if I had heard that Joseph Tom Mboya had been assassinated.

I got up, surprised and requested him to repeat the statement. He asked again if I knew that T.J. Mboya had been killed. I walked out of the hall to make further enquiry. Mr Mboya had been to Kisii High School the previous weekend and delivered a moving public lecture on Pan Africanism. He had spoken to us so eloquently that we were touched by our naivety regarding racial segregation. Also, Africa was one continent balkanized into tribal segments creating unnecessary antagonism. He further thundered that Kenya was one country but had been segmented into tribal cocoons which created undue animosity. This was an eye-opener to us young secondary school boys. To be told that he was now gone was one tall order to comprehend.

Soon after the confirmation from the radio news, games stopped and I could see students gathering in groups, talking in low tones. The rest was history.

Mboya's death was a blow to the nation as he was viewed as the potential future president of Kenya. There were major political repercussions soon after the announcements. I leant one thing: that everyone plans to win fame through the most uncouth manner. Some people will use every crude means necessary to advance themselves. All that one looks for are people's manoeuvres, their actions in the past and what one might expect in the future; you then realize that everyone is after power and to get it or keep it, one must destroy others. That is human nature and there seems to be no shame.

The students mourned the departure of a great statesman. I did not understand the whole reason for this inhuman act. It was only a few days before that we had been threatened by an agriculture teacher that he could shoot us like Martin Luther King. It started to dawn on me that killing of persons was not confined to one place but was worldwide. I recalled that there were many such shoot-outs only that we did not comprehend their motives and the consequences.

My final days in Kisii High School were eventful. We used to have house parties with our house masters. These get-togethers were great bonding activities which I later employed in my future administrative careers. Corporate bonding in many parts of the world is the current in-thing which has been perfected. During our house interactions, music, dancing, and food were plentiful. Final year students had special treats and were respected as they were now leaving the secondary school life for higher education or job placement.

The British teachers were leaving the country one by one as Kenyans took over from them. None of them was happy to leave the country. They looked sad due to the fact that their departure time was imminent. In fact, we students felt sympathetic towards the very good ones. They openly expressed their sadness at leaving Kisii High School, let alone Gusiiland, a place of natural beauty.

We bid them farewell one by one. The lucky ones were absorbed in the Directorate of Education, in the Ministry of Education, Nairobi. Others were posted to junior secondary schools while a good number opted to stay in Kenya and take up local citizenship. To date, I have witnessed various foreign nationals under the same predicament. Construction engineers, volunteers, NGO staff, technical experts and many more have fallen into this trap, never to return to their native countries and have opted to stay to perhaps get married here and become naturalized. Indeed, this country has its own unique beauty which many of us do not appreciate.

Reflections

My understanding of culture, people, behaviour, and future careers started to surface when I was a young boy. I learnt at an early stage that obedience pays; timely accomplishment of assignments provides other opportunities for one to do more. I also learnt that arguments wasted time if you knew the result of the debate. Creating harmony with others reduced anxiety and enabled us to accomplish more. I never wanted confrontations or a situation where I would be subjected to ridicule. I needed to be on the lead at all times where possible, whether in class, debates, errands and risks. I never shied away from difficult tasks. As a young person, I wanted to tell the truth and be told the same.

I considered all pupils the same and equal in my class no matter their age and status. That was why I could report a teacher's wife for punishment. This was innocent self-punishment, and suicidal at the worst. I thought I was doing great by being impartial in my monitoring duties. I assumed that everybody older than me was always right and deserved the highest respect. Hierarchy to me was an automatic command and had to be respected. I later learnt that it could be abused.

Leadership was accorded by others and I never asked for it, but was appointed to it. Humility overrides all other virtues, but being humble is not a reflection of weakness but respect. I had plenty of mentors and role models. My brothers, uncles and teachers all portrayed respectable images to emulate. I feared being the odd one out. They all appreciated good boys and girls and were ready to compliment us.

During my early years, I learnt that there is nothing more encouraging than praises and compliments. Just words like, "You are a nice boy", "keep up the good work", or "top of tops", a comment I once received from my English essay lady teacher, made me work even harder than the previous time. I recall the good and not-so-good remarks which made me improve my work. There were very constructive comments about my endeavours. Remarks that I rushed at issues and made hasty judgements still haunt me.

I indeed accepted the challenge but have continuously made every effort, over the years, to overcome it. I always tried to overcome the rushed decisions I made in order to reduce the impatience I had about the apparent slower thinkers. This is a handicap which I will always work on to be in line with others. I, however, find that quick action and risk-taking may be beneficial in certain situations.

Primary school days had their glorious moments. The mere fact that I could still recall virtually all my Standard Eight classmates gave me the power to identify and recall my future university students by name.

The harsh realities of class competition for position one in examinations surfaced later on in the survival of the fittest – Darwin's theory of natural competition. The actual long walking distance from Nyamagesa Village to Ibacho Primary School for a period of eight years became a norm to me and I never made any fuss about it. Today I can persevere to the ultimate in school work, field research, laboratory experiments and any other task which may need long hours of endurance to complete.

At an early stage we used to play and at the same time fight. I recall several times, especially end of term, when older boys would literally beat us before we parted ways. In fact we had some kind of gang of youths who protected one another during the end of term. In my case, I identified myself with some older boys who would protect me in case a fight broke out. But elders from the hills were always on alert and we reported the cantankerous boys for disciplinary action. Those small fights, however, hardened us for future survival. We developed survival tactics. Although I was a young leader, it gave me the courage to lead and make reasonable judgement. I had respect for school rules. I knew the hierarchy and my responsibilities were specific. My roles could not clash with those of the prefects.

My teenage life in Kisii High School was more of an exploratory maturing man compounded with heavy academic responsibilities. I realized very early that I had to work even harder to match up with my 100-odd classmates. I found sharper and more focused students from all over the country. I learnt to be humble, timely and obedient. My first year was challenging as I had to adjust to noisy dormitory life, scheduled food programmes, strict rules and regulations on aspects of boarding life. I, however, thoroughly enjoyed interacting and socializing with other communities. I had no choice but to be part of the Kisii School community. Old boys usually came to visit and give talks. The British teachers were regarded just like any other local staff and we had no inferiority complex.

We got used to their practice of strict deadlines. It was an advantage for me as I had always kept time. Stiff academic competition continued unabated. My belief was that I had to score high, a first grade and proceed to the next level or get a well-paying job. Kisii High School therefore succeeded in nurturing me socially and academically. I aimed as high as the other old boys. Great speakers like the late Tom Mboya inspired us. Great politicians whose names were being mentioned time and again became our absentee mentors even though we had never seen them. My tender age was slowly waning.

The debating society which I joined during my first year in school trained me to reason, think critically and have my facts at the finger tips. It also taught me to be tolerant to divergent views. I was further trained to be a good listener and give a chance to others to also be heard. This was the first place where I tolerated points of order from members, a speaker's power and all types of parliamentary procedures.

I must admit here that most of the members (99%) in the debating society were from the Luo community. I later learnt that the late Mzee Oginga Odinga and Tom Mboya had tremendous influence on the youths and on even older generations. I, however, enjoyed the debates, the voting, parliamentary procedure of crossing over and being in a majority or minority government. These debates would later assist me as a university administrator in terms of listening to everyone in case of disputes. I always tell graduate students to be tolerant to their colleagues whenever they present their research project.

My extra-curricular activities involved music, singing, SDA Group membership, plant identification and membership of the Young Farmers' Club. I participated in any other school activity that I could find.

I learnt one bitter lesson when the choir master discriminated against me. Fourteen students had prepared for the national music festival. This was through many hours of singing practice. I mean several hours. The choir pieces were one Luo song which I still sing now and an English one. Having spent virtually every evening singing in the halls, the choir master was told to select only twelve of us to travel to Kisumu, the provincial headquarters and hopefully to Nairobi for the ultimate musical festival.

To my utter dismay, the choir master, Mr Tom Oyieke, eliminated me and another boy! I actually saw blatant tribal discrimination because all the other singers were from one community. This was the very first glaring tribal bias against me. I dismissed the act, and later embarked on my school work. I never took up singing again. But, maybe, I can still do music.

However, I learnt one lesson from this: Indeed, tribal discrimination is a bad and destructive act. I made a major lifetime conviction that for every failure or disappointment, I would turn it around to boost me for better things. Negativity to me was a catalyst for greater heights. This has become my driving force despite any disappointments. A young person never forgets promises. They are debts which must be made good. Past memories of a young person never fade because they are a part of growth and development.

The developmental traits that I have listed are not exhaustive. There are many others like perseverance, confidence, and dedication which are essential. I did not change my habits and social life when I got to Kisii High School; rather I had to adjust to the new scenario. There were several boys in the school who had different philosophies of life. I slowly started realizing one important advice that our father kept on hammering into our heads as children.

He used to stress that the choices of friends we made at the time would determine our future lives. He further said that moral talents were God's gifts and intellectual talents were due to hard work. I started to realize the importance of choices in character, action, attitude and future career. I also realized that making any choice on any matter was a privilege and not a right. That choice could wreck or build me.

Our housemasters used to have special coffee meetings in the dormitories. Mr Green and his wife were our house heads. I took a religious study course taught by Mrs Green and was sure I could score a grade one in it. Indeed, I did attain a grade one in my final score. I liked the course because it exposed us to critical thinking as did the New Testament. I still debate the rationale of so many desert travels by the prophets and kings and question how they survived in those horrid deserts like the Negev!

The Greens would give us life skill tips to succeed in future. These have formed part of my life skills to date. Their talks enlightened me and made me aspire for more education after my fourth form. I kept on asking myself questions related to my future academic progression. Other than the usual character shaping, we were told time and again, I had my five personal traits that I adored. I called them five essentials to make me succeed in virtually any situation. I enumerate them as follows:

> First, I developed a positive attitude, which meant that everything I did was in good faith and would succeed in it. I knew that attitude was not necessarily taught in school, but depended on one's inner feelings. Positive attitude would always create a positive result and the converse was time. I therefore believed at this early stage that I would consider all things that I undertake positively and leisurely. In case I failed, I took that failure as a lesson.

> Second, I was a good speaker. I encouraged myself to be an efficient public speaker in any gathering; where possible to be able to speak clearly, forcefully, persuasively, and calmly in front of an audience. The size of the audience did not bother me. I would behave the same. This to me was one important skills I wanted to acquire which we were never taught. This trait would make me more comfortable, confident and saleable. Maybe, I would be able to sell any idea to anybody, be they products or ideologies. The skill would give me more opportunities for career advancement. I ended up delivering conference papers and speeches.

> Third is focus. Any success is dependent on effective action which in turn relies on the ability to focus one's attention where it is needed most. I believed in effective productivity habits with a strong sense of focus and discipline. There is no way one can stay on track without strict aim and vision. I knew one fact, either you manage yourself or get mismanaged.

> Fourth, self-discipline. This was a life trait which I considered most important. The ability for one to abstain from short-term leisure is so

difficult that we all become victims. School discipline and academic success were equal partners in that endeavour. I always paid special attention to instructions, especially those which had "dos and don'ts". I knew these could be short or long-term instructions and I decided to be a positive respondent on many of such directives. Self-discipline was very, very demanding. I convinced myself from primary school that discipline accompanied with humility and occasional smiles would get one out of trouble. I had always tried to be a disciplined individual wherever and whenever possible.

Fifth is time consciousness. This is a very critical factor which is unique. Many people take time for granted and forget that it goes by too quickly. I knew from the early ages that every second counts. This is a concern that I shall endeavour to explain in later chapters. Suffice it to say here that I never wanted to waste any time without either reading or doing something useful for myself. I knew that once a day went by, it could never be rewound. That date would be gone and gone forever. I put every minute into good use unless I was incapacitated. I considered day and night hours equally important in my work. Multi-tasking was my hobby.

I also had other additional beliefs. As a young boy, brought up in a Christian home, I always revered the Ten Commandments which we were taught in church and had to abide by them at all times. I made sure that I announced my Christian beliefs whenever I made a first encounter. I still do that today. I may not be the best of preachers, but my actions may tell it all. How did these values impact on me?

When I was about to complete Form Four, I kept on asking where I came from and what the future held for me. I convinced myself that the further I looked backwards, the further I was likely to look forward and succeed. This was a notion which was shared by many successful individuals in major successful corporations and business entrepreneurs. My parents were peasant farmers. Was joining them in that trade a worthwhile venture? The past shapes the future. I had to soldier on.

Goodbye to Kisii High School

The four joyous years that I spent in Kisii High School had to come to an end. There were obviously ups and downs during my days. I made friends, networked a lot and started to mature while studying. I bonded with the most hardworking boys and many cantankerous ones. I learnt how to keep clean shoes, neat stockings, a well-made bed, a clean mopped floor, shiny window panes and, most importantly, preparedness for house inspections and parades. These chores became part and parcel of my daily routine. As a prefect, however, I would be in charge of executing the said chores. Students loved my generosity in food rations and my liberality in meting out punishments. As a prefect, I did not make many enemies like some of my colleagues.

Time had come to have farewell parties and move on. We did our best in the final national examinations and had to wait for the results at home. The house masters organized elaborate parties for the leavers and made parting speeches. Unfortunately, we did not receive any presents when we left the compound because parents were not allowed to attend our parties. In any case, my parents would not have come. Parties were held the previous night before departure. We all packed our personal effects in the little wooden boxes, mainly the exercise books which we had accumulated over the four years.

I believed in one philosophy of life: Success breeds success. If I performed well in my Form Four level examinations, I would then be able to proceed ahead academically. In fact this was a major determinant of one's future career. I searched for a twenty shilling note bill which I had carefully tucked in my Bible for any emergency, found it and headed to the bus station on foot with my wooden box carried aloft.

The whole Kisii town bus station was filled with Kisii High School students proceeding home after closure. We were still in school uniforms. At this point in time, I started recalling the Darwinian Theory, the survival of the fittest, which we had learnt in the Biology class. There were very few buses to rural areas and those who pushed and shouted hardest got a chance to travel. It took me a while to get transport because of the scarcity of buses; those which were available were filled to capacity.

My ability to jostle for space was wanting. We had now joined the larger, unruly, disorganized community. I still behaved like a gentleman but realized that this would not get me far. I had to change tactics and jostle for space like the rest. I finally secured a seat to Nyamagesa Village via Keroka market.

This was at the end of November 1969 when the rainy season was at its peak. The road that bisected Nyamagesa is one mile away from my home and I had to climb up a steep wet hill to get there. It rained on me and I got wet, tired and exhausted, but happily got home at about 6 pm.

I had left my home a novice and returned there a mature boy. My parents were happy to see me and offered me some local porridge made of finger millet to welcome me home. This was another milestone in my life. However, I was pondering what would happen next after the results.

Note

1. They included James Nyarunda, James Ongwae, the Governor of Kisii County, Benson Kangwana, George Omolo, Bornaventure Wendo, Elly Otieno. Osmerah, Charles Odera, Sospeter Arasa Nyagwansa, Samuel Kiplagat, John Wangai among others. I occasionally run into these gentlemen and share the events of old days. Mr Ongwae was duly elected as Governor, Kisii County during the 4 March 2013 General Elections in Kenya.

4

First Job Placement

After a brief stay with my parents in Nyamagesa, it became clear that I had to move on and seek employment. Availability of jobs in the 1960s and 1970s was not a problem. One had to perform well in the examinations and have a good attitude to secure a job. My parents and senior brothers advised me to go to Nairobi and join my older brother, Hezron Tirimba (deceased), who was already working there.

He was a very shrewd, risk-taking business entrepreneur. I say this because at his very early working life, he had already bought a house in Buruburu and was capable of accommodating several of us when we visited him in Nairobi. I, therefore, had a domicile. He was not married when I joined him for a job search. He was very happy that I joined him and would possibly get a job and share the payment of his loan for the house. With my arrival, he was also assured of security and company, not mentioning ready prepared meals when he got home.

I joined him just before Christmas of 1969, and spent the holidays with him and several relatives who were staying in other estates of Nairobi. They were all working in various organizations, but would meet during the weekends and share the latest news from home or just spend time together as family members.

I started applying for jobs in private organizations. The first application I made was to Barclays Bank, Queensway. My brother was working with the Ministry of Education and he advised me that banks were paying much better than the government. I took his advice and it was indeed true.

As I was waiting for the feedback from the bank, our results came out around January 1970 and I had scored a second class division in the Form Four examinations. I actually lacked one mark to get a first division. The results were encouraging since I qualified to apply for any jobs in the country or join Form Five which I did not. I was happy, I had attained the grade.

As I applied for the post at the bank, I had also put other applications for universities abroad. Barclays Bank called me for an interview in its Queensway Branch. We were thirty in number. We did both written and oral interviews in

Mathematics and English. A few of us passed and qualified for the posts. It took one week to get feedback. I later learnt that only five of us qualified out of the group of thirty. I did not know anybody in the group. There were two girls and three boys who joined the Queensway branch of the bank.

I was posted to the Intelligence Department which I later discovered was handling loan cases and creditworthy customers. We were four in the department, a pleasant and friendly head of section called Gurmit Singh, one Joel Wainaina and a lady secretary of Asian extraction. I was a fresh trainee from high school. Our duty was to recommend or refuse loan applicants after carrying out a thorough and credible investigation on individual customers' credit-worthiness. I enjoyed doing the turn-over and learnt basic lending criteria in banks. My brother 'H.T' as we called him was very happy that I had secured a well-paying job in the bank. Indeed, the salary was much higher than what he was getting in government. This was good news to friends and relatives. Banks were considered good employers with generous benefits to their employees.

A large proportion of the staff in Barclays Bank were British nationals. The headquarters where I joined was almost 95 percent British staff then. There were a few Asians and Kenyan nationals. Despite having gained independence, absorption of locals had not been fully implemented. Private companies owned by foreigners were still dominated by the British.

My immediate boss, Gurmit Singh liked my work. He relied on me and Joel Wainaina to source information from other banks and produce reports for onward action by one Mr Bird. Mr Bird was the bank's ultimate decision-maker in our department. We rarely interacted with him. I recall one morning when he directly called me on my extension line and demanded that I see him downstairs. He sounded angry and was clearly pursuing something. I first wondered why he called me directly rather than call my boss Mr Singh who was present.

I walked downstairs gently, knocked at his office door and went in. He requested me to sit down and pulled out a pack of turn-over papers which I had worked on for a customer. I thought he was going to commend me for a job well done. He told me point-blank that the negative recommendation I had given to a client was not proper as he personally knew that applicant. I replied that the client's financial standing could not qualify him for a very large loan that he requested. I further told him that I applied the rules as prescribed by the bank. Mr Bird, a short, stocky man thundered to me that I go upstairs and recast the figures to read positive and award the loan. He handed over the papers to me. I looked at them momentarily and walked out of his office. I reflected back at my primary school days when I got a beating for reporting the late-coming of a teacher's wife. Was history repeating itself?

I had, indeed, done the necessary work, given my verdict. Little did I know that other powers would overrule documented results! I went upstairs, relooked at the documents, briefed Gurmit Singh and wanted him to handle the case. I recast the figures but they still turned out negative. I then told Singh to handle the case in its entirety. I learnt later on that the applicant got the loan, had problems repaying it and court proceedings were instituted to make good for the loan he had borrowed. My technical advice had been, however, ignored.

At one time I thought of reporting the incident to Mr Peter Nyakiamo, who was the first African to be appointed a bank branch manager in Kenya. He was therefore my mentor, and I looked upon him as my advisor. Bank managers at that time commanded a lot of respect and the whites respected him because he was a workaholic. Would he have protected me from Mr Bird? I knew he could not! He probably had enough headaches with the expatriate-dominated bank. In any case, the then young Africans working there considered him a loner and kept aloof from his community primarily because he did not assist in the negotiations of their salaries and other benefits.

During that time, as alluded earlier, most of the bank workers were either British or Asian. Peter Nyakiamo was among the very few Kenyans who had risen to a managerial level of a foreign-owned bank in Kenya. My instincts then were that for an African to rise to that level in a racial working environment, he must have been exceptional and hardworking. I therefore wanted to emulate him as my mentor. He later became a Minister in Former President Moi's government.

My late brother, Samuel Mose, had worked with him in the Kisii branch of Barclays Bank and used to tell us about Peter Nyakiamo, the good gentleman from the Suba community. He trained my brother and they became great friends. It was a coincidence that I had a chance to work with him in the Queensway Branch, the Head Office, Nairobi, several years later. He liked me and wanted me to take up banking as my career.

I started seeing injustices and recalled my primary school case of a teacher's wife coming late to class and not being punished. I also recalled my high school experiences of being trustworthy and fair to all. But I further recalled George Orwell's book, *Animal Farm*, which states that "all animals are equal but some animals are more equal than others". This was the typical colonial mentality then. I was never surprised later on to learn that the gentleman was told to leave the country after the expiry of his tenure in the bank.

My stay in the bank was not long. I had been trained in several departments and had become competent to work in any of them. My departure was a loss to the bank. New employees had to be taken to Limuru Barclays Bank Training Centre for bonding, training and induction. I remember a group of twenty trainees from the Nairobi branches assembling there for the week's residential training. We were being inducted to life examples and real exercises encountered in all banks. The trainers were older, experienced bankers, managers and senior accountants who had worked for longer periods than us.

Limuru, which is about thirty kilometres north-west of Nairobi, is one of the coldest places in Kenya. Temperatures would drop to 2°C in the cold months of July to September. I recall the very cold nights we experienced despite small fires lit in our rooms for warmth. Torrential rains would make the place even more chilly than normal. I thoroughly enjoyed the training, networked with other branch staff and shared our varied high school experiences. We were being prepared to take over strategic positions like managers, senior accountants and section heads as the banks were systematically replacing British nationals who were quitting their posts and had opted to reject Kenyan citizenship for the British one.

My close friends later became managers and senior bankers not only in Barclays, but other banks as well. I had the opportunity to network with all the local banks because of the nature of my work. We, the local workers, had issues with top management – despite good competitive salaries. We had a bankers' union which negotiated for better and equitable terms with the British and other nationals.

My typical working days and hours were precise. My brother and I could get up in the morning, have breakfast and by 7 am board a Kenya Bus to downtown Nairobi, a distance of about ten kilometres. There was no big transport hassle then. As people increased in the city, transport became a nightmare, which is evident to this day.

In the 1970s, however, the prices of commodities were lower than the subsequent years. We used to have a well-balanced lunch in town and nice dinner at home. Suffice it to say that the cost of living, transport, and other commodities kept on increasing every year. The current city of Nairobi is nowhere comparable to the Nairobi of 1960s and 1970s.

I did not stay in the bank long enough to utilize my training expertise that I had gained from Limuru Barclays Bank Conference Centre. I got a scholarship which was secured by my elder brother, Joel Onami, who was studying in the USA, to go for further education. I had applied to a few universities in the USA and UK but got admitted to Rutgers University in Brunswick, New Jersey.

It was now a family decision for me to either continue working with the bank or leave for further studies. My brother H.T. was of the opinion that I should stick around and develop a banking career. My other family members thought it wise for me to quit and seek further education. The two groups were both right. It was my turn and privilege to decide.

I said earlier that it was a privilege to make decisions and choices. It was my turn to think. Any decisions I would make had a life-time implication on my future career progression. I returned to Nyamagesa Village to talk to my parents, younger brother Amenya and sister Grace. We agreed that I would quit the bank and proceed for higher education. Adequate consultations were made on this matter with the final decision being mine.

One thing I learnt in the bank was that the employer sucks workers' blood. It was work and hard work at all time. Cashiers and tellers had to be alert at all times, serve customers diligently and be accurate to avoid discrepancies in their transactions. They arrived early and left late to ensure that the day's transactions balanced. Every minute counts when it comes to financial management and profit-making. Yet the pay was not commensurate with the work.

I returned to Nairobi and told my relatives and brother that I had decided to go for further studies in agriculture and environmental sciences. I went to my section head, Mr Singh, and told him the same. At my age, I considered myself young not to make banking my ultimate career. All my other colleagues were proceeding to higher education in UK, USA, Australia, India, and other countries. My brothers had also left for higher education abroad. It was clear that graduates did better socio-economically than Form Four or Form Six leavers. One stagnated career-wise if one did not progress academically. It is interesting to note that several people are now yearning for a second and third degree merely for promotions, prestige and higher earnings. That is the trend now in Kenya and many other countries.

My decision to quit the bank and proceed for higher education was a wise one. My hard work at Kisii High School and study technique there would pay dividends later in my university. The leadership traits and persistence added value to my future academic and research plans.

I wrote a formal resignation letter through my boss, Gurmit Singh, in which I explained why I wanted to quit. He was not amused. He had trusted and liked my work. He tried to persuade me to stay, but I kindly refused and told him that education took the first priority in a young nation like Kenya. The country needed expertise at all levels of development. He had no choice but to forward my letter to the human resources department. When my colleagues heard that I was leaving, they were not happy with the management. I mentioned to them that Barclays made immense profits through their dedicated work, and that they needed much more pay than what they were being given.

Later on, I learnt that there was a major bank strike which saw tremendous salary increase and other benefits. I kept in contact with my former colleagues but learnt that my section head, Mr G. Singh, had also left for the UK for some other assignment but not for the bank. Others also left for better opportunities.

My one-year stint in the bank was an eye opener. I was trained on various roles in public relations, confidentiality, book-keeping, business acumen, and most importantly, trust and confidence. I added these attributes to my earlier list during my high school days. I convinced everybody that I would study, make a career in agriculture and environment, with special emphasis on weed science and return home to work.

I had heard a lot about students who went abroad but never returned. My brother, Onami, saw my desire to continue with education and gave me support to accomplish my goals. Meanwhile, my colleagues at the bank formed a strong union to fight for their rights. The Kenya Bankers Association, of which I was one of the initiators, was registered soon after my departure to negotiate for better terms and conditions of service. I had made friends with many of the staff and leaving them was saddening.

My Exit from Barclays Bank

Queensway Branch was considered the elite and headquarters of Barclays Bank. All the printing work was done there and decisions which affected other branches were carried out here. It is in the Central Business District of Nairobi. Staff who were posted there considered themselves more privileged than those in other branches.

I had an opportunity to say goodbye to them with a promise that I would return. We exchanged pleasantries, remembered the days when we trained in Limuru and agreed to push for better terms of service. A few of my customers were also invited to bid me farewell. I recall one staff telling me that she could also follow the same trend that I took, and go for further studies. I learnt later that Jessica left the bank for further studies in the UK.

I left Queensway Branch with a small balance in my account. Most of it was utilized for ledger fee. I never got a penny from that meagre saving! It was a great pleasure to work with this prestigious bank in a section which had a lot of implications for Kenya's future industrialization agenda. Business tycoons used to borrow money from the bank for capital and other developments.

Although I did not make any tangible mark in my first formal employment, I made friends, clashed with colonial remnants, planned my time well, made far-reaching financial implications on customers, displayed my work ethics and finally learnt how to manage my finances. I matured a little bit while working in this bank and had the privilege of decision-making. It was a worthwhile engagement which tested my responsibilities at my developmental stage. This was not a wasted effort.

The short experience would later on manifest in my future managerial assignments. I left the bank in December 1970. I worked for 16 days in December before I flew out to the USA and was able to earn my full month's salary before my departure. That made approximately one year of service.

I maintained my bank account in Queensway which had no balance but later opened an account in Westlands Branch. I occasionally visit the midtown branch and admire the same old wooden parquet floor upstairs where I worked. This was my first official job after my secondary school and I very much enjoyed working with the group of people of three races, the British, Asians and fellow Africans.

The flight to the United States of America

All paper work for my trip had been completed by my brother, Joel Onami, who was in New York then. He sent an admission letter, an air ticket and details of how to get to him once I arrived in New York City. I got my passport and visa, bought a few clothing with my last salary and booked the travel date.

There was no internet or cell phone then for fast communication. We communicated through surface mail and the occasional trunk call. I confirmed the travel dates using the Trans World Airlines (TWA). The airline had offices in Nairobi and flew direct to John F. Kennedy (JFK) Airport, with stops in Murtala Mohammed Airport, Lagos, Nigeria and Monrovia, Liberia. This was an excellent flight since I did not have to change planes en route. I had not been in any aircraft before. You can imagine my anxiety!

I had been told of heights, speeds, turbulence, vomiting, toilets, clouds and the blue skies. This was exciting for a young boy from rural Kisii, travelling abroad. I neither feared heights nor aircraft speeds. My brother H.T. made sure that all the paper work was in order. He and some of our friends took me to the airport. My sister, Grace Kemuma, and younger brother, Amenya, also came to see me off at the Embakasi Airport. We took a Kenya Bus to Airport and we talked all the way.

My family members bade me goodbye and moved to the visitors' waving bay where they had a full view of the aircraft I boarded. They waved to me from the waving bay as I climbed the stairs of the airplane. I reciprocated. We were many in the plane. This was a day flight which was scheduled to leave at 4 pm with a first stop in Lagos, Nigeria. I had now joined another world, another community, another class of people I did not know. I did not know how to behave with the many passengers aboard.

The airplane's huge doors were firmly locked after all the passengers got in. The flight crew guided us to the seats. I had a window seat and the normal flight formalities started. The medium of communication through the inbuilt address system was in English. Occasionally, I could not follow the instructions but would ask. The passengers were mainly from the USA and West African countries.

I sat next to a white American couple whom I later learnt were from Denver, Colorado, the Rocky Mountain State, as they told me. Good personality pays. I struck a conversation with them and discovered they had come to Kenya as tourists and for game hunting.

They later told me several bad and good things about the USA. They hinted on racial discrimination and were frank in telling me that I could encounter its ugly face in some parts of the US. I quickly recalled the Kisii High School shooting threat by one American Peace Corp. I learnt a lot from our interaction. When they leant that I was destined for New York City, they even narrated to me in detail the good and the bad of the city.

New York City is considered one of the most sophisticated cities in the world in terms of people, class, affluence, entertainment, transportation and socio-economics. It compares with other major world cities like London, Tokyo, Johannesburg, Buenos Aires, Berlin and even Canberra in terms of lifestyles. That was the picture that was depicted of this great city. Mr and Mrs Meredith first told me that they actually had never been in NYC, but had heard good and terrible stories about it. It was a great city to visit but not to live in, they quipped.

New York City is fast and the people there are pushy, they stated. It has a population of about 12.0 million people. Many are rough and do not care about anybody or anything. They survive on fast foods and would do anything to make a dollar. There are many socio-economic problems which affect the black people and unlucky white ones. So I was to watch out for the company that I chose.

They, however, said that they had heard good things about the city. The city was the greatest in terms of job availability, movies, theatres on Broadway Avenue and great players. Times Square was in the heart of Manhattan. It was one of the busiest sections of NYC. It housed all types of commercial, social and transport systems/amenities. Very well-to-do people came here for various reasons. It had many great universities with world reputation. Many people longed to spend their summer time here because of numerous forms of entertainment.

It was home to world sports, boxing, world re-known performers and, most importantly, it had all nationalities from all over the world. It also had the best transportation system by bus, yellow taxis and subways. All world celebrities like movie stars, musicians, actors, ended up here. It was one of the most expensive cities in the world. Mr Meredith concluded his long talk by advising me to be careful about whom I chose as close friends once I joined the university with many other international students. In the past, New York City had a negative reputation due to robbery and drugs which were always highlighted in the media. It was at this point that the residents decided to create a positive image and called it the Big Apple. Positive praises began to promote its image, and this worked. Ultimely, the city had become the world's trading destination.

I was grateful for the free orientation, information and advice. By the time we finished our interaction, we were almost landing at Murtala Mohammed Airport in Lagos, Nigeria. We had dinner and beverages as we talked. The stopover in Murtala Mohammed Airport was for 45 minutes and we were not allowed to get out of the plane. I stretched a bit inside, peeped through the window, and saw the beautiful Lagos skyline. I had earlier watched the landscape and Atlantic Ocean from my window seat. The route from Nairobi to Lagos cut across several tropical countries of Africa and flew over Lake Victoria, the second largest lake in the world.

I was able to view the dense tropical forests of the Congo, Cameroon, and Central African Republic. I confirmed most of what I had learnt during geography lessons. Luckily, it was not cloudy during the flight and we gained time going westwards. The pilot kept on briefing us on the major landmarks as we got into each country's airspace.

The Nigerians and some other nationals disembarked at Murtala Mohammed Airport and a few more people joined us. It was now getting darker and the flight from Lagos to Monrovia was at night. What was amazing to me was that the sun-downer appeared never to set like in my home country where there would be sudden nightfall. I learnt that we were gaining time as we proceeded further westwards. The plane touched down in Monrovia at about 8 pm and we had another 45 minutes stopover. I do not recall seeing anything great here, except a small airport with a few light posts. The pilots updated us all along the flight routes except when passengers were asleep.

New York City was to be our next landing point after Monrovia, Liberia. We took off at about 9 pm and had to fly over the Atlantic Ocean for nine non-stop hours to John Fitzgerald Kennedy (JFK) Airport. This was the longest leg over the Atlantic Ocean. The captain told us just that and warned of possible turbulence. He talked of inter-tropical convergence zone as the main cause for the possible unstable flight. We were therefore advised to have our seatbelts on at all times while seated and avoid moving about in the cabin. Again, my geography lessons became useful when I heard of the tropical convergence zone.

This leg of the journey was long and many of us fell asleep aboard. There was, indeed, fierce turbulence midway the ocean. I must admit that they were rough and scary. We also encountered quite a number of these air pockets which made our flight uncomfortable. I cannot recall what I saw, but one continuous blue sky with a full moon and stars afar. I had a window seat throughout my trip and could see the outside outlines.

It was one boring continuous trip. As daybreak approached, we saw some islands as we approached our final destination, NYC. We landed safely at John F. Kennedy International Airport at about 1 pm. My friends had to connect to a local flight for Denver, Colorado. We exchanged addresses and they gave me their phone numbers for future contact. We bade each other goodbye. We all disembarked and had to clear as we went through immigration formalities. I hoped that my brother, Onami, would be waiting for me as earlier promised. I cleared through customs and immigration, collected my small luggage which was full of Kenyan dailies, The Standard, Nation and Taifaleo newspapers for my brother and his close friend, the late Joseph Magucha.

I stepped out on a cold, windy afternoon on December 1970. I had been warned of the cold weather. Luckily, my brother had brought a heavy, warm winter coat for me. They received me at the exit and we left the airport to his house in the Bronx. I looked outside and noticed graders scrapping snow from the highway. Joel and Magucha were excited to see a young Kenyan boy who would join them in the USA to study.

They were very inquisitive about the political situation and the latest happenings in Kenya. We kept on talking as we drove through traffic, but little did they know that I was more interested in seeing the landscape of NYC, the snow, buses, cars, than listening to their questions. My concentration was on sizing up the skyscrapers. I was sleepy but had to open my eyes and talk to two excited gentlemen. I compared the New York skyscrapers to the tall buildings in Nairobi; the comparison was not fair.

We arrived at my brother's apartment, 2011 Morris Avenue Apartment 7 C in the Bronx, one of the New York City boroughs (districts). Eleese, my brother's wife, welcomed me and I was sure she must have wondered how I survived the long trip from Kenya, East Africa, where she had never been before. We talked briefly.

My young nephew Terrence Mogaka was only two years old and we could not communicate. He was a young playful boy, who later became my best friend, associate and entertainer. He stared at me each time I spoke.

My sister-in-law was expecting her second baby, Thandi Moraa, and so she was careful regarding what she ate and how she carried herself in the apartment. She received me with respect and love as the young brother-in-law from Kenya. Her respect and admiration for my hard work and independence remain to this day. She must have had a brief from Onami. I therefore had a comfortable stay and became very helpful indeed as I minded my young nephew and later my new-born niece, Thandi Moraa. The two became my long-term buddies as I always minded them, walked them to a nearby kindergarten and we watched cartoons together.

I was not well-versed with the political events of Kenya though Joel, Magucha, his late wife Mary Nyaboke, and brother-in-law, Evans Keengwe, expected me to be a political expert. They seemed to be more informed in this matter than me despite their long stay in the USA. How did the group expect a young high school leaver to be so politically charged that I would narrate the day-to-day events concerning President Jomo Kenyatta, who later died in August 1978? As a local boy in Nairobi, all I did was to buy the dailies then read them casually. There might have been several political problems then, but I could not decipher the worst from the best in the government.

All I knew then was that we had a president who headed the Kenya African National Union (KANU), several members of parliament, and a very strong opposition led by the late Jaramongi Oginga Odinga. He was the Chairman of Kenya African Democratic Union (KADU). The two parties were so powerful that the country was divided alongside party beliefs. One party, KANU, was affiliated to the western bloc and the other, KADU, to the Eastern bloc. More important at that time, also, was the fact that Kenya was an independent and sovereign state. Period. That was all I cared about.

My brother wanted me to explain in detail the latest happenings in East Africa and Africa as a whole. Those were the days when names like Kenneth Kaunda, Julius Nyerere, Milton Obote, Jomo Kenyatta, Haile Sellasie, Kwame Nkrumah Mobutu Sese Seko, Muammar Gaddafi, Annuar Sadat, Joshua Nkomo, Samora Machel were household names on the African continent. I was aware of the struggle for independence, Pan-Africanism and the various movements across Africa. I knew that several African countries were getting their independence at different times.

I learnt one thing which remains a fact today: the Diaspora considered themselves more politically knowledgeable than those who resided on the African continent. They occasionally despised information of the local residents of their countries of origin. They considered themselves such know-it-alls that they could judge from afar and prescribe the best solutions for a country. I thought they had plenty of time and meeting venues to assemble and analyse issues, which was a fact. There was nothing wrong with their ability to critique the developments in a given country, but they were also obliged to join their brethren in building their nations by giving tangible solutions.

They used to be African students gathering in several cities attacking various African presidents for their wrong deeds. But they would hardly come up with concrete suggestions as to the way forward!

As mentioned earlier, I arrived in the USA in winter, December 1970. I was told that it was one of the coldest months of the season. February was equally cold with freezing temperatures. I was further informed by my friends that March came like a lion and faded away like a lamb. It literally meant that March began with fierce snow storms, cold wind drifts and torrential rains. But it slowly calmed down as spring set in. I suffered the harsh, cold snow drifts whenever I was out. I will always recall this statement which was made by one of my colleagues as he was making us prepare for a harsher March month. He was also preparing us to acclimatize to the new environment.

After a few months of my stay in New York, I was advised to seek a part-time job as I waited to start college. Universities in the US commence in late August or September. A few accept students in March. I therefore had ample time to catch up on the slangs and twangs of the American accent. I must admit that I had difficulties following their accent, and I still do. They also had difficulties following my African accent; I imagine they still do. They actually thought that we were talking like the British toys that they used to watch on television. Perhaps that is why they would take keen interest in the British accent from foreign students. Anyway, we simply tolerated each other when it came to the wide differences in accents.

I was fast in learning directions, roads, avenues and blocks. I could walk my nephew and niece around several blocks and get them home safely. I slowly gained courage to walk alone to the Bronx Zoo or to make telephone calls to a few familiar contacts. One thing I was told and advised against was being mugged or robbed by some street boys. I could be hurt in the process. I was advised to keep away from rowdy crowds and watch out for any strangers who would attempt to confront me.

The boroughs of Bronx, Manhattan and Brooklyn were the most notorious in drug peddling and robbery; I was advised. But this is true in all major world cities. In fact New York City was an excellent place to live in once someone got used to it.

I recalled my friends' lecture on the plane: New Yorkers are Pushers. The pace of work was just too fast for an average *rural* Kenyan. New Yorkers got agitated if one kept on saying "pardon me?". You were supposed to understand and follow conversations instantly. Period! Nobody wasted time on anybody. It was, indeed, survival of those who hassled most. I had no choice but to be fast enough and keep time and pace. My high school experience regarding time and adhering to deadlines started to bear fruits. I had to start preparing to begin college.

I was admitted at Rutgers University, New Brunswick, New Jersey, about 150 kilometres south of New York City. The transportation system in the USA is of high class, and the traffic laws work. My sister-in-law, Eleese, coached me on how to behave as a college student. They purchased for me the necessary personal effects and I was booked in one of the halls of residence in the campus. I shared a room with a first-year white student from Queens, one of the five boroughs of NYC.

After about six months living with my brother and his family, it was time to call it quits and start college life alone. I was now aware of my expectations, reactions to racial differences, keeping time and, most importantly, being on the good side of the law. I bought a meal card for use. Eateries were available at every corner or store one went to.

Junk food was and still is the most popular choice. I secured a part-time job as a receptionist in the Students' Centre. It was interesting to see the reactions of other students who found a typical African boy on a counter serving them. I worked for four hours daily, three days a week, during my free class periods. I was fast in networking and I got to know several students who became very helpful to me, and we actually formed study groups. They also assisted me in adjusting to the college and American life as I started to settle in Rutgers University.

5

University Education and the Foundations of Academic Leadership

I joined Rutgers University, College of Agriculture and Environmental Sciences (CAES) in 1970. As new intakes, we were all welcomed, given orientation tours and made to listen to pretty speeches by a number of senior students and administrators. I recall vividly being introduced to my academic advisor, Tom Concanon. He was the equivalent of a deputy registrar in Kenya's university system. Mr Concanon was my advisor for all the undergraduate courses that I took until I graduated. This is not the case in our universities, hence poor student-staff relationships and poor responses to the alumni calls.

Our campus, College of Agriculture and Environmental Sciences (CAES) was adjacent to a girls' boarding college, Douglas. One could not tell the boundaries of the two campuses. The main campus was spread over New Brunswick town, a university city, which relied almost entirely on the student population for business.

My advisor, Tom, gave me a very detailed talk on the academic courses to take, life at the campus, dining halls and, most importantly, my conduct as a foreign student. He was going to be my academic advisor for all my undergraduate courses. He had over fifty students to advise and always had time for us. Many of our universities lack this service. This may be part of the cause of the falling academic standards.

We became such good friends that my social or academic issues were readily tackled by him. In addition to this service, all foreign students were again introduced to the International Students' Centre. The centre handled issues concerning all foreign students from all over the world and we used to have monthly meetings in addition to special tea parties for networking. We spent most of our weekends hanging out at the International House.

My academic dean was basically involved in course advice and selection. He would also hold evening get-togethers for us mostly at the end of the semesters, holidays and long weekends. This interaction made life easier for the foreign students who had no families to visit during such breaks. This kind of arrangement is not available in most African universities. It was therefore easy to settle at Rutgers than in NYC because of student interactions. This was one of several extra-curricular activities.

I chose first-year courses which were common for all first years. They included Chemistry, Biology, Calculus, Physics and a History unit. We formed study groups and assisted each other in homework and practicals. I was familiar with group study and I had no difficulties adjusting to the exercise. The halls of residence were bigger and were fitted with larger cubicles than those in Kisii High School. The food we got in the dining halls constituted of different menus depending on the day of the week. I adjusted to the menu and strictly followed the timetable.

First-year courses were conducted in Douglas Campus by staff from both colleges. Some lecture halls were so big that lecturers were forced to use public address systems and mounted wall screens to deliver lectures. There were very well-arranged tutorials which were conducted by graduate students, mostly those pursuing their Masters and PhD degrees. I remember that close-circuit television sets were used for lectures. This was in the 1970s when these provisions were considered very advanced. The main duties of tutorial fellows were to review what the main professors had taught earlier in the day or week and add value to the lectures. Students were divided into groups of 20 or 30 at most. We gained a lot from this arrangement which was compulsory for all students.

Although this teaching arrangement also obtained in our local universities, it did not work due to lack of tutorial fellows. The difference in delivery of lectures in Kenyan universities and that in the US was the manner of conduct. The system of conducting tutorials by engaging graduate students could not be overemphasized. There was a semblance of the same in the early 1980s but the practice did not continue. This system trained young professionals early, as they progressed in the pursuit of their higher degrees.

If there were any academic gains I made during my undergraduate courses, it was when I was taught by a tutorial fellow. They reviewed what was taught by their professors and allowed for free interactions and critical thinking which lacked during large lecture meetings. I was able to ask questions because of the small numbers of my group. We were normally 20 to 30 in number. Under normal circumstances, a typical class could have 150 to 300 or more students in a lecture hall. Marking of examinations would be a nightmare for lecturers just like it is in Kenya.

In my former university, and I am sure in others in the US and the UK as well, tutorial fellows assisted in conducting practical lessons. The scenario in Kenyan universities is different. The numbers are high, lecturers are overworked and there are no tutorial fellows to assist. Few practical lessons are conducted as prescribed in the syllabus and students rarely have an opportunity to interrogate their professors. There is little, if any, interaction between staff and students during the delivery of a lecture. This is the exact opposite in some Western universities. The situation in our Kenyan and many African universities is hampered by lack of finances and soaring numbers of students. We adequately plan for the future generations. Students therefore attend lectures and want to get out of the lecture halls as soon the lecture period is over. Many do not show up altogether.

The Kenyan students, therefore, develop other ways of bonding and not through tutorials. They meet in games, dances, churches, mosques, choirs and other campus activities. This is great but there is the need to meet in subject clusters as this will improve the quality of education. My undergraduate and graduate training exposed me to life-long learning experience. I knew how to relate to my peers engaged in scholarly discourses, and appreciated other people's views during academic debates. We bonded well and made long-lasting linkages through this arrangement.

I recall that post-graduate students were provided with study cubicles which were labelled with their names. We also had consultation hours visibly placed on the doors for easy accessibility to your supervisors or student colleagues. We regrettably lack this kind of arrangement in most Africa universities. I hope that someday, these anomalies will be addressed to allow for more peer consultations!

My first year was mainly an academic exploration. I had no problems with class work since I started scoring good grades. My classes had very few black students, less than 0.5 per cent. In some courses, I could be the only black African student. Rutgers University during my time was dominated primarily by white students and a few African-Americans.

I experienced an incident which evoked some ugly memories. I took an Animal Science course with a Ghanaian student, Charles Mensah. The lecturer, whom I recall very well, Dr Ralph Mitchell, taught the course. We were about 100 students. During the final-term examinations, Charles and I scored the highest score of over 70, an 'A' mark.

The lecturer had the audacity to lower both our grades to 68 marks so that we did not get a grade 'A' in our transcripts. The answers were displayed and it was evident we got most of them right. Charles and I took him on and argued for our rightful grade. Charles was very vocal and I joined him. The Dean of the College finally intervened and we got our correct grades. Only two of us were blacks in the class and we considered this act racist. Charles continued taking courses in Animal Sciences, while I started specializing in Plant Sciences.

There were other small racial incidents that I encountered, but no so pronounced since human rights activists had created enough awareness on the rights of black people. From what I was told, the State of New Jersey was not affected by major racial uprisings like other southern state. However, there were pockets of racial discrimination.

I resolved to concentrate on my studies and make good grades to allow me join graduate school. At this point, I had set my mind on how to attain good marks, strive for a Master's Degree and finally attain a PhD. My undergraduate courses were more biased towards Plant and Environmental Sciences than Arts or Animal Science courses. The courses I took had many practical and tangible results.

Rutgers University had most of the good world-class renowned plant scientists. Some of the professors who taught us had been involved in crop science programmes across the world. They had worked in maize programmes in CIMMYT in Mexico, rice programmes which brought in green revolution at International Rice Research Institute (IRRI) and the International Institute of Tropical Agriculture (IITA) in Ibadan, Nigeria. I therefore had enough success stories to emulate from these exposures.

My first, second and third years of study went on smoothly despite encountering some hurdles along the way. Nothing, however, deterred my resolve to do well in my papers. I used to visit my brother and his wife in the Bronx, NY, every weekend and holidays. I moved to a special residential house for Agriculture students who performed well and were accorded the privilege of community residence.

Community Living at Helyar House, New Brunswick, New Jersey, USA

Universities worldwide recognize talented and needy students and accord them certain privileges. My desire for quality education made me score good grades and enabled me to be considered a special needy student to be accommodated in Helyar House. The 40-room one-storey building was donated by Rutgers alumni for special needy students who performed well.

I applied to join the group there and my advisor recommended me. I was given a room where students did not pay rent but were allocated some campus responsibilities. We were also totally responsible for all the food, clean-ups and purchase of household equipment. It was a well-organized, self-help students' club. I compared it to the one we formed during my primary school days, Nyamagesa Educational Club (NEC), back in Kisii, Kenya.

The communal living in Helyar House exposed me to the American culture and gave me an opportunity to save money to pay fees and use for other expenses. Most importantly, I had an opportunity to secure campus part-time employment given only to students. All the 40 of us were engaged in part-time work as research assistants, library attendants, in the dining halls and even fieldwork.

One had to do well in class for one to be given part-time employment. We did not pay for the accommodation and so the money we got was used to purchase food and other dormitory expenses.

I was able to save for other needs like books and clothing. I was grateful to the house master, who was also my lecturer, in one of the Environmental Science courses. Incidentally, Helyar House had alumni trustees who oversaw the smooth community existence. We also had a management committee which was responsible for various house needs. I was a member of one of the committees of the house.

The house was an all-male dormitory at the time, but I understand that later it accommodated both male and female students.

My part-time job was as a laboratory assistant in the entomology building. I was assigned to an Indian graduate student called Yesu Das. He was also a tutorial fellow in the university. Both of us were responsible to Dr Gupta, who was an entomology lecturer and researcher at Rutgers University. My work, therefore, was overseen by Yesu Das under the general guidance of Dr Gupta, who authorized our weekly pay. It was easy to work with both of them. Incidentally, our payments came from research grants and the college. My primary responsibility was to work in a vivarium rearing cockroaches for research.

Research on Cockroaches

A vivarium is a special laboratory for insects' culture. It was about three kilometres down the road from Helyar House. I worked during my free hours, weekends, holidays and in the evenings. I was paid US$2 per hour without any benefits, of course. My main responsibility was to ensure that the cockroaches lived well, got water and food pellets to survive and be ready for experimentation. The insects were for research and had to be reared specifically for this purpose.

Dr Gupta had received a state grant to investigate the best chemicals to use in the control of German cockroaches which were a menace in American kitchens. Nobody liked them. And Gupta's task was to discover, develop, research on the best insecticide to kill the roaches. I found this engagement exciting and was able to acquire experimentation designs and methodologies as I worked and earned my small salary. Mr Yesu Das, a PhD candidate, was also assisting Gupta in data collection and analyses. There were two other American students with whom we alternated during the weekends, but they accumulated fewer hours than me.

Dr Gupta liked me because I was reliable and very time-conscious. I never missed to attend to my duties, and reported to him or Yesu Das any incidents which could affect our experiments. I knew that the research we were conducting was beneficial to me and my bosses. I considered the research as my own pride and any good results would be equally beneficial to me. I knew I would be quoted in the acknowledgement page of any publication which would emanate from our work.

Cockroaches are very sensitive insects; if they lack water they will die through dehydration. I learnt a lot about insect biology during my involvement. We all know that cockroaches are the ultimate survivors with enough evolutionary tricks. Research shows that they have lived for over 350 million years and have completely adapted to the human species (*The New York Times*, 24 May 2013).

The nature of adaptation is impressive even for such an ancient, ineradicable lineage. Their feeding habits are strange. They can switch their internal chemistry as an evolutionary mechanism to avoid a poison. Their quick evolutionary change in behaviour offers controlling mechanisms a big challenge to researchers and the multi-billion-dollar pest control industry. It is through these evolutionary trends that developing chemical control baits becomes difficult to synthesize in the research laboratories. My work was to spray poison baits to kill cockroaches. The industry would always develop new compounds for insects and humans too. Cockroaches had the capacity to sense a poison and evade it. They have no taste buds but have taste hairs on many parts of their bodies to detect a killer poison.

My question here is whether mosquitoes have similar characteristics like cocroaches! They are uncontrolled and keep on causing continuous human suffering all over the world. They become resistant to drugs so quickly and one wonders how they acquired their superiority in evolutionary trends. They may be comparable in their survival traits to cockroaches. For example, studies have shown that mosquitoes change their behaviour; they don't rest on the walls that are treated with insecticides. This is an interesting comparison.

In the race for world domination or just survival for the fittest, cockroaches have scored the highest in adaptive capacity. They are superior to human beings, scientists claim. What is interesting is that cockroaches have been around since the time of dinosaurs, they can live for almost a month without food, and two weeks without water. Female cockroaches mate only once and stay pregnant for life. A cockroach can live for one week without its head, it can hold its breath for up to 40 minutes; and can also run up to three miles an hour non-stop (*The Standard*, June 14, 2013). Cockroaches are most active in the darkness. That is why they are so difficult to control.

Perhaps a comparison between the human behaviour and that of cockroaches may be appropriate here. Human beings commit most crimes at night. They also deal with matters of secrecy at night. This is the part of the day when all systems slow down. This is true for both humans and cockroaches.

Cockroaches come out during the night and cause havoc in the kitchens and food storage stores, whereas man waits until the late hours of the night to effect and or cause mischief of all types, notwithstanding that some other human decisions are also made at night. What is noteworthy is that both animates are nocturnal in causing havoc. This may, however, not be true to legalized human engagement or activities. Is there any psychological semblance in the nocturnal behaviour of roaches?

My exposure to laboratory research was the beginning of dedicated discipline. This was where I had to persevere for long hours and late nights to study the behaviour, eating habits, and survival mechanisms of roaches. Basically, my role was to apply basic research techniques to determine how long cockroaches could survive if they were not provided with certain foods and adequate conditions as needed.

I counted the number of nymphs one ovipositor could hatch, the duration they took to mature from one instar to another and conducive environmental conditions. I discovered that small cockroaches go through several metamorphosis stages before they reached adult life. Cockroaches survive under the most difficult environmental situations. They adjust to weather changes and seek appropriate niches for hiding and survival. They can survive on very little water and food droppings. They do not like light, and come out of the crevices only when it was dark and quiet.

They are intelligent insects which look for the smallest secure inaccessible places to hide. I was able to determine the cockroach sexes early usually during the second metamorphosis stage. This is within a few hours of their hatching. It was easy for me to give the approximate age of the young cockroach. All these research procedures were significant in allowing my colleagues and me to determine the best stage and method to kill them. I had to consider several variables before we designed an experiment.

The consideration for the provision of water, food pellets, optimum temperatures, lighting regimes, hiding crevices, noise, metamorphosis stages, and the sex dictated the kind of research we had to conduct. I thoroughly enjoyed sitting in the vivarium for long hours to observe the behaviour of these small insects which were considered the oldest in the world. I was able to discover that ovipositors (egg-cases) were deposited by female cockroaches elsewhere and the 'egg' hatched by itself, giving an average of 20-30 small nymphs. I was hardened in research procedures and patience during this whole period. I was also making a few dollars during this learning engagement.

The lessons learnt from my laboratory experiments were trust, patience, accurate data collection, data analysis and appropriate environmental conditions for cockroaches to survive. My two supervisors from India were so open with me that any publication which came out included my name. We carried field trials in several kitchens, smoking out cockroaches as we applied chemicals to control them. We were able to analyse the results of the chemicals which were potentially effective in killing cockroaches. We made conclusions on the efficacy of the compounds and forwarded our findings to the Drugs and Poisons Administration for onward consideration and registration.

The USA Food and Drug Administration is responsible for ensuring that any pesticides registered for use has been tested over and over again for its efficacy and environmental safety. Multiple testing sites are recommended and results are compared across the country. We wrote papers and made conclusions citing appropriate chemicals safe for use in the control of the cockroaches in American kitchens. The insects were certainly a menace and survived on dirt and squalid conditions. The cockroaches could shun all baits and adapt to the most complicated survival tactics.

Field and Laboratory Research

After my undergraduate course at Rutgers, I was enrolled for my Master's Degree course in Vocational Technical Agriculture in the Graduate School of Education. I had sat for a four-hour Graduate Records Examination which I passed well and was registered for the Master's degree course. I opted for course work and comprehensive examinations rather than thesis writing. The choice was risky, but shorter. My other friends opted for coursework and thesis which would normally take 2 to 3 years. My option took one and half years and I obtained my Master of Science degree. I had to put in extra hours and hard work to compensate for the thesis. My two great professors, Dr Charles Drawbaugh and William Smith (now deceased), supervised me until I completed the course.

We used to have day and evening lectures. I did my practicals for reports and fieldwork survey during the weekends and a few days of the week. Education courses were many and included tests and measurements, statistics, pedagogy, advanced courses in lecture designs and delivery, among others. I enjoyed the units and found them useful even after the course. I also took a few agricultural engineering courses, crop sciences and environment. I considered this approach all-encompassing and covering areas which I later found useful in my teaching career.

After completing the courses, I was expected to sit for a comprehensive examination which took four hours. All materials I had covered during the coursework were examined. I recall that we were four and locked in a room to answer all types of searching questions on topics in taught courses. It took a while to get the results but one of my supervisors called me to his office and informed me that I had done well and had passed. I was very happy that I had risked opting for the courses and examination. There were two failures from our group and the consequence was that they had to wait for the whole year to re-take the comprehensives or quit altogether.

During my BSc courses, I took several units in Crop and Environment Sciences. They gave me an excellent grounding for the Masters and PhD degrees. I was good in Crop Science units and we used to conduct greenhouse and field experiments. This was important for my future career progression. I started to consider enrolling in the PhD programme. The Master's in Vocational Technical Agriculture exposed me to communication skills and public engagement, extension service and teaching.

A combined training in plant sciences, soils, environment and botany made me feel competent in enrolling for a PhD. I applied for it and asked Dr Richard Ilnicki (now deceased) if I could work under him. I saw him once and he never commented much. He told me to go and return to see him another day. He was a well-known weed scientist who received a lot of research grants from chemical industries. He was a strict member of staff and worked only with those who could withstand his short temper as I was advised by my classmates. He was, however, a very clean-hearted professor who loved his students. He was particularly good in working with foreign students.

After a week, I went back to his office and reminded him that I desired to work under him for my PhD programme. It was at this time that I realized he wanted to know whether I was serious or not. He already had two PhD students with him, a Japanese and a Mexican. I would be the third.

My persistence finally bore fruits and he took me under his wing for my PhD. We had a lengthy discussion whereby he advised me on what to expect. He gave me a run-down on the need for research, the research grant entitled to me as a graduate student and the working conditions in the Department of Crops and Soil Science. For one to graduate with a PhD, one had to take some taught courses amounting to 48 hours, conduct research at the same time and write a dissertation. I had to pass the course-work first, conduct field trials, collect data to write a dissertation and defend it. All the work was to be done concurrently.

This was where time planning and management was vital. My stipend was for a period of three and not more than four years. As a foreign student, I had to meet all other obligations to legitimize my stay in the USA. We had two sites for carrying out field experiments: Adelphi, and on the campus. Adelphi was one hour away from my campus, Cook College, the then CAES. The bulk of my research was conducted here.

I therefore had classes for either two or three days and fieldwork for the remaining days of the week. Dr Ilnicki had a team comprising myself, three technicians (Jan Somody, Cathleen Napoli and Kathleeen Smowensky), Carol Napoli an MSc student, and two other PhD students who were a year ahead of me. We always hired summer help to assist in field design and data collection. In total, my professor had about eight to ten workers specifically assigned for Weed Science Section at any given time.

In addition to my field experiments designed for writing a thesis, we also had several herbicide trials which were paid for testing their efficacy. These were very important because all our stipends were paid through them. I remember getting new products from companies like Monsanto, Dow Chemicals, Stauffer, Ciba Geigy, Elanco, and Diamond Shamrock for evaluation. Each company expected a well-researched protocol to allow or disqualify it for registration. We were doing these experiments concurrently with other agricultural universities so that the results could be compared.

We had to be very careful in the manner we carried out the research. The findings were supposed to be comparable to those from other research institutions. We could then go to various weed science conferences and report the findings on the new product. Those national and international weed science conferences enabled me to publish several papers and later write a book on weed identification. It is therefore extremely vital to stress here the importance of research at graduate- level training. This is what we gravely lack in our African universities.

The conferences were held in several USA cities and they usually lasted for one week. That is how we became exposed in research and were able to interact in academic circles. We were also able to network and make local and international contacts.

My professor and I published several papers each year which was a credit to his and my curriculum vitae. Universities like Cornell, University of Florida, Texas

A&M, Virginia Tech, Pennsylvania State, among others, collaborated in research and student programmes.

Professor Ilnicki believed in one philosophy: work very hard during the week and then take it easy during weekends. Summer time was the most productive season for our experiments. We worked for five days, Mondays to Fridays, analysed data half-day on Saturday and all day on Sunday. The professor would invite us to his house for a weekend or we would go to the Atlantic City Beach where he owned a beach house. Life as a graduate student was the most important period for me in academic development. The brief experience I had gained in the vivarium was useful.

Prof Ilnicki's family knew all of us, the graduate students, and his caring wife, Helen, was very receptive. She would occasionally bring packed lunches to the field where we worked.

My second supervisor, Dr Roger Locandro, was equally pleasant. He also used to invite us to his house for weekends and public holidays. His family was always ready to receive me and share Kenyan tea or coffee. He taught us a course on Medicinal Herbs and Plants.

Another lecturer Dr. Jorge Berkowitz, took me for several environmental courses and covered important topics which included: Environmental water and land pollution, Chemicals and electronic pollution, Federal Laws of the USA on environmental pollution, The role of the Environmental Pollution Agency (EPA), Littering penalties, and Public participation on the protection and enhancement of natural resources. I found these courses useful when I joined NEMA and kept on referring to them.

What were the merits and demerits of field and laboratory research? The comparison between the two research methodologies was interesting. Basically, the laboratory research was continuous and persistent. The researcher had to endure long hours of sitting or standing to observe and collect data. One had to prepare to sit and observe experiments for as long as it took to get results. The process could at times be tiring and straining. One got fatigued but eventually became happy with the unfolding results.

Fieldwork is fun, laborious and very unpredictable at times. The researcher walks about, observes experiments, records data and can get results within one season. Fieldwork is usually outside in the open and no one can predict the weather patterns once an experiment is laid out. A researcher can lose data and either start afresh or discontinue the work altogether.

Once the results are obtained, the researcher becomes contented and continues to analyse and compare data. All experiments are unpredictable at times; it is like intelligent gambling. The hypotheses one makes should lead the researcher into making predictable conclusions as he or she conducts the experiment.

Researchers are generally reserved individuals. They think about their work and wonder if their hypotheses would proove true or false. The work I did during my undergraduate courses built on to the subsequent programmes. The many hours I spent in the cockroach laboratory gave me early confidence of collecting and analysing biological data. The mistakes I made in the vivarium became lessons for the fieldwork research trials that I later undertook. There is nothing small when it comes to practical training. I have come to believe strongly that persistence in any work usually pays off.

My interaction with industries, other students and professors across the US gave me confidence in paper writing, slides presentation, and eventually interviewing techniques. Graduate School was the ultimate academic challenge that any student ever enjoyed. It was tough, demanding, tormenting, and called for unparalleled endurance. One could easily give up and call it a day if one could not get along with one's supervisor. I once wrote an article on this subject which is reproduced elsewhere. I highlighted the route to achieving a PhD and how one became a professor of a university.

After collecting data for my PhD dissertation for three years, I had to compile and analyse it. I worked on the competition between soybeans and *pennisetum purpureum* using various chemical and manual control methods. Prior to starting the field experiments, I had to take some courses as part of the requirement for the PhD degree. They were high-level courses which were taught by various specialized lecturers on different topics. Most of my courses covered chemical reactions of herbicides in the environment, herbicide metabolism, environmental pollution and degradation, herbicide formulations, plant physiology, weed identification/taxonomy research methodology and statistical analyses.

I also did a course in computer science to help me analyse my data using COBOL programme then. The courses were common for PhD standard requirements for all students. I remember spending many hours in the library for references after lectures.

Farm Produce: After every harvesting season and having taken all the data needed from the experiment, any products which were considered safe for consumption were stored and used by us. We took home all that we could carry. Our experiments tested new products from all types of vegetables, field crops and horticultural crops grown in New Jersey. We could therefore harvest baskets and gunny bags of edible foods for our friends and own use.

Some of the foods needed deep freezers which I did not have. I instead donated them to my neighbours and friends. In some cases, I made juices and tomato puree from the excess quantities. I could readily store the finished products. We enjoyed the end of summer harvesting season and I was able to deliver products such as sweet corn, tomatoes, onions, water melons and strawberries to my brother and friends in New York City whenever I went there during the weekends. My friends still remind me of my generosity whenever we meet. It also reminds me of the time I was a kitchen/dining hall prefect at Kisii High School in the late 1960s.

All Masters and PhD students had their special study cubicles. The carrels were equipped with lighting lamps, sockets for calculators and book holders. Our offices had thirty such study facilities and a large coffee area. We used to meet there to discuss research progress and assist one another in social or academic matters. It was in the common room where we planned for seminars, prepared slides and practiced presentations. This was usually done in the evenings. I probably used the graduate room more often than anyone else. Our lecturers knew where to get us in case they needed us.

In Kenyan universities, a professor would call or leave a note for the graduate student to make an appointment. They generally do not have a designated meeting place. We also used to assist the lecturers in preparing for the practicals. It is very difficult to trace graduate students in many African universities since they do not have a central meeting place. At this level, social gathering and networking helped us to scout for jobs, discuss research progress and plan for any forthcoming conferences. We need more of these in developing nations to keep pace with the rest of the world.

Lately, however, many graduate schools in Africa are equipping their graduate students with the latest available laptops. Kenyatta University in Kenya, for example, is at an advanced stage of constructing a Graduate School block with all the modern amenities. This will enable closer supervision and interaction between the lecturers and students.

Seminars were conducted every week, usually on Mondays between 10 am and 12 noon. This was a pre-scheduled engagement and all of us had to present our research findings. I remember delivering over twelve presentations during my MSc and PhD training. They covered areas of my academic interest, guided by my supervisor. My best approach in delivery was to practice alone before presentation. My supervisor always made time for me to rehearse before him or with other graduate students. He timed me, took note of my mispronunciations and advised me on the delivery rate and I adjusted accordingly. This was the best practice I ever went through.

The practice became handy when I was a Vice-Chancellor at JKUAT in 1994 to 2003, charged with the responsibility of delivering several speeches in front of dignitaries, staff and students particularly during the graduation ceremonies. I indeed cherished my supervisors' constructive criticisms. I knew very well that some lecturers never accorded time to their students. My main supervisor at Rutgers made all his graduate students do presentations in front of him in a large theatre and he would sit at the far end taking notes. The practice was compulsory for all of his students. It enabled me become a confident public speaker, cautious of my pace in speech delivery and word pronunciations.

Travels within the USA

As students, we used to plan events. The events involved all of us, whether foreigners or Americans. I became popular in planning events like games, sports, public debates and even weekend retreats. Most people travelled during the summer with their families and we could be left behind taking care of field experiments or tending experimental animals in the farm.

I recall a trip we made to St. Paul's City, Minnesota, with my brother Onami and his family. We had been invited to attend the wedding of our sister-in-law, Barbara, and decided to drive there to enjoy the scenery. We left New York City on a Thursday afternoon to attend the wedding on Saturday morning. St. Paul's City is close to 3,000 kilometres north-west of New York City.

The two of us considered ourselves good drivers who could move across the States within 18 to 24 hours. This was during the hot summer month of August 1976. We had a rather beaten up but hardy car. My brother and I drove in turns all night without any rest but for short breaks. My sister-in-law, Eleese, and the three kids, Mogaka, Moraa and Ongeri, just took it easy. Our first enquiry about the distance remaining to get to Minnesota made the sheriff we asked wonder whether we knew where we were going!

We had done an eighth of the trip by midnight of Thursday. I appreciated the road signs and highway patrols. We drove through the Great Plains, seeing fields of wheat and maize throughout the trip. I honestly wished we had flown. Indeed, we debated on flying out of JKF but decided against it just to do some sightseeing across the Midwest plains. We drove through South side Chicago and had a glimpse of a great sprawling city with numerous skyscrapers.

It never occurred to us that the 46th US President, Barrack Obama, would come from that city, Chicago. We arrived at St. Paul's City in Minnesota on the morning of Saturday when the wedding was about to take place. I hit the bed, so did Joel and Eleese. We were woken up to find the groom and bridegroom ready to match. We took a quick shower and joined them.

We witnessed excellent wedding vows, attended the reception and had a memorable Saturday afternoon. Deep down in me, I was worried about the trip back to New York City. The long arduous, monotonous drive across several states in the Midwest bothered me. We had to hit the road back having had a very brief tour of St. Paul's City and its environs. Indeed, we were in for another long drive. This demanded determination and perseverance of the highest order. We actually set out for the East Coast on Sunday morning with an aim of reaching NYC the following day, Monday. I had to be in the field to join my team and our professor to harvest potatoes in an experiment where we were testing pre- emergence herbicides for weed control.

I vividly remember telling Joel that I had to attend to this important duty as I had to collect data for my paper. We had several flat tires which would be repaired and replaced along the highway. The delay for my fieldwork was imminent; I had no alternative but to stop at a petrol station, gain courage and make a call to my professor that I would be reporting to work late. I knew my professor so well that I could predict his reaction to such a message. The phone went through and I told him that I would be late. He quickly asked where I was and I told him I was around Ohio State, near a university he had attended.

He asked me what I was doing there. I explained and he told me to take my time and report the following day. This was a big relief to me.

We arrived in New York State via George Washington Bridge and went straight to my brother's apartment. The vehicle we used was all beaten beyond repair. The exhaust pipe was gone. Most tyres were worn out and the car itself was falling apart. We later gave it to the junkyard for recycling.

I took my car from Joel's place and drove to New Brunswick on Monday evening to attend to my field work on Tuesday. It was a hot August month when field crops were being harvested. I just could not comprehend why we drove such a long distance for such a short visit! My sister-in-law and Joel were on vacation then. I would have taken a few days off, if not for the pressure of work.

I made one conclusion after the trip; where there is a will there is a way. We were able to attend my sister-in-law Yvonne Barbara's wedding and returned to the Bronx safely, having covered over two thousand kilometres across the north-west states.

I recalled a book I read during my secondary school days, Around the World in Eighty Days by Jules Verne, published in 1873. In this book, several characters like Phileas Fogg, Jean Passepartout went around the world in actually eighty days using every means necessary to meet the deadline. Phileas Fogg bet his entire fortune that he could go around the world in eighty days with no special arrangement. He narrowly missed making it.

When I reflect on the trip we made to the Mid-West and compare it with those I made later in other countries, I see a big difference. I made extensive travels in Japan during my visits there. I also travelled a few times to France and the UK on various occasions.

Public transport by trains in Japan is so modern that one cannot compare it to other developed countries. One can comfortably traverse the whole of Japan within a short period and meet deadlines. The Japanese Bullet trains are so precise in timing, and reliable that mass transport system is so superb and convenient for everybody. A comparison between the US mass transit systems and that of Japan finds the former slow and at times unreliable. The transport system in the UK is comparable to that of the USA mass transport.

Timeliness in mass communication is so significant for the country's development. Our trip to Minnesota was by road and took three agonizing days. I later learned that students drove across the states for leisure and sightseeing, but they planned better. After a few years, a friend and I drove about the same distance in a new car and made several stops en route. It was my first time of seeing the Niagara Falls. We had made elaborate travel arrangements before setting off. This was my second long-distance trip while studying in the US. It was more leisurely than the earlier one.

Student Life and Entertainment

Life for students and young faculty staff is fast-moving. The greatest activities are witnessed during spring and summer months. Students plan and save money for the various activities. I recall many occasions when we teamed up to attend big concerts in various cities and parks. Cities on the East Coast are closely connected. We could move from New Brunswick to Princeton to Philadelphia over a weekend. Or move north-east to Jersey City, New York, Boston and spend the weekend there. Parks and public entertainment places would be fully packed with social activities.

I recall a group of us travelling to Atlantic City when it was first opened for gambling. All types of world celebrities attended the opening ceremony. We saw the world's renowned musicians, actors, actresses, singers and entertainers from all over the US. From Atlantic City, there are several beautiful coastal towns with various attractions. We also visited Asbury Park for live music performance by Cool and the Gang band. In another town were the Bee-Gees, also performing, and the Commodores in another city. Other weekends were spent in New York City where everybody wanted to visit. People particularly longed to tour Broadway, Times Square, China Town, Apollo Theatre, Central Manhattan Park, the UN Centre and many theatres spread all over the city.

One place I frequented after reading Malcolm X's *Autobiography* was the Apollo Theatre in Harlem where he used to meet the Black American Activists. The theatre is centrally located in Harlem which has a high African-American concentration. I enjoyed attending and listening to all sorts of speeches and plays primarily performed by the Black Americans. Nobody could notice that I was an African and a Kenyan national until I spoke.

Entertainment during summer in Western countries is so popular that everybody would save for it. People travel and spend their vacations in other cities. Local tourism for the young people is very popular in the USA. One cannot compare it with that of Kenya. It is cheap and affordable for every class of workers. Those who travel to Kenya are well-to-do senior citizens. I learnt one lesson about the youths in the USA and how they spend their days. They plan according to seasons rather than days. They save to accomplish their travel plans. They know what they want to do in a particular season yet I planned mine in terms of days and weeks!

In Africa, Kenya in particular, planning for certain activities is difficult because of less pronounced seasons. Our focus is somewhat limited and short-lived. People tend to assume a continuum in fulfilling all activities. They do not regard small monthly variations in the weather as important and hence time is not a factor in planning. 'It can be done tomorrow', as we normally say. We therefore waste a lot of time as if it was static.

Time management is not a priority to many of us. Deadlines are set but hardly met. How then can we ever move forward progressively, either as individuals or as a country if we have no sense of time management? My student colleagues always told me their plans for the following season or year. I adopted that notion and also planned for a year or seasons. College activities were obviously pre-determined but extra-curricular ones were not. The freedom of planning and the privilege of decision-making makes a young American college graduate more independent and mature than a typical Kenyan BSc graduate. That has been my conclusion.

Graduation Ceremonies and the Deepening of my Intellectual Horizons

Rutgers University is one of the oldest institutions of higher learning in the USA. It attracts prominent people during its functions. During my first degree graduation ceremony, we were honoured to have one of the most vocal US senators, Ted Kennedy, as a guest speaker. The Kennedy family had respected politicians and the fact that Ted accepted to officiate at our graduation ceremony was an honour. Grandiose preparation were made for his visit, and the President of the University, Dr Edward Bloustein, a lawyer by training, would not leave anything to chance. Security details were upped; rehearsals perfected and general university cleanliness was enhanced.

As usual, parking became a nightmare! Senator Kennedy landed at our campus with a chopper. He delivered a speech on equal opportunity to Americans which I still recall. He was straightforward in telling the Americans that everybody there was struggling and had an equal probability in becoming a great guy. Nobody was born with fortunes, but fortunes were struggled for. Survival of individuals depended on how hard and shrewdly one worked. He told us graduating students that the US was open to new ideas because research made that country what it was. The government would continue to inject research money into this important sector. He gave examples of research undertakings which Rutgers University had successfully accomplished. He further advised us not to stop at the first degree as this was the beginning of scholarly work. He completed his 25 minutes speech by encouraging the institution to enhance equal opportunities to all the Americans regardless of race or creed. The graduation ceremony was attended by all types of dignitaries from across the country.

The speech had emphasized research, academic continuity, the country's support for research funding and equality to all. This opened my mind for greater things ahead. Unfortunately, I lost my copy of this inspiring speech which had added value to my academic endeavours. Senator Ted Kennedy was a great, intelligent speaker. He intermingled with all the graduating students of my class and we eventually had lunch together in an open lawn next to the Passion Paddle, a clean water swamp on campus. I had a rare opportunity to shake his hand and exchange a few pleasantries.

He asked me where I came from and I quickly said Kenya. He was quick to comment on the great safari opportunities that exist in the country. The interaction lasted for about a minute and I returned to my seat.

After two years, during my Master's graduation ceremony, a Rutgers Alumnus of Medicine delivered yet another great speech. Unfortunately, I cannot recall his name. The gentleman was working as a medical doctor in California and he had flown in to grace the occasion. He was one of the best surgeons in the State of California. I remember the title of his speech was 'Integrity and Racial Equality'. This was a powerful speech which covered the values of human beings notwithstanding colour, race, creed, origin, sex, and religious affiliation.

The doctor took the pain of defining integrity and even referred to the *New Webster Dictionary*: that integrity is an unimpaired moral principle; honesty, soundness and the quality of being whole or undivided, original perfect state. He further expounded on this topic by citing other researchers who argued that integrity 'was like the weather everybody talks about but nobody knows what to do about it'. He further elaborated on what the concept embodied. When he referred to integrity, he had something very simple and very specific in mind. He stressed on respect for values and people.

Integrity requires three steps as pointed out by other writers: (1) discerning what is right and what is wrong; (2) acting on what you have discerned even at personal cost; and (3) saying openly that you are acting on your understanding of right from wrong" Stephen L.Carter, quoted from *The Daily Nation* of February, 25th 2013 by Rasna Warah. The writer concluded that integrity is about having the courage of convictions and the willingness to act and speak on behalf of what is right. This was the kind of courage which many people lack. Most people lack the integrity required to make far-reaching decisions for the benefit of an institution or country. Instead, they are short of ideology or conviction.

The speaker during my second graduation ceremony had encountered racial discrimination in the US as a medical doctor. Although he was a well-to-do white doctor, he was not ready to discriminate awarding practicing certificates to the blacks who had qualified to practice in the profession. His colleagues in the medical fraternity had unfairly refused to register qualified doctors from minority communities and he vowed to change such racial prejudices. He fought the war and succeeded in being fair to all as long as they met the required qualifications.

A quick reflection on the two most powerful speeches during my academic career taught me several lessons later in my career. First as a vice-chancellor, I had to prepare well for graduation speeches and pick topics relevant to the nation, audience and university community; and second, as a head of an institution, I had to encourage the young graduating students to proceed further because first and second degrees were not enough in the competitive world. I keep on stressing the same to my three sons and daughter.

The late senator's speech on equality and survival of the fittest made me realize that working hard and choosing a career in life was dictated by the quality and demand of the discipline. Each country had enough resources to cater for all if accessed and shared equitably. In his speech, he also stressed on integrity as a prerequisite for a fair and just nation. Without integrity, the people would suffer, and class conflicts would persist.

Corruption and other vices become endemic in communities. The divide between the rich and poor is widened and societies live in fear, development is hampered with and countries are left poor and experience slow industrialization.

I used these pieces of advice later in my academic and administrative roles. The speeches manifested well during students' agitations, riots, and admissions. The Kenyan Joint Admissions Board (JAB) was so well planned and fair in its roles that as its chair in 1998, I was proud to participate in fair admission exercises. Integrity therefore played a major role in my management of the administrative duties which I undertook as I carried on in my academic pursuits.

I always stress to my students that research, data collection and reporting are crucial stages which demand integrity. There are no shortcuts to good work. The many universities we now have in Africa, and especially East Africa, have to ensure that quality training is enshrined in their philosophy and their staff is of high integrity responsible for sustaining quality services. My summation on integrity is about the prize one pays for doing what is right and justifiable to society as courageously as possible, no matter the consequences. This virtue is rare in our midst.

The convocation speeches were educative and I had the opportunity to listen to several powerful special guests on other fora. I was able to translate them to my future management skills. My academic achievements and administrative duties borrowed a lot from the two speakers. I later realized that any formal speeches I made must have a component of integrity and fair play as far as Kenyan tribes were concerned. Tribalism and lack of honesty are cancers in our society today.

Choice of a Guest Speaker

How do we choose the guest speaker for a specific graduation ceremony in an African university? Who selects the speakers? Is it the council, the vice-chancellor, top management, senate, faculty members or the graduating class? This is controversial debate which has affected the speakers-to-be in some western universities. Commencement speeches are usually free when they are delivered on campuses.

That is referred to as freedom of speech in the academia. The notable speakers who are invited have the freedom to raise controversial issues which affect society. Cases have been cited in US universities where students from over 10 campuses protested against speakers invited to commencement events. The reason was simple, that the rejected individuals were people of little or no integrity.

One of those rejected was a World Bank personality whom students accused through face book as the architect of the Iraq war and a war criminal. Another renowned neurosurgeon and a conservationism icon dropped out as a commencement speaker in a great US university after students protested against comments he had made lumping together homosexuality, paedophilia and bestiality.

A similar case involved the US President Barrack Obama as a subject of controversy. He had been invited to deliver a speech at Morehouse College as a guest speaker. The college had earlier invited a Philadelphia pastor, Rev Kevin Johnson, to speak the day before Mr Obama. The pastor, who had written an article in *The Philadelphia Tribune* criticizing Mr Obama for not appointing African-Americans to cabinet positions refused to attend the commencement as he wanted to be the sole speaker during the event.

He was advised that the university wanted to provide a broad spectrum of views from both speakers. (*The New York Times*, Sunday, May 12, 2013). Another case was reported in Ottawa pitting one conservative commentator, Ms Ann Coulter. She made a comment about Muslims which angered about 2,000 students who demonstrated against her appearance. She cancelled her appearance.

Cases of environmental and human rights abuse have also featured on individuals' standing on the matter. Students have even demonstrated against the recipients of honorary degrees. The protests have taken various forms and have been largely used to portray one's image, standing in society and integrity.

Whose graduation ceremony is it? Those graduating at that point in time own the show. They should have a say on who can be their chief speaker during this occasion. Currently, students are not involved in the selection of speakers during graduation ceremonies. A mechanism can be found on how to engage students.

It is the students' commencement, not the staff, not the management. Many African universities may wish to consider mainstreaming integrity when choosing dignitaries for various university functions and awards. Universities are meant to be bastions of open-mindedness and free speech. They should not therefore deny the students their right to hear controversial opinions and draw their own conclusions about those opinions. The aim of these controversies and protests is to have a hand in choosing the guests

My PhD Defence (Viva)

My weed control research was complete when I harvested my last season crop and collected data on my experimental treatments. I analysed data and made informed conclusions. I wrote the dissertation and submitted it for examination. My main supervisor and two other examiners read the thesis and wrote reports to enable me attend an oral defence examination.

In the US, the public is allowed to attend any viva and ask questions. I was aware of this openness and prepared well in case a question came from a non-scientific stakeholder. My board of examiners included Dr Richard Ilnicki, Dr Cecil Still, Dr Roger Locandro and Dr Motto. These examiners had taught me at some stage during my Master's and PhD courses. Dr Still taught me Plant Biochemistry, Dr Locandro taught me a course in Plant Ecology and Dr Ilnicki, my supervisor, took me through Chemical Weed Control and Herbicide Metabolism units. I was therefore familiar with the members of the Board of Examiners. I prepared well for my defence which commenced at 9 am ended at 12 noon. I was asked several questions some of which touched on the Kenyan economy and tourism. I was able to answer most of the questions although some of them were outside my research mandate.

After about three hours of grilling, I was informed by the main supervisor that I had done well and was to be awarded a doctorate degree in Weed Science. I was elated. There had been cases of repeats or outright failures. I called my brother and his wife Eleese to tell them the good news. They too were most grateful and congratulated me. I became Dr. Ratemo Waya Michieka on the day I passed my defence which was on 8 July 1978.

I recall that date vividly because Dr Cecil Still's wife, Delores, gave me a wristwatch inscribed with my name and the date which I keep as a memento. This was a moment to remember and it seemed to be the beginning of good tidings. My fellow PhD students were also preparing for their defences. My friends organized a party for me and I had to plan for my next move.

Several companies whose chemicals we were testing wanted me to join them for employment, but I turned down the offers due to several reasons. I was not convinced that I wanted to stay in the USA after my studies. USA is a great place for one to work in and make a career, but future advancement could be limited no matter how hard one works. There was still that stigma of a black boy, a Kenyan boy, an African boy.

I was very much welcome to work and stay in the US, but opportunities elsewhere in Africa were brighter than those in the USA. I kept on thinking and considering whether to apply and continue my stay or seek other openings abroad. The latter prevailed.

6

Growth as an Academic Leader and Researcher

My first entry into research and academic life after completion of the PhD degree was not in Kenya, but in Ibadan, Nigeria, West Africa. I got a job at the International Institute of Tropical Agriculture (IITA), Ibadan. I saw an advert in an international journal seeking postgraduate students from reputable universities. I applied for the post whose duration was going to be two years. Those days, in 1978, letters were sent by air unlike nowadays when e-mail or tele-conferencing are used to transmit such important information.

After about a month, I received positive feedback that I was to join IITA in Ibadan, Nigeria, as a postdoctoral fellow. A Mr Reeves was in charge of post-docs and he was prompt in replying to my request. I accepted the offer and was ready to move to Nigeria and continue to add value to my PhD training. My position as a post-doctoral fellow was in Crop Protection, specifically in Weed Research in the humid and sub-humid tropics.

I received discouragement from colleagues but my brother encouraged me to proceed. I kept my other brothers in Kenya informed of my next plan after the USA. My professor also wanted me to stay in the USA and work with local companies or join assistant lectureship. After some soul-searching, I settled on IITA.

I invited my friends and a few of my lecturers for a low-key farewell party. The dean of students, Dr Richard Merritt, and his wife, Peggy, were so appreciative of me and gave me some nice reading materials as souvenirs. I still have them. My professor, Ilnicki, and his wife, Helen, gave me a small plaque which I still have. Others who joined us were Drs Cecil Still and Roger Locandro with their wives, Delores and Marylyn, respectively.

I got proper briefing and advice on Nigerian lifestyles. I was told about rampant corruption, ethnic wars and all sorts of discouraging information. But I had resolved to move on. All I knew about Ibadan was that it was one of the oldest cities in Western Nigeria and was inhabited by the Yoruba people. I also knew of the University of Ibadan, the oldest colonial institution of Nigeria. I further knew that the Ibos were the most industrious people who had staged a civil war because they wanted their own country, Biafra, to secede from the Federal Government of Nigeria. To the north were the powerful Hausas who were pastoralists, great traders and rulers. I was vaguely aware of Nigerian culture having read Chinua Achebe's novel, *Things Fall Apart*.

Details of my travel itinerary arrived and baggage allowance was two bags of 30 kilogrammes each. I had not accumulated much except a few books and personal effects. I actually had less weight than allowed.

I relocated from New Brunswick, the university campus, and stayed with my brother and his wife Eleese in New York City for a while before I set out for Nigeria. I was booked to depart from JFK airport to Murtala Mohammed Airport in Lagos. This was the same route I had used to travel to the States much earlier. My brother and his wife dropped me at the airport for my journey to Lagos. I made sure that all documentation was complete including my visa.

The salary was modest and the allowances were generous for a single man. A car was provided, two-bedroom apartment, medical allowances and insurance, free water and electricity. We were paid in US dollars and then would change into naira whose rate then was higher than the dollar. During my time in Nigeria, the naira was more powerful in exchange rate than the US dollar. I had my account in the First National Bank in New York where my salary was deposited. It was a good arrangement since this allowed me to save a substantial amount of money for the two-year period I was in Nigeria.

When I arrived in Ibadan, I found a few post-docs who were already working here. They used to come at different times for various programmes. As soon as I arrived in IITA, Ibadan, I joined others who were on orientation programme and I was introduced to my supervisor, Dr Okezie Akobundu, whom I had met briefly during our Weed Science conferences in the States. He had heard of me and was ready to receive me as his post-doc. This was a good beginning.

I was his number two in the ranks in the Weed Science Section but he was also supervising two PhD students, Ray Unamma and Steve Utullu. We became very close as I settled down. He was very helpful in my settling down. Essentially, we were a team of three scientists with Dr Okezie Akobundu as our leader. We also had technicians, a section driver and field workers. We worked as a team, became very close and the group assisted me to learn the lifestyles of the Nigerian people.

It did not take me long to get down to work. I became a team player and guided the two PhD students during their research, assisting Dr Akobundu. They both worked on weed control in yams and cassava, two major tuber crops in Nigeria. My work was to assist Dr Akobundu in discharging the section's mandate.

He heavily relied on me and we developed an excellent working relationship. Our section was under Farming Systems which comprised soil science, entomology, legumes and irrigation. The overall mandate of IITA was to conduct research in the tropics and sub-humid tropics of Africa. Similar centres were scattered all over the world and were mandated with specific objectives in order to increase food production. Sister centres were unique in their setups but overall mandates differed. Directors-general were the bosses of the centres. Little did I know that later, I would become a director-general of an institute, the National Environment Management Authority (NEMA) in Kenya.

Our role was to conduct weed research in selected crops of the tropics in a number of states of Nigeria. The major crops were cassava, yam, rice, cowpeas, maize and beans. The three of us worked as a team to find out the best weed control methods for the crops. Ray and Steve were specific on their research pursuits and I assisted them having just graduated in the same discipline.

The trials were spread across Nigeria and I recall having trials in Ikenne in Ogun State (the home of the late Obafemi Awolowo), Mbiri, Umenede, Ahmadu Bello University in Kaduna, Nsukka and Ilorin. I would travel to these sites to set up field trials, collect data and analyse it. I recall meeting all types of researchers during this time and we would discuss the best ways to combat aggressive weed species. The most common weeds included *Eupatorium odoratum, nutgrass, Imperata cylindrica, pennisetum purpureum, talinum triangulare, and couch grass,* just to name but a few. The weeds affected all crops in arable and waste lands. We had to use many types of control management measures including herbicides for selected crops.

Tropical rains in south-western Nigeria were a major concern during the trials. I carried out extensive work on minimum tillage trials using the already registered herbicides like glyphosate and paraquat, among others. Dr Akobundu and I published papers from the trials and attended a number of scientific conferences in Nigeria and outside the country.

I forged linkages with many university professors from the University of Ibadan, University if Ife now Obafemi Awolowo University, University of Lagos, Ahmadu Bello University and University of Nigeria, Nsukka. Comparative trials were important for decision-making and future trials. I recall scientists like Prof. Lagoke from Ahmadu Bello University (ABU) who loved to work on weed control in onions. I learnt a lot from his work although onions are not a major crop in southern Nigeria, the region being too wet and humid. Onions do not do well under such conditions.

I was also involved in organizing scientific conferences in weed research. I remember presenting papers during the Nigerian Weed Science conferences in Lagos and Ibadan. I found these meetings refreshing and reminded me of the ones we used to have in North Eastern Weed Science Society, USA. I did not have any problems presenting and defending the papers. I enjoyed the interactions because they equipped me with wide experience in the scientific world.

As I was conducting countrywide research in Nigeria, I encountered very interesting and different cultures in the various states that I visited. Each trip that I made gave me a different perspective on the rich cultural diversity that is unique to Nigeria. Whenever I was in Ondo State, for example, I had to know if there had been a dead personality who demanded to be buried accompanied by other bodies. In Ondo State, the story goes, when an important person died, he had to be buried along with others who were usually caught unawares, killed and used as such. Normally, foreigners fell victim. I was told to watch for bald-shaven persons, an indication of a dead personality.

To this end, I had a driver and an excellent travel guard, Samuel Olu, who was always with me as a guard and an advisor. I spent a lot of time with him and learnt a lot from him. He later passed on from cirrhosis because of excessive illicit alcohol consumption.

Another incident was when we visited an Igbo home and we were welcomed in the traditional way of Igbo culture. As we arrived at some elder's homestead he would welcome us in the compound, and request us to sit down and bitter berries would be brought to us to eat. Kola nuts were also provided for us to break. I did exactly that, having read Chinua Achebe's novel, *Things Fall Apart*. I was at ease with this procedure and enjoyed chewing the kola nuts although they were a bit bitter and quite sharp on the tongue. It was only after this ceremony that one was allowed to talk or declare one's mission.

This tradition was still revered and commanded respect, discipline and hierarchy. This practice cut across several Nigerian states and it was an excellent culture of perpetuating a sense of community among the people. This is part of the discipline that many nations lack.

The most surprising event was when I was taken to a fortune teller in a small town of Shangamwe, a few kilometres from Ibadan on the Ibadan-Benin highway. I was taken by a Nigerian friend and this reminded me of my maternal uncle, Birari, who would perform similar functions.

My friend and I went to this fortune teller, who was a man. He lived in a low lying area, almost similar to a cave. We had to bend to get into his house. We were requested by his handler to carry some 50 kobo (less than one US dollar). We were ushered into his small waiting room, told to hold a live cockerel, and he slaughtered the chicken as we held it. He lifted the chicken and let the blood drain on to a small goatskin which was placed in the centre of the room.

He then let the chicken down, brought a box which was filled with all types of paraphernalia. He poured the contents onto the floor next to the blood to see which item fell closer or farther from the chicken blood. The man told us that we would be great people in our careers. I was sure he must have been told by his handlers that I was a foreigner, a suspicion he confirmed. He said so many things as he concentrated on looking at his paraphernalia. Each item meant different things to him. We did not talk, but his aide explained to us the meaning and interpretation of the rooster, blood and the items in the box.

We both gave five more naira and crawled out of the shrine. By the way, all this time we were here, we had to kneel on the floor and face the fortune teller. Again, I had witnessed my uncle perform the same tricks when I was a young boy in Kisii in the early 1960s.

Another visit to the Alaafin of Oyo was equally exciting. During my stay at IITA, the spiritual Chief of Oyo State resided in Oyo City, a few kilometres north of Ibadan. He was the chief and spiritual leader of the people. I decided to visit him. He was so famous in the area and entertained many dignitaries that came by to visit him in his parlour.

I drove into his expansive compound and declared my intentions after identifying myself. I was specific about the intentions of my visit and explained that I had heard of him and his generosity to his subjects and wanted to just meet him. It was a Saturday afternoon and his workers were not very busy. They allowed me in. He had an estate to himself. There were several houses built around a large field of about two hectares with the rooms facing each other and having a central meeting parlour.

The centre was open and was used for activities like plays, prayers, meetings among others. His main house was strategically positioned and he was able to monitor all activities within the expansive area. He could also monitor visitors as they came in. He had a number of workers and handlers, wives and many children. I got a chance to go to his parlour, told him who I was, a researcher scientist from Kenya and working in Nigeria. He spoke perfect English; hence communication was not a problem. I actually broke a kolanut with him.

He explained to me the significance of the ceremony as we sipped a glass of palm wine. We exchanged pleasantries and he took me on a guided tour of the compound. There were several well-thatched round huts with one central meeting place. The lawn was well maintained and a clean compound with nice flowers and dome palms surrounding the periphery fence. The chief looked young in his expensive Yoruba attire of 'Shola and Bola'.

I was later told that he had six wives and several concubines. There were several young boys and girls walking about the compound. I returned to IITA and told my colleagues about my visit to the Alaafin of Oyo. Nigerian chiefs command power, respect and authority. This is noticeable even with the present-day chiefs. Many of my friends from Nigeria are chiefs' sons and they command respect and must be addressed so: Chief, Dr, and Prof, Sir, so and so. I find this form of salutation appropriate and sound within the African setup. We all need some titles and forms of salutation.

Post-doctoral fellows had a lot of activities in common. We used to go out together, conduct seminars in groups, and play games together. Dr Lewis Jackai was my closest colleague with whom I shared a lot in common. We used to travel together and go shopping in downtown Ibadan. He was a meticulous scientist who carried out his research in entomology with ardour.

There were certain behaviours which mesmerized me, especially driving on highways. Despite the many positive attributes that Nigeria is known for, such as being the most populous country in Africa, driving in any of the big cities is a nightmare. While driving in Lagos, Abuja, Ibadan, Benin, Onitsha, Sokoto and even small towns, drivers do not give way. A motorist would rather spend a night at a junction rather than give way to a fellow motorist!

Due to inconsiderate motorists, I once slept in Ore City, one of the most densely forested parts of Nigeria. A bridge had given way due to heavy tropical rains. This is a common occurrence in many parts of the world, including Kenya. Luckily, a small section of the bridge could be used to cross the huge river but one vehicle at a time. This small section of the bridge would allow only one vehicle to cross and proceed on. None of those who were in front of us were willing to take turns to go through. There was no such a thing as passing through in turns.

As the morning approached, I noticed one very saddening event which had occurred during the Biafran war. Virtually every house had bullet holes occasioned by the war. The City of Ore was the epicentre during the war, I was told. The Federal Military and the rebels converged there and occasioned the fiercest battle in Nigerian history! Lives were lost, properties damaged and all systems collapsed. That was the end of the secession. I marvelled at the gaping holes on the roof-tops and walls and imagined the number of civilians that lost their lives during the battle!

My driver, a Yoruba man was very eloquent in explaining to me the whole episode despite the fact that I had already read it in the papers. As we drove away from Ore City, I saw some young men and women going about their business in the dilapidated shops. I left Ore with the sad memories of a battle-field.

The reckless driving culture on Kenyan roads is a worrisome affair. The matatu menace is real. The speed limit is not observed both in Kenya and Nigeria. A strange observation in Nigeria, for instance, was the manner in which public transport moved. On a highway, a brave driver would complain that his foot was tired of pressing the gas pedal. He would get out, pick a boulder or some heavy object and place it on the pedal to keep the vehicle moving as he only steered the wheel. The pedal was pushed flat and on motion for several kilometres before the driver would remove the weight from the accelerator; so the story goes.

This may be an exaggeration perhaps, but it demonstrates the serious dangers associated with public transportation systems in Nigeria. I cannot remember how many accidents we witnessed on Nigerian motorways. The traffic police were not bothered as they were busy being bribed. Public transport vehicles were christened damfo, which meant '*going to heaven*' or one-way journey. Some long-distance buses were, however, driven by responsible people who behaved better than town touts.

I once travelled to the East and crossed the River Niger at Onitsha City, the home of Dr Nnamdi Azikiwe, a renowned politician. I saw very modern buses with all amenities necessary for distant travellers. This was a total contrast to what I witnessed in the south-west of Nigeria.

Work Environment at IITA

During my engagement as a doctoral fellow in Ibadan, there were five international agricultural research centres (CGIAR) then. IITA was one of them. Its mandate was to conduct research for the humid and sub-humid tropics. There were several scientists here who were in-charge of various research protocols.

IITA had various working categories, scientists, technical staff, postdocs, administrators and support staff. A scientist's retention in the institute depended on the work output and satisfactory performance before the contract was renewed. The post-docs were on the normal two-year contracts. If one wanted, one could request to be absorbed once the two-year duration lapsed. The absorption depended on many considerations.

I do recall that we were about eight post-docs in different research programmes. We used to have our own meetings to compare the working conditions in the institute. Several of us wanted to be absorbed here. A few of us were not keen on settling here. I was the only Kenyan post-doc.

The research centres were run by a Board of Trustees who were appointed from all over the world. They used to meet once a year or when need arose. Post-docs were known to the trustees and they used to meet us separately at times. In our group, there was a number of Europeans, Asians, Australians and Americans. We worked as a team.

Rifts

Strictly speaking, a post-doctoral fellow is one who has been awarded a PhD degree from a recognized university. The purpose of working as a post-doc is to further excel in one's area of expertise before one is fully engaged in one's career. Depending on who one knew, one could be categorized as post-docs yet have only a Master's degree.

This revelation bothered many of us because the beneficiaries were mainly from the developed nations. No African or Asian post-doc had a Master's degree. We all had PhD degrees. We once cornered the director general, Dr William Gamble, about this discrimination and he was unable to respond. We wanted to know the criteria used to allow persons of lower grade and brand them post-doc fellows.

Mr Reeves was in charge of recruiting the post-docs. The notion that some fellows amongst us did not deserve the title bothered us. In fact, these young scientists were so well protected that we did not know what kind of research they were engaged in. They even had better packages than the African post-docs. We felt short-changed and therefore dissatisfied.

At one point, we asked African scientists who were members of the Board of Trustees and they were also not able to answer. This caused further discontent. The young fellows were entitled to more rights and privileges than what the African post-doctorates had. A few post-docs who had the PhD degree and were from other developed countries were also concerned about this disparity and sided with us. They equally questioned the rationale of equality without comparable academic credentials.

Notwithstanding this anomaly, I continued working for the organization with conviction. My colleagues kept on complaining about the discrepancy but got no satisfactory explanation.

I noticed one clear operating machination. There was selective conversion of some 'loyal' post-docs to scientists' posts. Those who asked least and never questioned the status quo were awarded full contracts on the expiration of their period. This silenced a number of us. The problem arose when the budget was slashed and not many would be absorbed. One thing I knew for sure was that they needed weed scientists and I would be retained. My supervisor had put a strong case for my retention at the expiration of my two-year term. My stand against injustice, however, remained solid and unwavering.

We requested for a meeting with the trustees one afternoon after their normal morning board deliberations. The meetings of IITA trustees were made public through a newsletter from the communication office. I knew a few of the trustees: Prof. Reuben Olembo, my Kenyan mentor, Prof. Bunting and an Asian whose name I cannot remember. I talked to each one of them separately and later on requested if all the post-docs could meet them for a short consultation. The chairman agreed and asked the director general to attend the meeting. He was actually worried and started to ask me questions and the agenda of the meeting. He was also an employee of the Board of Trustees. The post-docs under my general guidance drafted a memorandum of complaints to hand over to the trustees.

I planned the meeting in such a way that a colleague was to read the memorandum and hand it over to the chairman. I did not consider it appropriate to lead the team and then read the statement. I was only instrumental in arranging the afternoon meeting. They agreed and one of us presented the grievances. It was a jovial meeting but hard-hitting facts were presented. The director general was very understanding and took the criticism positively, and even made changes later. Those of us who wanted to stay were then given a chance; they benefited from the short consultative meeting we had with the trustees. Whatever changes came thereafter were for the good of future post-docs.

Prof. Reuben Olembo and Prof. Bunting were sympathetic about the events at the institute and later on made recommendations which enabled it to acquire more donor funding. Staff at IITA knew that I was behind the negotiations and appreciated it. I, however, did not stay to benefit from the negotiations.

A few weeks later the DG, Dr William Gamble and his wife came to my residence and had a cup of tea with me. We had a pleasant relationship thereafter before they left for another assignment in Europe. The experience had made me resolve never to condone mediocrity because it never builds individuals or nations; instead it destroys nationhood. It pays to point out the wrong happenings.

Culture, Music and Lifestyles

Nigeria's culture is unchallenged, uncorrupted and intact. The traditional attire still persists. The respect with which elders are saluted is impressive. The chiefs, queens, kings, princes, princesses, the royalty and village elders are all held in the highest esteem. Other than some western culture which has taken root through the movie and music industry, Nigerian rural populations still maintain their indigenous and traditional habits. Their belief in *juju*, and juju music is still pronounced. The local artistes perform in both villages and cities but, of course, the youths now tend to go western.

During my stay there, musicians like Sunny Ade and Fela Akintokun used to perform in hotels like Premier in Ibadan. The playing of drums to signify something new was popular, particularly during the harmattan when the weather would be cold and the atmosphere dusty. Late afternoon drum sounds and impromptu preaching was rampant, particularly in the south-western parts of Nigeria. Freedom of worship and association was allowed and I had a chance to visit most churches and prayer meetings during the weekend.

IITA had only one mosque where I occasionally joined our Muslim brothers for Friday prayers. A Seventh Day Adventist Church was near Oyo town, a short distance away. Several branches were established later. Many traditional gatherings converged for worship. When I last visited Nigeria in 2004 and

2006, I was amazed at the number and names of churches that I witnessed in the country. They were all being advertised on large billboards. I took the liberty of counting and recording the various denominations between Lagos and Ibadan. The number reached close to 100.

Religious sects in Nigeria are a big, booming business. This trend seems to have hit countries like Kenya. I saw the biggest and well-raised tents which could easily accommodate 10,000 people. Every village had a church or a section of a religious outfit. The City of Nairobi in Kenya is quickly catching up on these money-making ventures in the name of religion!

The Pride of Nigerians

Despite the negative publicity that Nigerians receive, they are very patriotic individuals who love and cherish their country. Individuals can differ but the philosophy of living in the most populous country in Africa overrides the differences.

I have had a chance to work closely with Nigerians in many international meetings. They come out as very patriotic people. It is through this patriotism that I noticed each of their states has an additional nickname. All states in Nigeria have second names, just like the USA. I took the liberty of recording a few as follows:

- Abuja City – Centre of Unity
- Oyo State – Pace Setter
- Rivers State – Treasure Base
- Ondo Stat – Sunshine State
- Osun State – State of the Living Spring
- Kwara State – State of Harmony
- Anambra State – Home for All/ Light of the Nation
- Kano State – Centre of Commerce
- Borno State – Home of Peace
- Sokoto State – Seat of the Caliphate
- Yobe – The Young Shall Grow

The naming of states shows a sense of pride and ownership by the inhabitants. Although the idea was borrowed from the United States, the philosophy is commendable and patriotic. Nigerians are united when it comes to protecting their country and national interests. I remember this characteristic during international environmental meetings organized by UNEP. They will push for their rights without relinquishing! They say it like it is and do not care the consequences thereafter!

I actually admire their boldness and love for their country, culture and shrewdness notwithstanding other vices. Many African countries do not display as much patriotism as Nigerians do; except perhaps during games and sports.

IITA, an Interactive Meeting Centre, Ibadan, Nigeria

The mandate of IITA was to carry out research, train scientists and conduct seminars relevant to the humid and sub-humid tropics. Many scientists from various countries used to meet here for either conferences, training, short-term courses or annual research reviews. I had a chance to meet several Kenyan scientists, who frequented the centre. I remember meeting Prof. Reuben Olembo (now deceased) who became very close to me and regarded me as his student.

Prof. Olembo was the first person to prompt me to return home, Kenya. He was emphatic on the request and since he was one of the trustees, I welcomed the idea. I knew he would not support my stay at IITA.

Other scientists whom I met at various conferences included Prof. Chris Karue, who was the Dean of the Faculty of Agriculture in the University of Nairobi, Prof. Shellemiah Keya and Prof. Daniel Mukunya who were also members of staff in the faculty. All these good friends encouraged me to return home and take a lecturing position at the University of Nairobi. In fact I remember Prof. Christopher Karue taking my CV to the then Chairman of Crop Science, Prof. David N. Ngugi.

Prof. Ngugi got in touch with me through a letter asking me when I would be in Kenya next. He also informed another scientist in the department, Prof. Daniel M. Mukunya to follow up my case. My area of expertise, Weed Science falls under Crop Protection which was headed then by Prof. Mukunya. Meanwhile, Prof. Olembo kept on checking on me whenever he came to Ibadan. He was working with UNEP in Nairobi.

I travelled home to Kenya for a short break during Christmas holidays in 1979. I made a point of calling on all those people whom I had met in Ibadan. I went to Prof. Ngugi's office and had an interesting talk on the university setup and courses taught in the Faculty of Agriculture. I later saw Prof. Mukunya who spent a lot of time telling me about the University of Nairobi, the only one in Kenya at the time since 1971. He spoke so proudly of it that he wanted me to join his section of Crop Protection.

He was very frank and he told me that despite all the praises, staff salaries and remuneration packages were pathetic. He never minced his words. He, however, defended the salaries by comparing them with those of the Civil Service staff which were below what university lecturers earned. A good number of lecturers were expatriates who served on contract. They had very generous packages but had to leave for their countries at some point in the future. Getting a local lecturer was more cost-effective financially than keeping several foreign nationals. This would mean having a stable faculty manned by local permanent staff.

I returned to Ibadan and continued with my work. Deep down in my mind I was considering joining the University of Nairobi despite the low wages. In any case, I was not ready to haggle with IITA administrative staff for permanent employment or extension of contract. I collected my field data from research stations, analysed it and made relevant conclusions regarding the investigations. I wrote a few publishable academic papers on our work.

On a Friday afternoon in January 1980, I received a letter from the Chair of the Department of Crop Science, Prof. David N. Ngugi, informing me that I had been offered a position as a lecturer and that a Mr Solomon Karanja would get in touch with me.

To be a lecturer in the University of Nairobi, one had to have a PhD and possibly some publications. I had both. Within a few days I received the registrar's letter of offer and my starting salary scale, I recall, was K£2,774 per annum plus other allowances. The package was not attractive, and so I decided to sleep on the idea for a while.

I shared the contents of the letter with my close friends, Drs Amare Getahun and Lewis Jackai. They both advised me to make a positive decision and start a career. I also wrote to my brothers who were working with the Ministry of Agriculture.

They advised me to take up the offer and join them in Kenya. Meanwhile, my older brother in the USA, Joel Onami, had also advised me to proceed home and get a career job. But the decision would entirely be mine and it was my right and privilege to make it. I had to seriously consider this option since it involved a big difference in terms of financial rewards.

I had also been approached by Monsanto Company of the USA to work for them in Kenya and South Africa. They called me for an interview in Brussels, Belgium, and I performed well. I was required to man their Nairobi and Johannesburg offices on 50/50 time allocation. I refused the offer because of two reasons: Shuttling to the south and living in Kenya would not serve best of my interests. I could not settle down and make a career and start a family under such circumstances.

Secondly, South Africa was going through the most trying period of apartheid and racial discrimination. I did not want to encounter racial prejudices again which would certainly remind me of the encounters in USA. Little did I know that I would encounter the ugly face of apartheid on my way to the Comoros Islands later. Kenya got her independence in 1963 and so I was convinced that I would settle better here than in South Africa, despite the very good salary package Monsanto offered me. Again, the very final decision was going to be entirely mine.

Relocation into Academic Progression

I replied to the registrar's letter on 30 January 1980. I had spent enough time making a decision, which was in the affirmative. I had been told of the kind of work I would do at the University of Nairobi, which would include lectures, research and service to the community in form of extension work. I was told that all the units which were taught by a one Dutch expatriate, Mr Van Eijeten, would be handed over to me. These were 'Weed Science Courses' and 'Introduction to Crop Production'. I was also told that there were Master's research programmes which had just started and I could start supervising the MSc students.

I was aware of the tasks and started to internalize my roles and demands. I was, however, sure of the basic teaching and research needs ahead of me. I had done extensive fieldwork and collected data to use for advancing my mandate; I was, therefore, qualified to teach as I had relevant courses during my MSc and PhD in research and teaching methods. My MSc course was on Vocational Technical Education from the Graduate School of Education in Rutgers University in the USA. I had taken units in Tests, Measurements and Evaluation of students which became very useful in my teaching and grading techniques.

I wrote a letter of resignation through my section head and personally handed it over to Dr Bede Okigbo, who was the deputy director for Administration at IITA in Ibadan. He looked at me with surprise and placed it on the table. He said this: "So Ratemo, you plan to go back to your home country, Kenya? What has not pleased you here in Nigeria?" I responded to him politely that I loved Nigeria and my work. However, I had not planned on staying in Nigeria permanently. Instead, I had hoped to secure a permanent job in Kenya after my post-doc. I, however, assured him I would be back in Nigeria someday.

Prof. Okigbo was a competent administrator and very friendly to the IITA staff. He was a well-respected scientist and scholar from Umuahia City in Abia State, Eastern Nigeria. He released me and promised to act on the same. I now had the liberty of telling all my friends that I had resigned and was headed for Kenya within two months from January 1980. This was going to be the beginning of my long-term career in the academic world.

My supervisor, Dr O. Akobundu, was taken aback when I informed him of my last day at work despite having forwarded the letter. We were always very close as we planned the experiments and wrote the results together. He had hoped to retain me at the institute after my two-year contract and was working on that. I had learnt a lot from him. His wife, Regina, was equally surprised that I had decided to leave IITA. We had become close and they always advised me where to stay when I was in the East as both of them came from Umuahia. Dr Akobundu was busy writing his book on 'Weed Control in the Humid and Sub-humid Tropics'. Since I was his assistant in the weed science section, he delegated some responsibilities to me, and devoted most of his time to writing the book. I guided the research agenda and was able to defend any issues pertaining to weed research.

I learnt from him the techniques he used in writing his book and I later wrote a similar one on the *Weeds of East Africa* with a translation in Kiswahili. My stay in Nigeria had, therefore, been academically and financially rewarding. I kept a good network of scientists whom I met at IITA.

I planned my schedule. I had to sell a few personal effects that I had accumulated and gave some out to my workers who were very kind to me. We had worked as a team and colleagues and had had some memorable times together. I recall an incident with a snake in the bush as we set out experiments. A python crawled into the car driver's seat and coiled itself on it. It was a hot afternoon and the only shady place was inside the car. The driver had left the door of the car open to allow for air circulation. They readily killed it and roasted it for a meal. Snakes are consumed in Nigeria and my guys had an easy task of getting a ready meal inside the car.

Another incident occurred when I was spraying a herbicide at Ikenne field station. I encountered another medium-sized snake which charged at me. As I ran away from it, my technician, Morana, gave it a chase, killed it, made a fire and roasted it for a meal.

On our final meal together these were some of the memories we reminisced about. The workers loved working with me because I used to take them to the field for research, hence they made extra allowances. I had also campaigned for them to get extra pay whenever we got a new project for efficacy herbicide trials.

My Preparations for Departure

I was supposed to report at the University of Nairobi on 30 March 1980 and start lectures the following week. I banked my money in a New York Bank and changed some to naira for use during my last few days in Ibadan. I cleared my bills and closed the Nigerian bank account. I made sure that all my documentation was in order, including my International Driving Licence which would help me drive around Nairobi once I arrived there.

I knew that the world of academia was different from a research institute and so I prepared all my academic credentials and the CV for presentation to then then Vice-Chancellor, Prof. Joseph Maina Mungai. One assignment I accomplished efficiently was the preparation of my lecture notes. I had been given the course title and topics which were mainly in Weed Science. I utilized my free evenings and weekends so well that when I left Ibadan, I had fully prepared course outlines and the accompanying lecture materials in order under every title.

I consider myself an average but persistent writer. I utilized all the available latest international journals to prepare my lecture notes. IITA, in Ibadan, had excellent library collections. Mrs Braide, the documentation manager, was so kind to me and she gave me several articles on Weed Science in the tropics. She took the trouble to source articles from other sister institutes like International Rice Research Institute (IRRI) CIMMYT and ICARDA and FAO.

I shipped several hard copies to Kenya and made sure that all data on my research slides were documented, arranged chronologically for each circumstance and shipped. My friends had arranged a farewell party for me. They made good remarks about me, my interactions with the post-docs and other staff, my research accomplishments and my love for the Nigerian cultures. I thanked them for their hospitality, assistance and promised to visit them once in a while.

Dr Amare Getahun, a close confidant and former mate in Rutgers University remarked that I should go home to Kenya and start a family. I considered him an older brother. I used to stay in his house whenever he travelled out with his family. His wife, Elizabeth, saw me as a young boy ready to discover new academic frontiers in my home country, Kenya.

They gave me some mementos which I still keep and are in touch even after they returned to their home country, Ethiopia. I bade all staff at IITA good-bye and planned to fly to Nairobi on 28 March 1980 through my familiar Murtala Mohammed Airport, Lagos. I got the ticket ready to fly by Nigerian Airways.

I was dropped at the airport by an institute vehicle with my two large boxes, and a briefcase stuffed with my certificates and lecture notes. The Nigerian Airways destined for Nairobi took off at night and landed at the Jomo Kenyatta International Airport in the morning of 28 March 1980. I was ready to embark on my lifetime academic career with zeal at the age of 30.

Academic Life and Leadership at the University of Nairobi

I arrived at the JKIA and there was transport ready for me. Mr Stephen Mwita had a Volkswagen kombi and had come with an assistant registrar who organized for hotel accommodation at the Milimani Hotel in Nairobi. There were no readily available houses for staff then. I collected my luggage and headed for the hotel.

The University of Nairobi was the only institution of higher learning in Kenya then. The staff and student numbers were small compared to the current ones. The assistant registrar discussed various issues as we drove through Nairobi with hardly any traffic to slow us down. As I was being booked at the hotel, I asked the registrar what time lectures started; he told me 8.00 o'clock in the morning.

I knew classes were to resume on 30 March 1980. That was why I had prepared my lectures in advance and got my last pay from IITA in March before I took over my teaching position at the University of Nairobi. It was an excellent idea since I did not have to prepare for every lecture. I had prepared adequately for my two units. The MSc and BSc courses I taught covered 45 hours each per term. I only used half of the materials I prepared.

I made plans to be receiving IITA journals monthly to update my lecture notes and also kept abreast of the research progress there. Messrs Steven Utulu and Ray Unamma eventually completed their PhDs and sent me their graduation photos. I was impressed with their work which I had partly supervised. I wrote back and congratulated them. I was proud of their achievements as I had played a role in their research programmes.

The beginnings of the 1980s were, however, turbulent times both for the country and the University of Nairobi. Students and staff were engaged in national politics. The government detained rebellious staff and student activists were expelled from the university.

Hence, I got into a situation where academic programmes were carried out in a hostile environment with frequent closures. There was interference with curricula, frequent staff exodus for fear of imprisonment or detention. I do recall that there were restrictions on what we taught, the books we used or even research we conducted. This affected staff in the humanities and social sciences. Guest speakers to the university were vetted and at times rejected. The promotions of academic staff were based on political sycophancy. This widespread fear and intimidation adversely affected the quality of university education and consequently inhibited the future development of the country.

Student leaders were regularly expelled and the Academic Staff Union was banned in 1980. The August 1982 coup d'etat attempt forced the government to tighten its grip over the university. The coup attempt resulted in the longest closure in the history of the university. The government closed it down soon after the coup was thwarted and did not reopen it until October 1983--a period of fourteen months. It was clear that the university administration was being manipulated against their will.

The coup attempt and the subsequent closure of the university had far-reaching implications at the institutional and personal levels. A backlog of students waiting to join the university emerged. It is due to this backlog that we had to introduce a double intake of students in 1987, severely straining the facilities. I chaired the powerful committee which surveyed the appropriate campuses to establish the facilities available to admit extra students. I was directly affected in these circumstances. Many bright students missed lifetime opportunities; lecturers had to double-teach and ensure quality was sustained.

Little research was conducted then and the general public looked at university lecturers and students as rebels. As a member of the academic community, I supported the course of the staff and student unions, and the issues they were articulating. The staff and students were not political in the sense they were perceived by society, but sought continuous improvement in academic standards. You cannot talk about standards without addressing issues of working environment, books, remuneration, infrastructure and trained personnel. This was not politics in the sense peddled by the political establishment and perceived by society.

At a personal level, my ambition was to excel in academic work; hence I engaged little with the politics then. I was in a familiar territory, Nairobi, Kenya, and I started speaking Kiswahili, a language I had missed, with students and staff. I did not need to acclimatize in my own country.

I was picked from the hotel the following day and dropped at the Vice-Chancellor's office, first to report to him, and secondly, to hand in copies of my certificates and curriculum vitae. Prof. Mungai was one of the finest men that I ever interacted with. I arrived at his office at about 8.30 am and he saw me as he walked past the reception. He called me in and we started talking. My first impression of the man lasted in my memory for the duration I was with him. He laughed as he talked, cracked jokes about Nigeria, about the coup d'état and even asked me how I had survived for two years. He asked me about the University of Ibadan Teaching Hospital but

I did not know much about it. I had visited the campus a few times but could not delve into academic programmes there. I, however, knew the main campus pretty well.

I found Prof. Mungai a down-to-earth administrator, respectful and responsive. We became very close later on as I matured in the system. The hearty discussion we had that morning of 30 March served as my interview. He called the academic registrar, Mr Solom Karanja, who took my papers and was instructed to give me a formal letter of appointment with all the details for a lecturer's post. From then on, Joseph referred to me as the Weed Science man from Ibadan. He was so proud of me that he used to assign some duties like chairing disciplinary cases, junior promotion reviews and student complaints.

When I reflect back on my earlier experiences, I realize that humility is a core trait in public relations. Later as an administrator, I always recalled Mungai's humble reception of a young person who wanted to develop a career. I employed similar conduct in all my future interactions with my staff, students and the public. I learnt the art of being obedient, receptive and accommodating to my staff and visitors. I had first-hand knowledge in dealing with very sympathetic cases of minor thefts due to poverty and long-awaited promotions. I made appropriate decisions which many times favoured the junior staff.

I left his office and went to my campus where I would work for over 33 years – my domicile, despite taking leave of absence to work for other institutes. I took copies of my testimonials again to my department chairman, Prof. David N. Ngugi, who became my immediate boss for close to ten years, as a chair and dean.

David received me very well and congratulated me on joining his department. I spent quite some time with him and he was glad that I was ready to take over courses which for a long time were taught by a Dutch lecturer, Van Eijten. Despite having no appointment for a meeting with him, Prof. Ngugi, took the trouble to give me an orientation of the university and its role to society and government. We had tea together. I was also introduced to a long-serving secretary, Ms Jane Mbugua (now deceased). A good number of the staff I met in Kabete Campus were at the rank of assistant lecturers who were pursuing their PhD degrees. Many had also gone out of the country to pursue their PhD degrees. I was therefore senior to them in terms of hierarchy.

Prof. Ngugi again scrutinized my CV and appreciated the publications I had presented. All of them covered various aspects of Weed Science both in the USA and Nigeria. I was happy that both the VC and Chair recognized my efforts in the academic world. Since I knew we would be meeting more often, I asked him, if he so wished, to assign me another staff to take me round. I figured that he was a busy chairman. He declined and instead took me around the campus himself. I thought this was, a good gesture.

We had a quick tour of the campus; saw the greenhouses, field trials, laboratories and a herbarium. He finally showed me to my office and gave me the keys. Apparently, he had reserved one office for me because I kept on assuring him about my joining his department soonest. He made one comment as we were parting that morning to the effect that the other courses I took in environmental sciences would be of use in the future. His words were realized about 15 years later when I started teaching several units on environmental sciences.

He left me with a technician whom I would work with in the Weed Science Section, Mr Francis Kiinjanjui, a former plant curator with the East African Community in Muguga, Kenya. Mr Kinyanjui followed instructions and settled me in my new office which was much smaller than the one I had at IITA. There was a desk, some reference books and an old chair.

The office was in the Faculty of Agriculture and faced nice greenery and beautiful flowers. This was where I spent most of the time teaching and conducting research. After my administrative tour, I returned to the same office in 2006 and shared it with my student, Dr Safary Ariga, whom I had taught and supervised for his MSc and partially PhD. There was also Ms Parin Kurji, who was using the same office as a temporary working station. The number of staff had increased and taken over offices.

Sharing an office was a humiliating and debasing act which showed the poor management of both the chairman of the department and dean of the faculty and relevant administrative organs. It meant they never cared for their returning colleagues just as much as they did not care for students' facilities and welfare. The non-caring attitude of those in authority was a manifestation of the declining services and academic standards in many public universities.

Kabete Campus is about twelve kilometres from the main University of Nairobi Campus. It is to the west of Nairobi City and enjoys a cool climate all year round. The campus is located on a semi-plateau terrain overlooking Nairobi, Mount Kenya, the Aberdares and Kiambu Shopping Valley.

The greenery which surrounds the buildings makes it the finest of campuses to work in. The two faculties, Agriculture and Veterinary Medicine Sciences are the oldest in the University of Nairobi. The master plan of the campus has catered for a lot of greenery when new structures are being planned. The Wangari Maathai Institute of Environment and Peace is located in the College of Agriculture and Veterinary Sciences (CAVS). This makes the place a perfect environment to work in within the suburbs of Nairobi. Those who work here may not appreciate this scenery until they visit other campuses.

I have spent close to half of my life working in this pristine environment. The beautiful golf course adjacent to the campus offers the largest free open land with well-manicured lawns for teeing. A walk through the 18-hole goal course makes one rejuvenated. Staff enjoy playing golf when they are free. I am a member of the club.

Upper Kabete Campus is located in a high prime land where livestock and crops research takes place. The Faculty of Veterinary Medicine is the oldest in East Africa and trains students from the whole region and afar. On my return to Kabete I quickly settled down and started teaching two groups of students: BSc and MSc. I started to question myself what else I could do besides teaching and carrying out research. My departure from Ibadan, Nigeria, was so that I could contribute to Kenya in scientific development and advancement. I knew I would do well in my duties and so why I started to plan. I had a few free days each week when I did not have classes.

I had to devote my free time to conducting research trials in and outside the campus and, at the same time, offer free counselling and advice to young students. I also decided to join local and regional professional organizations. This worked. I applied for research funds from two organizations: the National

Council for Science and Technology, now called the National Council for Science, Technology and Innovations (NACOSTI), and the International Development Research Centre (IDRC), a Canadian-based research funding agency.

Each organization applied different methods of awarding research grants. The requirements varied from project to project and discipline. The local one, NACOSTI, made me write several elaborate proposals and it funded my research request with KES.11, 000.00 (about US$1,000). The chief executive then, Mr Starkeys Muturi, ordered me to go and conduct a survey on the occurrence and control of *Oxalis latifolia,* HBK in Kenya. He figured the amount would be enough. I thanked him and did exactly what he directed me to do.

Funding for research was not a priority in Kenya then, hence the limited innovations! The exercise was nowhere close to what I was used to in the USA and Nigeria. I did the work with the aim of applying for more funding. I wanted to keep and continue with my research tempo and habit. I quickly realized that if I did not maintain my objectives, I would easily drift into other social activities which would not benefit me in the long run.

The Canadian funding agency, IDRC, was more generous. In 1982, I wrote a detailed research proposal on minimum tillage for maize production. I had done similar work before in the USA but on wheat. An officer from IDRC, Roger Kirkby, went through the proposal with me, explaining the Canadian requirements in a fundable research proposal. I included all the items necessary to conduct the experiment for five years.

Luckily for me, I had written similar proposals before and had received funding. We had also been taught proposal writing in college. I got handsome funding which included, among other things, a double-cabin pick-up for ferrying research tools. The research covered several stations in Kenya. The vehicle was for use by my collaborating researchers at the site, but I allowed the department to use it also. This was my first research breakthrough in the early 1980s. Just like in Nigeria, I carried out trials across the country.

The trials involved growing the maize crop in partially tilled fields and we used selected herbicides to control the weeds and raise the maize crop under the least disturbed soil cultivation. This would reduce soil erosion in fragile areas. During my fieldwork, I was exposed to many challenges which were unique to Kenya. Weed species differed from location to location. Rain patterns posed a challenge as they could fail in some seasons and farmers would therefore lose a whole season's crop. I learnt the hard way to understand and appreciate the natural climatic changes from one location to another, recognized the variations and had to adapt accordingly. The research grant impacted a lot of staff in the faculty. Staff started applying for funds and a good number of them secured grants.

I trained two MSc students under the project. There were other programs within the Crop Science Department, but of lesser magnitude than the one I secured.

History repeats itself. During my field trials as I was busy spraying my experimental plots in Nyanza, I encountered a huge python in a bush coiled around on the grass ready to attack me! I was quick to retreat and called my helpers to untie the knapsack sprayer I was using so that I could run away. I recall vividly that sunny morning. Unlike Ibadan, Nigeria, where they eat snakes, my staff got scared and we abandoned spraying the site until it was slashed and cleared of the bush. I developed a phobia for snakes and whenever I go near a bushy area, I always look carefully and carry a stick in case a snake appears from somewhere.

As usual, the data I collected was analysed to find out the effectiveness of the treatments. The programme was funded for eight years and I was able to present papers in the Weed Science Society of Eastern Africa (WSSEA) which I was also instrumental in its formation. I was secretary and later chairman of the professional society which existed for about fifteen years and fizzled out when I was appointed Deputy Principal to the Jomo Kenyatta University College of Agriculture and Technology (JKUCAT), the then University College of Kenyatta University. I was bothered that such a regional society which covered several countries could just die off like that.

We met biannually and present proceedings. I was the chief editor for the proceedings. I also helped design the logo and recruit members. The countries included: Uganda, Tanzania, Ethiopia, Malawi, Zambia, Zimbabwe and Kenya. We shared a lot in common and meetings rotated from one country to another. The last one I attended was in Kampala and was opened by the then Minister for Agriculture. He praised the society for the good work we were doing. I was also secretary and later chairman of the Kenya Agricultural Teachers Association (KATA). The association was meant to promote agriculture in Kenya through teaching at all levels of education.

The association brought all agricultural teachers together, both secondary and universities. I recall senior members, like Prof. Shellemiah Keya, being a very active chairman of the association. Dr Nathan Kathuri served as secretary for a long time and made sure that the proceedings were produced on time. Kenya Institute of Education gained a lot from the annual meetings we used to have. The secondary school syllabus drew a lot of material from our proceedings.

The last memorable meeting we had was held at Kenya's Mombasa Beach Hotel. We had invited the Vice-Chancellor of the University of Nairobi, Prof. Joseph Mungai, to be the Guest of Honour. He accepted the invitation and I was to arrange for his travelling logistics. When I asked him how he wanted to travel, he suggested that we use a train. Mombasa to Nairobi railway transport was very good those days; I saw no reason for us to take a bus. The rest of the team went by road; Prof. Mungai and I were booked in the first class coach to travel overnight. We had planned the travel together and he carried books to read and edit.

This was the time when I knew the type of scholar Prof. Joseph Maina Mungai was. We discussed all sorts of things the whole night. The professor told me stories of his parents, how he grew up, walked to school, witnessed the resurgence of Mau Mau fighters, his academic progress to the highest. He explained how he studied under some difficult situations before qualifying to join Alliance High School.

He finally went to Alliance High School as a small boy from Muguga Primary School and eventually became the great medic and a vice-chancellor of the highest institution of learning. I got to know him well and deeply admired his intelligence, achievements and determination. He was a down-to-earth scholar who studied in Makerere University College and smuggled cadavers to Kenya to start a medical school.

That was the kind of man I interacted with for a reasonable period of time. In fact, mentors are never short to consult. Prof. Mungai was my mentor when I was a young lecturer and later a VC. I learnt humility and humour from him. I did not think that anyone in the University knew that we had travelled with the VC to Mombasa in the same coach for over 12 hours. I always reminded him of the trip whenever we met.

He stayed in the conference for a week and contributed a lot during the discussions. We travelled back by the same mode but, this time during the day. It was fun listening to the professor who told me the weirdest stories he could remember. He later wrote his autobiography and cited his past experiences.

As a vice-chancellor, Prof. Mungai witnessed some of the worst students' riots of the time. He was brave and very articulate in his speeches during functions like graduation ceremonies. I learnt a lot from this great man who steered the University of Nairobi through the most turbulent times in the 1980s. Little did I know that I would be a vice-chancellor later! My association with Prof. Mungai gave me some insights on how to run a university. He viewed me as his younger brother who needed his counsel to manage a university.

Later on, I worked with him at the Commission for University Education after his tenure as a vice-chancellor. If there is anyone I remember as my mentor, Prof. Mungai is the one.

Student Training

As a lecturer and researcher, my role was to train as I taught. I took this practice as the living Bible. I convinced myself that I had to train and supervise MSc and PhD candidates to completion. Even now, if I accept to undertake student supervision, I ensure that they complete their work successfully and on time.

One promise I made to my colleagues was that I had a role to mentor the youth and upcoming lecturers. During my tenure as a lecturer and later senior lecturer, the following students completed their PhDs.[1] I was sure of one thing during my supervision: my research agenda would proceed uninterrupted. I would keep up with the latest in science and technology and would publish research findings in the process. I also knew that the graduating students would run future universities and government ministries. I was not far from the truth.

I cannot at this point fail to emphasize the importance of research and teaching. One translates what one researches on into theory and practice. Continuity is therefore vital to keep abreast with new frontiers in science and innovations. I continued to receive newsletters from the local and international weed science societies to which I was a subscribing member.

We were very vibrant young scientists in the department. Many lecturers had projects for research. The number of students was small and each course had about 80 students. The quality of teaching and students' interactions were better than what we have now. Postgraduate students were also few, about ten per cohort. I trained more students during the 1980-1990 period than in the years 2000 to 2014.

The teaching load was heavier in later years. I knew more students by name and academic interest than I do now. Lecturing was carried out seriously and we had more time to prepare, plan, examine and deliver lectures than we have now. It was more prestigious to be a university don in the 1980s than it is now. The frequency of rioting was, however, more pronounced then than it is now. I will cover some of the causes and consequences in later chapters.

In 1985, I was promoted to the rank of senior lecturer at the university on the basis of teaching, research and publications. After three years, in 1988, a position of associate professor was advertised and I considered myself qualified and suitable candidate based on my qualifications. I had added more publications and students' supervision. I had further attracted research funds from the FAO on a project that I brought to the university. In addition, I had attended numerous conferences where I presented research papers. Two of us vied for the post, Prof. Kimani Waithaka and myself. We both passed the interview and were promoted to the position of associate professor.

After six years, in 1994, while serving as Principal of JKUCAT, the position of full professor was advertised in the University of Nairobi, where I was still a staff member on leave. I had by then consolidated all my research work and published in local and international journals. My FAO book on weeds had been published, and several other proceedings on the subject to which I was an editor. I had also represented the university in several conferences. I was fully involved in community service in Kenya and Eastern Africa as an administrator and an external examiner. I secured Commonwealth funds which enabled me to lead a team of scientists to conduct research in Malawi, and we came out with a publication.

My promotion to the position of full professor was not easy. I went through all the requirements and had to pass the interview, whose panel included selected professors of the university, with the chairperson of Council chairing. I consider those earlier interviews more demanding, rigorous and of higher standards. I recall my panel members as Prof. John Kokwaro, Fred Onyango (deceased), among others.

These days, requirements for promotion are often flouted and non-deserving cases get promoted while deserving ones are left in abeyance for long periods depending on one's networks and inclinations. There is also a lot of variation in the appointment of staff to senior positions, especially associate and full professors across the Universities, such that an individual qualifying as full professor in one university may not even qualify as a senior lecturer in another university.

This is because current practices, especially in the newly established Universities do not apply strict academic criteria for promotion. Some are politically influenced, while others lower the criteria to favour friends. The worst scenario is when members from one ethnic community are hired or promoted to a home institution without regard to their qualifications and academic track record. Chairpersons of Councils who are the appointing authorities are sometimes misled, and they allow or use the most ridiculous appointing criteria. Some are ignorant about university requirements and universities take advantage of this ignorance.

In extreme cases, some Vice-Chancellors, using political networks, become so powerful that Council members do not freely participate in Council discussions especially when they have divergent views. Suffice it to say that appointments to senior university academic and management positions in Kenya are still flawed and influenced by sycophancy, political and ethnic networks.

My Inaugural Lecture

Inaugural lectures are normally delivered by scholars on topical issues which they consider important. Having grown in academic ranks from Lecturer, Senior Lecturer, Associate Professor and Professor in the University of Nairobi, I decided to share my experiences through a public lecture. I had attended a number of such scholarly talks and admired them. These were academic discourses in one's field of study and they were the ultimate academic initiations. I chose to write on an environmental topic because it was a concern to many Kenyans who saw the rot in our environmental management. Despite my busy schedule and frequent travels, I was determined to accomplish my target.

My topic was "Environmental Degradation and Pollution: Let us Reverse the Trends" delivered on 9 September 2004. My area of expertise was Weed Science and Environmental Pollution. This was therefore an appropriate topic cover to my peers at the University of Nairobi and for the general public. It was easy for me to gather primary and secondary data since I was fully engaged in environmental matters. I had delivered several speeches during conferences and it was not difficult for me to compile data for my lecture. My idea was to bring to the attention of the public that we were destroying our natural resources and polluting our country at a speed which could render Kenya highly degraded.

Our rivers, lakes, wetlands, coastal marines, parks, urban areas and air were targets of pollution. Solid wastes were a major concern in the urban areas, especially in major cities where population was increasing day-by-day. I realized that we were losing our plant species and microorganism without a care for the future generations. Urban slums were on the increase as poverty continued to bite. A few people cared about the future, including some NGOs and a group of concerned individuals. My 30-page booklet defined pollution as "*the presence of contaminants in the environment in quantities, characteristics and duration such as to be injurious to human, animal and plant life or which unreasonably interfere with comfortable enjoyment of life*". I further defined degradation of land and ecosystems as "*the loss of productivity – qualitatively and quantitatively*". This could be due to mismanagement as a result of various human activities which include physical, biological as well as chemical processes.

I considered that scholars who may not be in the same discipline as mine wanted a simple and clear definition of the terms "pollution", "degradation" and "*polluter-pays-principle*". I cited very many live examples, both in Kenya and other parts of the world, which impact directly on the environment. My staff in NEMA, especially the Public Relations Office, went through the manuscript before I submitted it to the University of Nairobi Press to print the required 500 copies. I made sure there were no mistakes in the booklet.

I saw the Vice-Chancellor, Prof. Crispus M. Kiamba, who graciously accepted to officiate the event. Normally, the Vice-Chancellor would introduce the speakers. The University statute stipulates so. I was a University of Nairobi staff on leave of absence. This was a university function. He was delighted to host me. The public relations office of the University and that of NEMA worked together for publicity. They printed posters indicating the title, date, time and venue of my public lecture. It was also posted on the internet for wider publicity.

This was one lecture for which I prepared very well. I almost memorized the whole text to make sure that I did not make any mistakes during my speech delivery. I covered all areas which I thought the audience would be interested in. I knew well that issues on electronic waste, chemical disposal, raw sewer discharge and climate change would feature. I therefore did not limit my speech to the specific topic but covered a wider area than I had indicated. As usual, I went through my speech several times before the D-day.

The delivery was not like the graduation speeches which I would make and nobody would challenge me back. This was scholarly, high-level speech in front of the most learned group of people from Universities and the public. I had to excel. I knew this kind of lecture happened once in a scholar's life. I did not want to disappoint my peers and colleagues from NEMA. A good number of staff from UNEP also attended the lecture.

On the material day I went to the Vice-Chancellor's Office to gown and join other professors to escort me to the venue of the lecture. I must admit, I was a bit nervous. I recognized several renowned scholars of the University of Nairobi.[2]

After an elaborate introduction by my Vice-Chancellor, I was given one hour to deliver my speech. I remember it was 2.30 to 3.30 pm, a time which I adhered to. It was a hot afternoon and my gown, cap and hood did not allow for free air circulation around my body. I had dressed in a navy-blue suit. I later left the podium sweating. It was not an ordinary class lecture where I could talk for several hours without getting tired. It was a hard talk in front of scholars who would later on engage me in discussions and critique the talk.

Immediately after my speech, I got a standing ovation. Everybody present clapped their hands and I was relieved of the ùburden of talking on a subject that was so dear to me.

Two distinguished scholars took to the podium immediately after my speech. The first was Prof. David Wasao who I thought was going to throw a salvo at my inaugural speech. He instead called me up the stairs, raised his hand and shook mine in accolade for a well-executed talk. He was the chairman of the University of Nairobi Council and a renowned world academic. I was glad and my fears were assuaged. He actually said, "*Young man, well done and keep up the good work!*" These are seldom remarks from scholars.

The other scholar who was equally impressed with my talk was Prof. Canute Khamalla, who was my NEMA Board Chair. He had kept nodding his head during my talk. He looked straight into my eyes and thanked me. He was most touched by the practical examples I cited during my talk and wondered when I had gathered all the data I presented. He was visibly happy and proud of me. He told the audience that as his director general at NEMA, I was a man of class and substance who could drive the environmental agenda to higher heights. I was elated to get such good comments from my seniors. The 500 booklets were circulated and I never got extra copies for my friends. I could tell the happiness of my family members who had been equally nervous when I was speaking. My wife, Esther, later congratulated me and told me that I had made a great presentation.

The whole function lasted for 2 hours and by 4 pm the Vice-Chancellor and his entourage retired to his parlour for a cocktail reception. My wife Esther and children joined me at the reception where we stayed for one hour before going home. It was an excellent inaugural lecture and I felt satisfied to have successfully delivered it when I was 56 years old! This was my ultimate academic ascendancy and recognition. I later learnt that many upcoming lecturers were using my booklet for lectures. It had covered several aching environmental problems in Kenya which needed urgent attention.

I had seen myself through the ranks. I had experienced the odds and rough terrains during my services. I became a lecturer on 30 March 1980. I was then promoted to senior lecturer and subsequently an associate professor. I thought this was good enough, but I had another hurdle of being interviewed for the full professorial appointment. In all these promotions, proof of published work, research, student supervision, service to the community and fundraising were the prerequisites. Managerial skills were also considered then.

I then took on administrative roles. I was the chairman of the department, but was then taken out of the classroom to assist in the establishment of the 5th Kenyan public university, JKUAT. After a while, I was appointed the Deputy Principal (Academic) and later elevated to the Principal of the then University College. The elevation of the University College to full university status saw me appointed as the founding Vice-Chancellor of the new University, JKUAT. I was now fully converted from a lecturer, and researcher into an administrator, leader and advisor. I did not, however, give up my academic undertakings. I still taught and supervised MSc and PhD students as I carried on with my demanding administrative roles. I still published papers and my first Weed Science book of which I still have a few copies.

My administration acumen became more prominent when I was appointed to chair one of the largest agricultural research parastatals, KARI. There, I worked with dedication and acquired over 100 title deeds of parcels of land which had been grabbed by powerful persons in Kenya. These were large tracts of land meant for high-level, advanced research, especially plant and animal breeding. I became an enemy of a few but a darling to many. I was also appointed to assist in several other quasi-government bodies like the National Council for Science and Technology.

The establishment of African Institute for Capacity Development (AICAD) and revitalization of the Inter-University Council of East Africa (IUCEA) were other tasks that I took on and saw them succeed. I am proud of the two sister institutions of East Africa which are complementary in functions. All these undertakings involved negotiations, persuasion and fund-raising. As a manager, I had to utilize these skills. The programmes saw the continuous involvement of donors locally and internationally. Good networking and public relations played a major role in the establishment of the institutes. I utilized these strategies during my administrative tenure in NEMA.

The inaugural lecture culminated in the final achievement in the academic hierarchy. I had done my 13 years as Principal and Vice-Chancellor and I knew that there were no any other academic promotions. The only scholarly work left was teaching, research and publications. These have no end. One can carry on for as long as one is alert and capable of communication. Another role which I considered vital for me was service to the Kenyan community. I continuously helped those who needed my assistance and made appearances either in the press or public meetings.

I was awarded several accolades both locally and internationally. As a Principal, I was decorated, with State commendation, the Shining Star (SS) by the Head of State, President Arap Moi, which I considered a great affirmation of the contribution my work had made to Kenya's educational and societal development.

I was again, later on, honoured with another one by the same Head of State, the Elder of the Burning Spear (EBS). The honour was awarded to me when I became the first Vice–Chancellor of JKUAT. I very much appreciated the kind gesture from the Head of State who genuinely recognized the good work Kenyans were doing.

During the same period, the Japan International Cooperation Agency (JICA) also honoured me by awarding me a certificate for their best programme that I undertook. I had successfully supervised a number of JICA-sponsored projects. I was among the 21 prominent persons worldwide who were honoured by the president. I am still the patron of the Japanese Ex- Participants of Kenya (JEPAK).

In 2003, my Alma Mater, Rutgers University, USA, honoured me as a Distinguished Alumnus who had done exemplary academic and administrative services to my country, Kenya. During this occasion, my wife, Esther, and I joined world movie celebrities, Flockhart Hart and Harrison Ford in the event. Flockhart was my college-mate and she was also being honoured at the same time for her exemplary role as an actor. She was in the Mason Gross School of Performing Arts.

These accolades did not come easily; they carried a price tag of hard work, bravery and risk-taking. I had to make bold decisions and stand by them for the benefit of institutions. That was the prize I got for the work. The academic and leadership trails are not the same, but highly complementary.

The rise in hierarchy from lecturer to Vice-Chancellor follows a specific trend of events. I used a lot of past mistakes, errors, omissions, commissions and regrets to arrive at decisions. Did I allude earlier that institutions have no blood? They just drain without giving back. I also indicated earlier that institutions are images of their chief executives. Conversely, chief executives are seen in the same light as the institutions they run. The names evoke the true character of that head. We all know that fish starts to rot from the head, so do institutions. If a head is corrupt, so shall be the lower hierarchies and appendages. For a long time, JKUAT and NEMA became synonymous with my name. The influence and responsibility that I set forth laid firm foundations for the two institutions.

Notes

1. Dr Pihri Sibuga Kalunde now a Professor in Sokoine University, Morogoro, Tanzania; Dr David Kamweti, (deceased) a Private Consultant, Dr Safary Ariga, currently a lecturer at the University of Nairobi; Dr Leonard Wamocho, currently a lecturer and Dean of Students at Masinde Muliro University and Dr Wariara Kariuki, a lecturer at Jomo Kenyatta University of Agriculture and Technology. Others include Ben Kisielo Wanjala, James Ongusi Bango, Joseph Oryokot, Boniface Ita, Fridah Kiriri, Jowi, Olang'o, Lucas Ngonde, Mr. Balaa and James Baraza, just to name but a few.
2. Great academicians like Professor David Wasao, Prof. John Kokwaro, Prof. Canute Khamala (my NEMA Board Chairman), Prof. Shem Wandiga, Prof. Joseph Nyasani, Prof. Florida Karani, Prof. Daniel Mukunya, Prof. Lucia Omondi, Prof. Geoffrey Muriuki, Prof. David Ndetei and a host of lecturers, senior lecturers and the public.

References

Animal Farm, George Orwell, 1945 13. Around the world in Eighty Days, Jules Verne, 1872.
Degrees that are of little value, The Standard Xtra, July 8, 2013, page 1.
Did the British send Kisii Worrier's Head to London? Daily Nation, October 14, 2013. page 8.
Environmental Degradation and Pollution: Let us Reverse the Trends, Inaugural Lecture, Ratemo Michieka, 2004.
Integrity is about paying the price for doing what is right, not what is legal, Warah, Rasna, Daily Nation, February 25, 2013, page12.
Parents Lobby wants review of varsity exams, The Standard on Sunday, December 15, 2013, page 22.
Roaches shun sugary bait in adaptive survival tactics, The Standard, June 14, 2013, page 7. The Commencement Controversy, The New York Times, May 12, 2013, page 4
The Leadership Book, Charles J, Keating, 1982. Things Fall Apart, Chinua Achebe, 1958.
Varsity expansion must consider quality safeguards, The Standard,May 15, 2013, page 18.
Will Cockroaches Find Another Survival Trick: Laying Off the Sweets. The New York Times, May 24, 2013. page A 13.
World Bank Survey shows how teachers abscond classes, the Standard, July 20, 2013, page 6.

7

The Triple Balancing Act: Academics, Family and University Administration

Having settled as a lecturer, my next task was to get my life partner. I was advised by my parents that I needed a wife with whom I could live and work. They did not specify what kind of girl I should marry, but they hinted that I needed a wife from a reputable family and one who would relate to them respectfully. That statement was loaded. My other family members encouraged me to marry, settle down and raise a family. The idea was welcome and I started to search for an appropriate partner.

Gusii customs allow for formal and informal courting of a future partner. The customs also allow for relatives and close family members to do a search. Many African traditions use this method although the final decision is between the two persons.

I was introduced to my wife, Esther Nyabonyi, in 1980 by a group of relatives. I had heard of her and the information I got made me anxious to meet her. I sent emissaries and she did the same on learning about me. I was sure that many people knew about my interest and were closely watching my moves. The larger Gusiiland where both of us come from is not so large that one cannot know prominent people.

I knew through my church elders in Nyamagesa that my fiancée was the daughter of Mr Jackson Mokaya, a well-known teacher and pastor in the Gusii community who had done part of his missionary work in Nyaribari, my local home area. Her mother was one of the respected workers in many mission stations especially in Kamagambo in Kenya and Bugema in Uganda where Pastor Mokaya taught and evangelized. Her brothers, who were young students then, were also well-known in the community.

Esther was the only daughter of Pastor and Mrs Mokaya, and had eight brothers. My brothers-in-law are, in order of seniority: Stanley Memba, Samwel

Otora, Peter Ogera, Charles, Daniel Menge, Robert, Tom Migiro and Leornard Onduso. They are all professionals in diverse fields of study. Some reside in Kenya and others outside Kenya. We are always in touch and share a lot in common.

My family is large and spreads throughout the whole of Nyaribari District. My parents were also well-known in the community as staunch Christians who brought up their sons and daughter in a Christian way. My mother was instrumental in church-building and welfare matters of the community. She was a respected opinion leader in the area. Many pastors, especially during the camp meetings, stayed in our home. I would not be surprised if my father-in-law-to-be did spend some nights in our home.

Esther was teaching at Suneka Secondary School. I drove there and the headmaster was kind enough to allow me to see her. We spoke briefly and I left the school so that she could go on with her lessons.

I returned later to take her to Kisii Town for her transport home. I later made several visits mainly during the weekends to avoid cancellation of my lectures. I recall one Saturday afternoon that I went to her house; and her mother came home from church only to find a White 504 Peugeot parked outside. She was very polite to me, received me well, and then I left. It rained cats and dogs afterwards. Courting went on for over one year.

On 3 September 1981, Esther and I tied the knot followed by an elaborate party at Nyanchwa Secondary School in Kisii Town. Several people attended the wedding, including my close friends from my college. They came in the university bus. Friends like Prof. Kimani Waithaka, Eliud Gathuru, Moses Onim, Jane Mbugua and David Karanja attended our wedding.

September was a rainy period in Gusii. We were all soaked wet; cars got stuck deep in the red soils and all our wedding attire got muddy. It was the 'rainiest' day of the month. But we were comforted that those were God's blessings.

After the reception, driving to my home was even worse. We had to push vehicles to get home. It was a crowded, wet and blessed wedding. My parents were most excited to see their son get married. The rest is history, as we started to navigate through life as a married couple.

We requested for a transfer for my wife to Nairobi and it was granted. It was such a relief when my wife was posted to Mary Leaky Girls' Secondary School, next to my Kabete campus. I would drop her there and return to my office, just two kilometres away. Then we would go back together to Hurlingham where we stayed. It was a quiet neighbourhood with few people then.

I used to stay in a university three-bedroom flat on Argwings Kodhek Road, next to the Army Headquarters. University houses and apartments were so spacious and well-kept that all lecturers wanted to stay there. We were paying very little rent, basically for maintenance. The situation is different now.

This University housing arrangement proved to be academically rewarding as one had a convenient environment to pursue academic work. This situation had changed by the time I became a VC, with academic members of staff making their own housing arrangements in a manner that defeated the collegial academic culture.

Although the non-provision of university housing encouraged a few academics to own houses, in most cases, academics ended up staying in deplorable, slum-like conditions unsuitable for academic work and growth. I know of my colleagues who have lost students' academic documents under these circumstances. This has been caused by poor remuneration packages, particularly housing allowances.

We settled quickly in our flat and met several residents who were already staying there. Professors De-cock, Henry Mutoro (now DVC, Academic Affairs, University of Nairobi), Henry Maritim and Inonda Mwanje. We actually formed a soccer club. We used to meet, and the wives shared a lot in common. The twelve flats in two blocks had adequate space for sports, meetings and barbecue area. It was a lovely community of university dons.

On 11 March 1982, my wife delivered our lovely first son, Nyakundi M. Michieka. We thanked God and prayed for the baby as we got blessings from both our parents. Further blessings were witnessed when my brother David Ombogo's wife, Mary, gave birth to a baby boy, Samuel Goima Michieka, within the same week. They were born only a couple of days apart. The two were like twins as they grew up together.

Unfortunately, Samuel Goima Michieka, who was training as a medical doctor at the University of Nairobi passed away through a tragic road accident in Nairobi at the age of 22. The loss of our nephew, and the only son to Mary and David, devastated us so much that we are yet to come to terms with it.

Our son, Nyakundi, now a PhD holder, grew up with a strong desire to excel in his academic career. We later had a son, Dr Amenya M. Michieka, who was born on 11 May 1983, a daughter Kemunto Mong'ina, an advocate of the High Court, born on 20 July 1984 and Michieka Okioga, a medical doctor the last born on 22 July 1985. My wife, Esther, and I have given them the best parenting. They all attended Consolata Primary School in Westlands, Nairobi. My family life and the children's upbringing will be covered in another chapter in which I focus on my tenure as Principal and later Vice-Chancellor (JKUAT), followed by my tenure as Director-General (NEMA).

An Attempted Coup d'État

On 2 August 1982, barely two years after I started teaching and one year into marriage, the country became almost ungovernable. My stay in Nigeria was generally under military rule until Shehu Shagari took over as a civilian president in 1979. I did not expect our beloved country, Kenya, to fall under military rule. This was the least I expected.

One Saturday evening, on 2 August 1982, we all ate well and retired to bed. No one of us imagined that we could be awakened by gunshots all over Hurlingham. It was at about 4.30 am when we heard the gunshots. We thought it was a robbery somewhere in the neighbourhood. We lived opposite Kilimani Police Station and thought maybe an accident had occurred. That was it – a political accident. It was an attempted coup d'état.

We got up as more gunshots rent the air. The radio had a strange music playing and no news came through. I called my brother, David, using a landline and he told me that a friend had told him that the military was taking over the government. My colleagues in the Ambala Flats were all up by 6 am to find out what was happening. It was, indeed, an attempted coup d'état.

We monitored developments in the country on the radio, and heard veteran broadcaster, Leornard Mambo Mbotela, announcing that the government had been taken over by the military. Several military land rovers swept along Argwings Kodhek Road from the military headquarters. There was total confusion.

Our flats were less than half a kilometre from the army headquarters but we could not walk there to enquire. We also learnt that there were gunshots all over the city. We were all getting confirmation from many sources that the country was, indeed, under military rule. Further information trickled in that the University of Nairobi students and the air force had taken over the government. It was now dawning on all of us in the compound that we were under siege.

I knew that our little son, Nyakundi, had no more milk for the day. It was 7 am when only military land rovers were speeding along Argwings Kodhek Road. I quickly thought of a solution and decided to rush to the nearest petrol station which had a small shop at the present Yaya Shopping Centre to buy a few packets of milk. I took the car keys and drove as fast as I could.

I bought several packets of milk, two loaves of bread and butter. As I engaged the reverse gear in the panic of confusion, I hit a road kerb and dented my car rim. I nevertheless got the milk and bread for my son. I told my wife of the decision to get the milk first and explain later. In fact several military land rovers were now driving everywhere and advising people to stay indoors.

I was shouted at by the military personnel to go back to my house as all regular policemen and women had been ordered to stay away from duty. At least I braved to get the baby's milk but risked death. I considered my act brave even though it looked ill-advised. But, the fact was there was no milk for Nyakundi. My wife, Esther wondered how I had come back so fast, and alive!

Later at about 11.00 o'clock, Prof. Henry Mutoro and I walked to the shopping centre when we saw civilians moving hastily to various destinations. We walked amidst the crowd to the nearby city mortuary since everybody seemed to be heading there. At this time, there were fewer land rovers plying on our road. It was quiet but we could hear more gunshots in the down-town direction.

I later regretted going to the city mortuary! What we saw is beyond anybody's imagination. Hundreds of dead bodies were heaped all over the place one on top of another. It was unsightly, horrific, hideous, dreadful and utterly shocking! Bullet-riddled bodies were scattered everywhere.

Henry and I quickly left the place as more bodies were being hauled in. By midday on Sunday, total confusion prevailed. We went to our residences and found our colleagues standing outside, talking. We did not talk much as we tried to come to terms with what we had just seen.

Some radio announcements started to come over the air. As we stood outside, two air force helicopters flew past our flats towards Ngong Hills. We did not understand their motive, but assumed more fighting would go on between the air force personnel and the soldiers. We were told that the air force had staged the coup d'état but the army troops had crushed it; hence the several military land rovers on our roads at that time.

We retreated into our flats for more news. We were informed of the aborted coup d'état and that the air force personnel were being hunted alongside the university students who had supported the coup. The army, which was loyal to the government of the day, had restored some order and were busy flushing out the coup plotters.

My wife and I then became worried. We had several relatives in the university and their halls of residence were military targets for supposedly harbouring insurgents. My two brothers-in-law, Peter and Charles, were staying in the halls. There were no cell phones then. And we feared for our flats. It was known that Nairobi University staff stayed there and would certainly be an easy target for the army to come looking for rebels.

At six in the evening, my two brothers-in-law, Charles Mokaya and Dr Peter Mokaya, and a colleague, Dr Kambuni, came running into the compound for safety. They came in and narrated to us how their fellow students had been rounded up. Getting to our residence safely had not been easy. We were happy to see them safe and told them to lie low. We were informed that some people in the town centre were looting the stores. At the same time, shooting was going on unabated.

I do recall my late brother, H.T. Michieka, who had been caught in the fiasco, had to sleep hungry in the store in town to avoid being hit in cross-fire exchange. He was lucky to be alive and was rescued with several others after two days of total confusion. The coup d'état was unsuccessful and people slowly started to assess the damage.

Everything went quiet on Sunday evening. Normalcy resumed slowly and many people started going back to their residences. There were continuous announcements that all was now calm and the government was in control. But we were advised to tread carefully. The insurgents could charge at innocent citizens in their last attempts for survival. My brothers-in-law heard that students were being followed, searched and hunted down rigorously. Many may not have had anywhere to go.

A good number of students were put in police cells. The search continued for several days as the city slowly returned to normal. I recall one incident when the late Robert Shaw (of Starehe Boys' Centre) cornered a university land rover carrying students at the Roysambu roundabout. He stopped it at gunpoint, ordered every student to get out and lie down. Perhaps, they might have been using the vehicle to escape. They were later released as they had no weapons.

A Risky Trip to the University Dispensary

The country was assured of security and that Kenyans should carry on with their work. That was on Tuesday, 4 August 1982. Our son, Nyakundi, again was due for routine clinic check-up. I innocently told my wife that dates and times were important for children's clinic: *"Let us take the young baby for the normal check-up,"* I told her. We prepared the baby, got into my car and drove past State House Road to Nyerere Road headed for the University of Nairobi dispensary.

On arrival at Nyerere Road, opposite Serena Hotel, we saw a military truck full of armed soldiers pointing guns at any coming vehicle!! Nyerere Road is one-way and all vehicles were going towards that truck. They were searching for air force boys who might have been hidden in private cars. I had to raise the young baby high up, with my ID aloft and my wife had to explain the circumstances which made us leave our house.

We were told to make a u-turn and return to our flat as fast as was practically possible. I took a sigh of relief, rammed onto my accelerator and headed back home. That was the end of my son's clinic. It was my second risky venture within two days which I later considered silly on my part. Yet, our son's health was still paramount in my mind. The love for our children was and is still immeasurable.

After the attempted coup d'état, Kenya was never the same in many aspects. Prices of commodities shot up. There was a lot of distrust among people. Staff and students were seen as traitors and suspects, and detentions increased. All security machinery was always on the alert and then President Daniel Arap Moi made a lot of changes in his government.

It was during the 1980s and 1990s when public universities in Kenya recorded the highest number and most destructive riots in the country's history. As a lecturer, I continued to teach and conduct my research unabated as I had to account for every donor coin spent, despite the closures. I recall the Vice-Chancellor, Prof. Philip Mbithi, remarking that universities were like electrical switches which can be switched on and off instantly. He meant that we were opening and closing Universities so frequently that we had to be on the alert at all times.

University education became expensive, unpredictable and unpalatable to the nation. In fact, lecturers were viewed as the worst individuals to the nation and any crime that one was suspected of committing attracted immediate detention. There is a chapter on Kenyan University riots and their likely causes later in the book. We soldiered on despite the sudden unpredictable and long closures.

Whenever students rioted and universities closed, the terse statement to order closure was so brief and did not indicate when to open. The affected groups, the students and staff, were left in abeyance and were given no timelines to expect re-opening! Everybody was left in suspense!

Suddenly, Senate would be called by the VC and dates for re-opening would be proposed. Once the proposed dates were agreed upon, an announcement would be drafted for the media to advise all those concerned that lectures should resume on that date without fail. It did not matter where the staff or students were, they just had to report and resume lectures. The lecturers were treated like automatic switches which you could switch on and off at will. The one switching would assume there was always power and the bulbs or whatever had to provide light. What the lecturers suffered was a typical Kenyan way of approaching life. For the students, they always had to be alert for re-opening announcements and it was assumed that they had money to pay bus fare to travel to their campuses without fail.

University closures are the most expensive decisions that the VC and Senate undertake. It is an agonizing decision to make! That is why I never rushed to close JKUAT whenever there were problems. I opted for protracted negotiations with the affected groups.

Family Responsibilities

Not a single day did I miss attending my children's open days unless I was out of the country or engaged in an important function. Luckily, visitation days used to be during the weekends.

My thirteen years at the helm of JKUAT took their toll on my life. The demanding times during my mid-forties, at a time when many people want to consolidate their lives were spent bringing up two great institutions in my life. I fully balanced my roles as a husband, father and family man with those of being a principal and eventually a Vice-Chancellor. My family needed me full time and so did JKUAT. I also had to ensure that other duties as chairman in some parastatals were fully executed. I promised not to compromise my integrity. Above all, I continued lecturing and supervising graduate students – hence my eventual return to the University of Nairobi as a professor.

My family was very young at the time I was appointed a deputy principal to JKUAT. Our first child, Nyakundi, was eight; Nyakundi and Amenya were already attending Consolata Primary in Westlands and Kemunto was in a kindergarten at the same compound. My wife and I decided to relocate to a central place in Westlands instead of commuting from Kikuyu Township to Juja and back daily. We left our constructed house with a large compound in Kikuyu, Kiambu County and moved into a house on Rhaphta Road in Westlands which was not as spacious as ours. At the same time, my wife transferred to Ngara High School in Nairobi from Rungiri Secondary School in Kikuyu Township.

Our children continued attending the same school. My wife, Esther, and I were very particular in ensuring that all the requirements for classes were provided. She could prepare packed lunches for the three kids daily. Michieka Okioga later joined the same nursery school and we then had four children attending Consolata Primary School. We knew virtually all the teachers and the head teacher of our kids. I still meet them in shopping malls at Westgate in Nairobi and we make reference to those early days. They still remember us as dedicated parents who love their children.

Usually, my wife would drop them and proceed to her school, then the driver would drop them back at home. There were no traffic jams in the mid-1980s and early 1990s. Driving in Westlands was a pleasure. We used to get book lists for each child in November of each year. I would buy all the required school books during early December to avoid the after-Christmas rush. I also bought their uniforms in November to avoid the January rush and long queues.

I must confess here that I would stop every activity and spend any available money to ensure timely payments of school fees for the year. I did not like getting reminders on delayed fees payments. Equally important was the fact that none of my children missed classes. They were obedient and we always shared with them the social good of education.

My wife is a renowned teacher and she could always relate to them how wonderful it was to be educated. As a head of an institution, I could also demonstrate to them the beauty of being educated. For example, those who graduate wear long, beautiful gowns of academic power during the graduation ceremonies. They would hear me during the news and watch me on the television, especially during graduations. We continuously inspired them and they had role models. We used to tell them the beauty of being learned, educated and independent. Education can give you skills and opportunities.

As parents we promised ourselves one thing: That all year round, we had to attend every single academic career day and parents' day without fail. When I was young attending my primary school at Ibacho, none of my parents ever came for any school function. They were rural, agrarian church-goers who thought that it was not necessary to visit my school. They, however, encouraged us to work hard. But I do not recall them ever attending any school function for any of their children. I understood their attitude since all other kids during my time saw no parents in the compound. In fact, we had no formal parents' days except for during classroom construction. My father was always ready to take part in building classes for the school.

Back to our visiting days, Consolata Primary teachers were hard working and they always pointed out the strengths and weaknesses of each and every pupil. I used to be the first or second in the queue waiting to see the teachers. It may be instructive to point out here that to be able to see eight independent teachers for the eight subjects meant queuing eight times from 7 am to around 4 pm. We usually met on Saturdays and this meant that I would forego my Seventh Day church services. I thoroughly enjoyed interacting with the teachers who gradually became our very dear friends.

I took notes for every subject for each child and on returning home; I went through all the weaknesses and strengths of each child for every subject. I spent adequate time telling the kids what their teachers had said about them. This was very powerful in instilling discipline in them. Luckily, visits were arranged at different weekends and terms. Where necessary, I heavily reprimanded my children. I would explain and re-explain any weaknesses in specific subjects. I would also praise them in subjects where they had done well. I would then append my signature on their report cards. I still have all their end-of-term results for posterity. I would request my wife to continue with advice on the basis of on any serious observations or feedback from the teachers.

When my children joined secondary schools, the same academic and parents' days prevailed. Having done well in class 8, Nyakundi and Amenya joined Mang'u High School on Thika Road, five kilometres from my JKUAT campus. I knew that this proximity had some psychological implications for them. They saw my shadow next door, I was certain, and would therefore behave.

Furthermore, at this stage of their schooling, I found no reason to visit them unnecessarily. The principal of the school then was one of the finest disciplinarians who surely minded his work. He was Mr Paul Otula. I knew Paul and his brother, the great science writer Owuor Otula was my class-mate at Kisii High School in 1965. I knew them as good performers in whatever assignments they took on.

The principal was the national chairman of Kenya Basketball Association (KBA) and I knew that many of his students were good basketball players. The tall, strongly-built gentlemen exuded authority in Mang'u. The school was the pride of Central Province at the national level and beyond in terms of examination performance. Certainly, Mang'u High was one of the best schools in Kenya in terms of discipline, basketball and KCSE results. It rivalled Starehe Boys' and Alliance High. I used to visit the school on speech days and give career talks whenever I was invited by the principal. The school was 6 kilometres away from my JKUAT campus. I used to invite the boys for functions at JKUAT.

Nyakundi and Amenya finished their schools and were admitted into our local universities to pursue higher education in Engineering and Medicine respectively. Our daughter, Kemunto, joined Alliance Girls and later studied Law in Moi University, Eldoret. She is now an advocate of the High Court. Michieka Okioga joined Starehe

Boys' under the able directorship of Mr. Geoffrey Griffins (now deceased). Mr. Griffins, who started the centre, was a no-nonsense administrator. I knew him well and he used to invite me to give talks to his boys during career days.

Starehe Boys' Centre was one of the best-performing schools in Kenya for several years. It was an ardent academic competitor with Alliance High and Mang'u High schools during the Form Four (KCSE) examinations. Again, we visited these children every term for the four years of their secondary education. I never took a chance or excused myself unless I was out of the country. My wife would prepare some food and we would all share it at their grounds alongside several other parents. I still meet some of the parents that we used to interact with those days. These were my most relaxing days.

One big lesson I learnt was that children enjoyed whenever their parents visited them. They would then take their class work seriously because they knew that the teachers would monitor them closely and inform their parents of any misconduct. Boarding schools were particularly vulnerable to vices. As parents, it was possible to detect misfits early and correct them. There is no higher parental pride and esteem than school visits to well-performing kids.

The ultimate parental satisfaction was when we as a family attended Nyakundi's PhD graduation in Economics Engineering in West Virginia University, USA. Dr. Nyakundi was among the first Mechatronics students at JKUAT which had only 22 students then. Dr. Amenya's graduation in Dar-es-Salaam was equally superb. He graduated in Medicine and later continued to specialize in Paediatrics. It was a joy for us to also attend Kemunto's graduation at Moi University and later witness her admission to the Bar after her pupillage at the Kenya School of Law. She later graduated with a Master's degree in Business Administration from Baraton University in Eldoret.

We finally ended with the huge graduation ceremony at the University of Nairobi where Dr Michieka Okioga was amongst those who received their degrees in Medicine and proceeded to specialize in Gynaecology and Obstetrics. On average, we tried to ensure that our children got reasonable education which would make them competitive in Kenya and even beyond.

As parents, we are deeply grateful to their understanding and their love for us. We treated them as adults from their youth with no tolerance for any mischief. However, boys are always up to something and Nyakundi and Amenya were not any different. For example, they damaged my car despite having received good driving lessons. I condoned the error by putting them back onto the driving board. We thank God for having accorded us an understanding family.

Just like any youths, we allowed the young persons to go out and socialize with their friends, but my wife and I kept a close watch. The area we lived in had the highest concentration of night clubs and eateries. I occasionally monitored them in these joints.

My Role as a Vice-Chancellor

My students and staff at JKUAT also demanded similar attention as my family. My duties were to manage and guide them. As the fifth youngest public Kenyan University, JKUAT was seen as a privileged institution which received technical aid from JICA. It was viewed more like a Japanese college than a Kenyan one. Over time, we built an unparalleled reputation and the demand for admissions was overwhelming.

Many students who qualified for university admissions wanted to join JKUAT. My students nicknamed me Uncle Ratemo and I so happily responded. I had an open-door policy but was precise in my responses. I never entertained long stories from both staff and students. Most of my responses were often in the negative and rarely in the affirmative. If, indeed, I could not solve a problem I did not have to offer long justifications before arriving at a negative answer. It was, and still is, a sheer waste of time. 'No' is as good an answer as 'yes'; but that is something Kenyans are yet to learn.

There were several examples of students' unrest which solidified me as an administrator. I must confess here that JKUAT, like any other Kenyan university, had disciplinary problems. We dealt with many cases, some of which had bearing on the university's security and reputation. I had very good deputies who normally chaired such cases. We equally had staff problems which could also cause staff unrest. In both scenarios, our problems were not at the magnitude of the sister Universities.

In fact, JKUAT was considered the best university for students to enrol in. Despite a few insignificant threats by student leaders, we were on academic schedules for all the thirteen years I worked here. We graduated our students several months ahead of other public universities. This practice encouraged parents, guardians and sponsors to continue bringing their students to JKUAT.

Academics and higher education administrators often cite the struggles they go through to balance the demands of their real families and the wider family of the academic community. Ours was not different. From the onset, I had to learn to multi-task while ensuring that academic and administrative work was never carried home. Nothing went pending unless it was something that needed wider consultation. In this way I created adequate time for family and academic work, without interfering with the progress of the other. I also got strict with time management.

I planned my work ahead and accomplished tasks ahead of time. Many times I would make mistakes, but it was better than postponing and leaving work pending. I also advised my deputies to ensure that a day's work was completed, to avoid backlog. They understood and appreciated my demanding roles. I loved to delegate roles and functions to my deputies and anybody I considered would deliver. This was my way of assessing my staff without them knowing. Most of them performed well.

Nyakundi excelled in the course and graduated after five years. My family and several relatives attended his graduation ceremony in Juja. We later held a party at our residence and invited his friends to enjoy. He applied for an MSc course and was admitted at Stroudsburg University, USA, where he studied Natural Resource Economics. He graduated and later applied for a PhD at West Virginia University in the USA to pursue the same discipline.

He received his PhD in May 2013 in the same field. I attended his graduation which brought all the family members in the USA together in Morgantown, W.V., USA. It was one of those fine moments that remain a landmark in Dr Nyakundi's academic development.

He was supported by teaching assistant grants while pursuing his studies. In summary, my son Nyakundi is a fine, fast-moving and decisive young man. He has an excellent personality with a high sense of humour. He is an independent thinker. He is an accomplished and a trusted individual who is truly reliable. We shaped his character as parents. He is a post-doctoral fellow in the UWV.

Amenya, our second son, was also in Mang'u High School. Nyakundi was already there and Amenya settled quickly under the watchful eye of his older brother. He passed his Form Four well and was called to join Moi University, Eldoret, for a course he did not want to pursue. He turned down the offer and opted to go for Medicine in Hubert Kairuki University in Dar-es-Salaam, Tanzania. Amenya worked hard and graduated with a Bachelor of Medicine and Bachelor of Surgery (BMBS) after five years. He was a student leader during part of the time in the university. I recall meeting his Vice-Chancellor, Prof. Esther Mwaikambo who liked him very much and influenced him to specialize in Paediatrics.

Prof Mwaikambo is a paediatrician herself and was a member of the Governing Council of the IUCEA during my Chairmanship. Prof Esther Mwaikambo is a strong supporter of the East African Community. She wanted Amenya to stay and work in Dar-es-Salaam after his graduation. This was not, however, possible as we also needed him in Kenya.

I recall travelling with Amenya to the university for the first time and submitting the registration documents. We had forgotten his photographs in Nairobi on our dining room table. We almost had a fit as we did not know where we could take other photos to enable him register as the deadline was midday. We exchanged a few terse words and started looking for instant photo studios to have his picture taken. We finally succeeded.

Amenya's stay in Dar-es-Salaam was very beneficial to him. He became independent and mature as he interacted with several other students from the East African region and other nationalities. He became a very fluent Kiswahili speaker although he was already good at it. I usually took him to meetings in case I needed any assistance in translation.

We used to visit our son twice a year. My wife, Esther, and the other children would drive over 1,100 kilometres to Dar-es-Salaam through the most picturesque scenery imaginable. The route from Nairobi to Dar-es-Salaam through Arusha, Tengeru, Moshi, Pare Mountains, Mombo, Kigoma, Kongowea and Chalinze goes through Hills and valleys along the Eastern Rift Valley. We enjoyed the trips and learned a lot during our visits.

Driving to Dar es Salaam takes one through one of the most beautiful breath-breaking landscapes in the Eastern African region. Our trip used to take us through a wide range of ecosystems, ranging from the savannah grasslands through the moist mountain areas of Mt Kilimanjaro, into the coastal reefs and mangroves. We always longed to travel to Dar es Salaam to just enjoy the beautiful scenery.

Amenya's graduation was great. My family and sister-in-law, Isabella, attended it. Dr Salim Ahmed Salim was the Chancellor of Hubert Kairuki University and Guest Speaker during Amenya's graduation in 2006. He gave a good speech which reminded me of the speech by Ted Kennedy during my graduation at Rutgers University, USA in 1974.

Dr Salim talked about the role of a Tanzanian as a citizen. Ted Kennedy's speech had also covered the role of Americans in nation building. I thought the two speeches, given several years and miles apart, had similar themes and conclusions in shaping the youth for national development.

My wife, Esther, and my sister-in-law, Isabella, made the loudest Kenyan ululations as Amenya's name was called for the award of the degree. The congregation marvelled at the beautiful Kenyan sounds. Subsequently, all the names called thereafter received similar ululations. It was a hilarious moment. The Tanzanian crowd learnt fast. It was a thrilling moment to remember.

I was a proud father watching my son graduate. It was now my turn to celebrate and take pictures with my family. I was familiar with the proceedings since I had officiated in similar events in JKUAT. Witnessing a child's graduation is the ultimate pleasure to parents.

After the graduation ceremony, we went on a leisure boat ride to an island for lunch and supper. The Tanzanian coastline is a jewel to marvel at. We visited Bagamoyo and went through the very famous historic sites and graveyards of early missionaries. We later flew to Zanzibar Island for a retreat. We had a chance to visit the Clove Island, also known as the Stone City. It was at Stone City that slaves were packed in caves ready for shipment. We were saddened when we viewed that historic site. My wife, Esther, who is a historian, expounded to us the horrors of slave trade as we went around the caves. She further told us the role played by the early missionaries who built churches on the Island.

Amenya, my second son, is a kind-hearted and caring person. He is a paediatrician and is pursuing his MSc at the University of Nairobi in the same discipline. He meets his deadlines and wastes no time on trivialities. He respects family ties and enjoys visiting the family. Amenya is a medical students' leader

One may wonder what roles I played to my beloved family as I was building up persons and setting up institutions. I have deliberately dedicated this section to my wife, Esther, and our four children who supported me during some good and bad times. It is important to highlight my role as a husband, father and grandfather in the light of family upbringing.

My wife, Esther, is a professional teacher by training. She taught in many secondary schools across the country after graduating from the Kenyatta University College (KUC), a constituent college then of the University of Nairobi. She later pursued a Master's degree in Sociology at the University of Nairobi.

I have been a university professor involved in teaching, research and community service for 33 years. Both my wife and I, therefore, are seasoned teachers who have made tangible impact on the Kenyan youth. We both nurture and advise them at various stages of their development. We know the value of education and so we advise them to work hard. That is why both of us assist the bright but needy students. We have supported a number of them to attain university education. Many are now renowned professionals gainfully employed. We have four children, three sons and one daughter. We also have two lovely grandchildren, a granddaughter and grandson.

All our children started their nursery and primary education at Consolata School in Westlands. They passed their Standard Eight Examination and joined various secondary schools in Kenya. Nyakundi and Amenya attended Mang'u High School. I had prepared Nyakundi for the boarding life as this was the first time for him to leave the comfort of his parents. I remember him calling me one afternoon when I was in JKUAT that his nice 40 mm thick mattress was missing from his bed. It had been stolen by one of the boys and he could not trace it! I had not heard of such an occurrence in boarding schools. I had to abandon my work, travel to Thika Town and purchase another mattress.

I told my son about life in boarding schools and that he must be prepared to meet all types of characters from the whole country. I reported the incident to the principal, Mr Paul Otula, and he promised to carry out an inspection. Mr Otula was one of the finest and dedicated principals that I ever came across, besides Geoffrey Griffins of Starehe Boys' Centre. We were good friends and I used to give talks to his final-year students. Indeed, he found the mattress and I told him to take the necessary action since I had already bought a replacement.

Nyakundi sat for his Form Four examinations and qualified to join Universities in Kenya. He was admitted into JKUAT where I was as a Vice-Chancellor just before I left for NEMA. He was among the first 22 pioneer students who studied Mechatronics. This was a new course as a hybrid between Mechanical and Electronic Engineering. Prof. Ndirangu Kioni, the current Vice-Chancellor of Dedani Kimathi University, was the founding dean of the course and my son's academic advisor. He was very fond of the young man.

at the University of Nairobi. He is a disciplined young man who makes friends quickly. He is an early riser and enjoys nature.

Our third-born is a daughter, Kemunto, who joined Alliance Girls' High School after her Standard Eight Examinations. The four years she spent there were eventful as we always went to visit her during the parents' days. I do not recall any school visit that we missed. One memorable moment was when we were running late to visit her. I drove so fast that we took the shortest time to reach the compound before we were closed out. We made it!

Kemunto was quick to make friends and had a habit of visiting the children's homes and the elderly people. She joined a group of students who had similar interests in visiting the elderly and children. I remember that she would come home and take some of our house effects for donations. She still has a liking for children.

Kemunto joined Moi University in Eldoret for a law degree and graduated with an LLB in 2008. She continued her studies at the Kenya School of Law where she passed and eventually got admitted to the High Court of Kenya. She also graduated with an MBA course from Eastern African University, Baraton in 2013. She passed her examinations in litigations and is now a registered member of the prestigious group.

Kemunto enjoys her work and attends a number of charity organizations. She is a member of the Rotary Club, Westland Branch, Nairobi. She has two children, our beloved granddaughter, Shante Nyaboke and the grandson, Amali Ratemo.

Our last-born, Michieka, attended Starehe Boys' Centre after passing his Standard Eight. He had a choice of going to Alliance High School where he had been offered a Form One place or Starehe Boys' Centre. We opted for Starehe. I knew Geoffrey Griffins, the Director, well. We had interacted during my tenure as Vice-Chancellor at JKUAT and he was also the Director of National Youth Training Programme. I once travelled with him to Naivasha, for (ladies) and Gilgil (men) National Training Camps to address fresh recruits before they joined the Universities.

After Form Six, students had to go for a one-year mandatory training course to supposedly instil discipline in them so that they would hopefully behave better once they were admitted to their respective Universities. Mr Griffins was the Director of National Youth Service then. My university, JKUAT, also used to admit the highest number of students from Starehe Boys' Centre. I used to call Mr Griffins for consultations in case I had any problems from his students. He was glad to have my last-born son join him.

Before Michieka went to Starehe, he had vehemently resisted joining the Centre. The reason was that there was bullying in the school and he did not like boarding life! Unfortunately, or fortunately for us, we had taken the young boy to Strathmore for examinations to be admitted here in case he missed admission to other schools. He did very well and got admitted to two national schools instead. Strathmore was hence our alternative. The young man then wanted to join Strathmore and forfeit Starehe because the former was a day school.

I firmly told him that Starehe was our choice and made preparations for his entry into boarding life. I took time to explain to him the beauty of the school and promised to visit him once a month without fail. This we did. He actually obliged and the rest is history. We knew the effect of staying home and not reading or not doing his homework. We believed in boarding schools after Form One. I was a Form One student in Kisii School at his age and so was their mother in Kamagambo Secondary School. We both enjoyed our boarding school days at different times and places. None of us regretted that arrangement.

I recall the first time I took Michieka there. A Form Two Asian boy came to receive us. He was responsible for orientating my son. I was so impressed with the explanations he gave Michieka and how he took us round the campus. The young Asian boy was so articulate that I told my son to note and emulate him. I thought that all secondary schools should have this method of welcoming fresh Form Ones.

There was no bullying. This is what Mr Griffins had promised me from the very beginning. Michieka did well in his Form Four examinations and joined the University of Nairobi. He graduated after five years and opted to specialize in Obstetrics and Gynaecology. He is now a practising doctor at the Kathiani District Hospital in Machakos County.

Michieka Okioga is a likeable young man who loves his work. He is always conscious about anything he does. He is a meticulous young man who can at times take hours to tidy up his room or brush his shoes. He enjoys the company of mature people and relates well with all ages. He is a fanatic of Formula One car racing and would stop doing anything to watch the race at any time of the day or night. He has indoctrinated me.

We had one of the best graduation ceremonies for him in December 2010 at the University of Nairobi Graduation Square. Being the last to graduate in the family, we called our relatives home and had a quiet party in praise of the Good Lord for the role He had played in giving us such wonderful children. It would have been efforts in futility if I did not put my children in order.

The society would have questioned my role as a model for my own children. My wife and I are both teachers at various levels and we knew better. It was our turn to be happy. Later on, Dr Michieka Okioga graduated with an MBA degree in Management from Presbyterian University in 2013. He is a member of various organizations.

I remember an incident that I like to share to show how to discipline children without hurting them. Our children used to be given homework. I used to sign their books as directed by the class teachers. I did this without fail for all the children. My wife could also sign when I was out of town.

One day Nyakundi delayed in completing his homework and wanted to sleep early and I sensed this. I told him to do his homework but he was still reluctant and wanted to go to bed. So, I trick him without arousing his suspicion.

I left the sitting room unnoticed and slowly headed to his room. I put off the lights, went into his bed and pretended to sleep. Sure enough after about thirty minutes, he quickly walked in and wanted to sleep. As he pulled out the bed sheets and blankets, he squarely put his hand on my face! He jumped up and wondered whether it was a ghost he had touched or a human being! He knew it was me and quickly returned to the reading table with the broadest smile ever. He knew he had been tricked and vowed to be a good boy and do homework as required. The rest of the children heard of the story and dared not to be caught unawares.

Reflections

Parental care cannot be substituted. Early relationships always bear fruits. Children recall the most trivial events during their formative ages.

My wife and I gave our children the love, humility and discipline they needed. They recognized it and always expressed it through cards or email messages. We know they love us. We vouched to attend their parents' days without fail. We might have overdone it, but that was our choice. They respected us and obeyed us.

As they grew up and became young adults, we always reminded them that life is a continuum. It perpetuates. They will have a role to play similar to the one we played, but may be in a much different environment and circumstances. The best traits shall withstand the stiff competition.

We respected their friends and accorded them respect as young adults. Our children appreciated this gesture and their peers were equally appreciative.

The other vital role I played to all my children was to teach them how to drive. I made a big notebook on dos and don'ts while driving. I spent more time discussing the theory of driving than the actual practice. I personally took them on road tests every Saturday and Sunday afternoons. After every practical lesson, I would have a review session and correct all the mistakes we might have made on the road. My first sentence to them was: *"Assume all drivers on the road are mad; practice protective driving."*

They then went to driving schools to polish up before going for the licensing tests. My notebook is still in their possession and they always refer to it. That was my extracurricular activity.

We always told the children to excel in their own areas of expertise. They should not expect free gifts from anywhere but through hard and focused work. My wife, Esther, and I spend quality time advising our children and telling them to fear God, the Creator and also respect people, both the young and old. They had to be achievers in all aspects and strive for excellence. They are understanding, co-operative and achievers. They have not let us down and we are their living mentors.

My love for extended family members is strong. I have always believed that my position in society should have no effect on the nurturing and upbringing of my family. It should not result in the dispersal, decline or loss of the wider family touch. That would be inhuman and unnatural. Indeed, my positions were arbitrary and would come to an end someday. I used to tell them that university education can teach anybody skills and give them opportunity, but it can neither teach one sense nor give one understanding. Sense and understanding are produced within one's heart and love. I simply meant that humility cannot be taught, but practised.

The former President of Libya, Muammar Ghaddafi compared life to a tree. He gave an excellent analogy which I concur with. He stated that a man who is the head of the family should stand firm like a full-grown tree. If he breaks his family, it is like destruction of a plant, breaking of its branches, and fading of the blossoms and leaves. Any man or society in which the existence and unity of the family are threatened, in any circumstances, is similar to the fields whose plants are in danger of being swept away or threatened by drought, fire or withering away.

The blossoming garden or field is that the one whose plants grow, blossom, pollinate and root naturally. The same holds true of human society. The individual is therefore linked to the larger family of humankind like the leaf to the branch or the branch to the tree. They have no value or life if they are separated. The same is the case for the individual if he is separated from the family. That individual without a family has no value or social life.

Ghaddafi concluded his beautiful comparison by saying that if a human society reached the stage where man existed without a family, it would become a society of tramps without roots, like artificial plants.

This highlights the importance of intergeneration progression in any family or society. I made sure that I nurtured a family which would not regret my heavy responsibilities in society because I have always balanced my time well between roles and responsibilities, and have remained focused in life.

8

Engagements with Academic Leadership at the Grassroots

Under normal circumstances, the university procedure dictates that the VC appoints the chair of a department, whereas the dean was elected. For one to be appointed to chair a department, one needed to have been a senior lecturer with a record of academic publications and teaching experience.

The time of my appointment as chair of department, I had won two competitive research grants that brought some substantial amount of training grants to the University and department from IDRC and the NACOSTI. So my consideration to serve as chair was deserved. However, instead of the three years renewable once as chair, I served for two years, and then was appointed Deputy Principal in charge of academic affairs at JKUAT. This was a first, to be appointed deputy principal without serving as dean or completing my term as chair.

Given the culture of appointments to serve in senior university administrative positions during those days, one would say that my appointments were accelerated and politically favoured. But looking back, I think that the appointments were deserved and reflected the time and quality of my input in administration at the lower levels. For example, during my two years tenure as head of department, I was able to source and secure various Master's and PhD scholarships for staff development. I also served as secretary and later chairperson of the Weed Science Society for Eastern Africa. This helped in promoting the development of the discipline and young scientists in the department.

The society's biannual proceedings became a fertile training ground for junior academics in the department, Kenya and the region. I also strengthened the teaching of statistics and biometrics in the department. During my tenure as chair, the VC engaged me on several occasions to chair mediation and promotion committees for junior staff.

At no time did I ever harbour ambitions of serving in higher university administrative posts. My appointment as chair was in itself surprising. And despite the fact that I was qualified to be appointed, I do not know which other consideration the VC then had in mind in offering me the appointment as my focus was on teaching and research.

In the previous chapter, I have demonstrated how deeply I was engaged in research, student supervision and publications. I also thought of gradual progression through deanship, as the practice was then. So my eventual appointment as DP after serving only two years as head of department was more surprising and I wondered what the appointing authorities had taken into consideration.

The various committees I sat on or chaired gave me prominence. The decisions we made in such meetings were appreciated and carried the day at that time. There could be variations on the decisions I made by the higher authorities but none was nullified. I worked well with my heads of department, the deans, Deputy Vice-Chancellors and the Vice-Chancellor of the University of Nairobi. I occasionally acted for the departmental chairperson whenever he was away. Drs D.N. Ngugi and D.M. Mukunya were respectively my chairpersons when I was in Kabete. Deans Prof Karue, S. Keya and D.N. Ngugi worked with me.

In 1985, the University of Nairobi created six campuses: College of Agriculture and Veterinary Sciences (CAVS), College of Architecture and Engineering (CAE), College of Biological and Physical Sciences (CBPS), College of Education and External Studies (CEES), College of Health Sciences (CHS) and College of Humanities and Social Sciences (CHSS), each headed by a Principal.

The University appointed five Principals to head them. My first Principal in the CAVS was Prof. Geoffrey Ole Maloy, followed by Prof. Dominic Oduor-Okello and later on by Prof. Shellamiah Keya. I had an excellent working relationship with each one of them as I assisted in many students' advisory services. I had respect from my colleagues and the staff played a key role in quelling riots whenever they surfaced.

I played another demanding role in the final year students' field and industrial attachments. I chaired the committee which scouted for suitable students' placement. My predecessor, Mr Don Thomas, was very helpful in identifying large-scale farms. I worked well with him also.

The meetings which I chaired or where I was a member taught me several lessons. I learnt how to handle diverse situations. There were issues which needed immediate solutions and I had to have answers for them instantaneously, sometimes regardless of the quality of an answer. I learnt how to handle both staff and students on various matters. I was on a students' advisory committee, a mentoring group and a junior staff appraisal committee. Ms Esther Thiongo served as my secretary on the committees, especially for farm practice arrangements.

My first contestable position was to be a leader of a co-operative society called 'CHUNA'. The idea was hatched at Kabete Campus by a few staff members. We decided to form a society to enable us save a portion of our salaries and borrow despite the fact that I was qualified to be appointed, I do not know which other consideration the VC then had in mind in offering me the appointment as my focus was on teaching and research.

In the previous chapter, I have demonstrated how deeply I was engaged in research, student supervision and publications. I also thought of gradual progression through deanship, as the practice was then. So my eventual appointment as DP after serving only two years as head of department was more surprising and I wondered what the appointing authorities had taken into consideration.

The various committees I sat on or chaired gave me prominence. The decisions we made in such meetings were appreciated and carried the day at that time. There could be variations on the decisions I made by the higher authorities but none was nullified. I worked well with my heads of department, the deans, Deputy Vice-Chancellors and the Vice-Chancellor of the University of Nairobi. I occasionally acted for the departmental chairperson whenever he was away. Drs D.N. Ngugi and D.M. Mukunya were respectively my chairpersons when I was in Kabete. Deans Prof Karue, S. Keya and D.N. Ngugi worked with me.

In 1985, the University of Nairobi created six campuses: College of Agriculture and Veterinary Sciences (CAVS), College of Architecture and Engineering (CAE), College of Biological and Physical Sciences (CBPS), College of Education and External Studies (CEES), College of Health Sciences (CHS) and College of Humanities and Social Sciences (CHSS), each headed by a Principal.

The University appointed five Principals to head them. My first Principal in the CAVS was Prof. Geoffrey Ole Maloy, followed by Prof. Dominic Oduor-Okello and later on by Prof. Shellamiah Keya. I had an excellent working relationship with each one of them as I assisted in many students' advisory services. I had respect from my colleagues and the staff played a key role in quelling riots whenever they surfaced.

I played another demanding role in the final year students' field and industrial attachments. I chaired the committee which scouted for suitable students' placement. My predecessor, Mr Don Thomas, was very helpful in identifying large-scale farms. I worked well with him also.

The meetings which I chaired or where I was a member taught me several lessons. I learnt how to handle diverse situations. There were issues which needed immediate solutions and I had to have answers for them instantaneously, sometimes regardless of the quality of an answer. I learnt how to handle both staff and students on various matters. I was on a students' advisory committee, a mentoring group and a junior staff appraisal committee. Ms Esther Thiongo served as my secretary on the committees, especially for farm practice arrangements.

My first contestable position was to be a leader of a co-operative society called 'CHUNA'. The idea was hatched at Kabete Campus by a few staff members. We decided to form a society to enable us save a portion of our salaries and borrow money from the savings and repay it at a very low interest rate. A few of us steered the process and CHUNA was created.

I later campaigned for a slot on the supervisory committee to oversee the right operations in the finances. I won the elections. We were three on the committee: Profs Karega Mutahi, Henry Oruka Odero and I. My leadership traits in public affairs were pronounced during the meetings.

We were strict and demanded monthly statements showing staff income through contributions, loans given out to them and recurrent expenditure. This was my first financial management duty. Members were happy with our work but the main management committee was not too happy with our restrictions. In fact they took it that we were indeed supervising them. We followed the Societies Act and always reminded them that we were the watchdogs of members' contributions.

I was a member for three years and did not contest for the second time because of other more demanding roles. During our tenure no money was misappropriated. The three of us did not want to tarnish our names by being compromised in any way. We left the society sound financially. Later on, we learnt that our successors were sacked and surcharged due to loss of money. CHUNA is now one of the biggest and well-managed cooperative outfits in Kenya. In fact, it assisted me acquire land on which I built my residential house in Nairobi. Many University staff members have benefited a lot by getting loans which are paid back at the lowest interest rate in the market as compared to the commercial banks.

Departmental Chair

The role of Vice-Chancellors in most African universities is very clearly stipulated in the universities' acts and statutes. Among them is the appointment of staff to various university positions. The University of Nairobi statutes state that the vice-chancellor chooses the department chairman to head the department from amongst its (senior) staff. Under normal circumstances, he consults other senior members of the department as to the suitability of the candidate before making the appointment.

Normally, in many universities in Kenya, chairpersons of departments are appointed for a period of three years, renewable once. Chairpersons constitute University senators who make decisions for the University. They run the departments and are the academic and administrative heads of that important teaching cadre.

Among their roles is ensuring that quality teaching takes place in the departments, timetables are prepared on time, examination processes are adhered to and examination board meetings are held before the results are presented to the faculty or school boards. These are very vital organs of the university administrative hierarchy. They are, indeed, the small vice-chancellors of the university, representing the big office.

The most important consideration before one is appointed is that one must be a respected lecturer, who can perform the academic and administrative roles with integrity, is respected by colleagues and can work mutually with others for the growth and development of the department, and by extension, the University. It is important to note that the appointee deals with teaching staff, technicians, support staff of all levels, students and the public at large. The same person offers lectures and carries out research for the purpose of promotions within the university cadre. One must be a team player.

At the time I served as head of department, Nairobi then was the only public university. The tensions that are commonplace these days between management and grassroots academics were rarely noticeable, and as head of department, it was easy to navigate ones' administrative duties without compromising laid-down procedures. I was not micro-managed from the office of the VC, and ran the department fairly independently. Occasionally, I had problems with members of the teaching staff in the department regarding how they carried out their teaching duties, but we managed to resolve such issues within the department amicably, without involving the university administration.

Two issues, however, which confront heads of department nowadays is expanding Universities and the management of external examining processes. This is serious where there are so many part-time lecturers. These are necessary components of university academic processes and can pose challenges. However, external examiners in some cases would delay moderation of examinations and submission of reports to the detriment of students' progress. External examining is an important aspect of quality control in the universities and cannot be dispensed. The delay, therefore, slows down academic processes in the department and is one issue that I had to deal with as head of department from time to time.

Part-time lecturers would not, in most cases, submit examination and CATs on time. At times they would disappear with students' scripts; and as head of department, I had to go around tracing them. This was time-consuming and caused anxiety among students. Reflecting backwards what I had to go through as head of department, and these days when universities are increasingly relying on part-timers, I would make a strong recommendation that part-time lecturing be done away with. In some cases we have whole university departments largely run by part-time lecturers. Even if they were to be used, I would recommend that this be in very small rations and under the supervision of a senior permanent member of the academic staff. The problem is lack of funds to pay them promptly.

I always delivered my lectures in the mornings. I preferred free afternoons for preparations, students' appointments and research planning. I taught undergraduate and Master's classes in weed science and environmental courses. I also taught the Introduction to Crop Production to first years. I enjoyed teaching the young students who had just been admitted to the university and who expected a difference in the mode of teaching from that of high schools.

I made my stand and requirements clear from the first day of lectures that all students must be on time and meet my obligations in fulfilling the subject requirements. During the early and mid-1980s, there was an adequate supply of teaching materials and laboratory provisions. I had an excellent technician, the late Mr Francis Kinyanjui, a former East African Community Plant Taxonomist, who was very conscious of his work. Practicals were prepared on time and clear instructions for the students were always ready before it started.

One morning, I got a call that the Vice-Chancellor, Prof. Phillip Mbithi, wanted to see me. It was soon after my morning lecture. I wondered why he asked for me knowing that we always made appointments through the chair and deans' committee. I drove to the Main Campus which is about twelve kilometres away from Upper Kabete Campus where my department was located. I arrived there and requested his secretary to allow me see him. She was kind enough and made a call to him.

Prof. Mbithi walked out to the reception area and ushered me in with his usual broad smile. Prof Mbithi was a very interesting Vice-Chancellor. He knew many people and related well with them. As an agricultural extensionist-cum-sociologist, the professor knew the right method of persuading staff to support him. He called me by my first name as we sat down. He enquired about the faculty and department and whether the students were behaving well. Students' riots were so common in the 1980s and 1990s that lecturers were always alert. Any small misunderstanding between students and mainly the government would spark demonstrations which would always cause closure. He merely wanted to know whether all was calm at my campus. The VC had one of the finest networks of detecting trouble.

As we continued chatting over a cup of tea, he called his secretary, Catherine, and requested for a letter which was supposed to be delivered to me. It was my appointment letter as the chair. He gave me his signed letter, explained why he had called to personally deliver it to me. He wanted to know my reaction. He wore heavy rimmed spectacles and could occasionally peer piercingly through the thick lenses.

I had no prior indication whatsoever about this appointment. Apparently he had consulted the outgoing chairman then, Prof. David Ngugi and other staff regarding a suitable person to take over from him. This was the time when I was so busy collecting my field data for the IDRC project. I saw myself having two divided roles.

I honestly knew the work of the chair was demanding. Under some circumstances, one would refuse the appointment, but this had never happened due to the wide consultations carried out before such appointments were made. It could also be disrespectful to the head of the institution and may not augur well for the appointee. One thing I liked Prof. Mbithi for was his humility to call and tell me in person about the change. Other Vice-Chancellors might have sent a letter without giving prior indication.

In any case, I accepted the challenge and promised that I would do my best to run the department. My department had the highest number of teaching, technical staff and research projects. It is still the largest in the Faculty of Agriculture.

I left the VC's office and drove off to Kabete, called my wife, Esther, who was teaching in Rungiri Secondary School and told her of the appointment. She commended me and told me that it was part of my duties as prescribed in the terms of service. I did not ask what comes with the appointment but the professor had told me that there were some allowances I was entitled to. My sincere concern was not even the benefits, but about time and efficiency of running a department. My research work and lecturing were my prime concerns.

I had trials scattered all over Kenya, and saw the danger in conducting them. Field trials call for the researcher to diligently set them up, collect data, analyse and make conclusions appropriately. It is important to stress here the need for quality research and dedication. It is a very personal undertaking and conclusions made from it reflect the scientist's grasp of the knowledge matter. My technicians would assist but they were few. I relied a lot on graduate students.

Handling all these were my immediate challenges. I decided to soldier on. The many meetings could also interfere, as we were expected to be present. I, however, used to send my other staff to represent me, especially at Senate meetings.

I told my staff members in the department that I had been appointed to head Crop Science. In the 1980-1990s, the 1 o'clock news in Kenya would announce appointments. It could be from the Head of state, or any other high-ranking quarters. I heard my name announced as the new head of Crop Science. The University of Nairobi was the only public institution of higher learning in Kenya. Others were colleges. The then President Daniel Arap Moi was the Chancellor and Prof. Phillip Mbithi was his Vice-Chancellor, answerable directly to him. It was therefore prudent that he was continuously briefed on the university's matters at every stage.

The 1 o'clock news aired my appointment and I took over the mantle from Dr David N. Ngugi. The staff was excited and I had to carry on. While I had been at Prof Mbithi's office, I had asked him what my specific role would be as the head of department. He gave a short answer to the effect that one learns from other chairs because there was no formal training. He further urged me to excel in whatever I did and should always engage in wide consultations. That was it, as far as I can remember. Similar future appointments did not have terms of reference, not even a briefing! One was simply directed to assume a responsibility and learn on the job; period!

At a personal level, and having done my university studies in the United States where university governing bodies exercise a greater degree of autonomy, I found the practice where a Head of state as Chancellor has to know and announce who even becomes a head of department, a little too intrusive of the political establishment into the governance of the university even at lower levels. However, since that was the culture then, I carried on with my duties as head of department. I however found departmental head-ship challenging in that most of the time was taken up by mundane administrative duties and little time was left for any academic work and progression. The current three-year duration, renewable once, is therefore adequate as one needs to exit and concentrate on academic mobility.

In May 1988, I took over from Ngugi who was vying for deanship at the Faculty of Agriculture. He handed over the department to me and told me to follow up the most urgent matters like examinations and research protocols which were pending.

My Agenda

As a new chair, I had some ideas on what I could excel in. I wrote a manifesto for my guidance. There was no written strategy or plan for the department but I formulated one. I set myself some targets and called the teaching staff to advise on the same. I set up the following tasks having participated in the previous meetings and noted shortcomings:

> To immediately stop having any part-time lecturers, instead appoint permanent ones to take over teaching Biometrics and Statistics;
>
> Increase teaching staff in Crop Protection and initiate its full autonomy as a department; hence secede from the Department of Crop Science. Increase the number of research proposals in the department to attract funding; Assign more staff to mentor first-year students.

These were the urgent concerns as the status quo continued. I kept my dean posted and kept an open door policy with my staff. I actually met the four objectives within the short period I was chair. Biometrics and Statistics which was always taught by part-timers was now fully staffed as we had advertised posts and had recruited full-time staff.

Research proposals were increased in number after we held proposal writing seminars. The department of Crop Science became autonomous and I was given a vote to run it. It lasted for a few years and was later amalgamated back to Crop Science and Crop Protection for no apparent good seasons. The staff we had hired were absorbed in the original department of Crop Science.

As a chairperson, I used to attend and participate in the Faculty Board meetings, College Board meetings and Senate. We also used to hold seminars and workshops on the management of the departments. I recall the first Principal of my college, Prof. Geoffrey Ole Mayoi, telling us at one sitting that new chairmen had the duty to perform well since most of us were young entrants.

I was 37 years old then and my desire was to publish as many papers as possible. University promotions are permanently based on publications in addition to other duties. In fact the latter never count. The common saying in university corridors is "publish or perish". I therefore had to run the department, continue my research programmes, supervise postgraduate students, teach my normal allocated hours and, most importantly, ensure that my young family was catered for.

My wife was teaching and our first two sons were attending nursery and later primary school. I could drop them at various stations in the mornings and pick them later in the afternoons. I had to plan carefully. I was able to run my errands and perform my academic roles with ease. I learnt to delegate and follow through unless I was personally required to participate in meetings. This is a habit I acquired throughout my academic and administrative careers. But I was selective as what to delegate and what not to assign anyone.

My staff was supportive. We developed an excellent working relationship amongst ourselves and our department was a leading example in attracting research grants. I pushed for promotions where appropriate and advised staff to adhere to the respective promotion requirements for each cadre. I could not push any request until I was satisfied that the criteria had been met. Quality was my guiding principle in the discharge of my duties.

Director, Kenya Marine and Fisheries Board

Appointment as a director of Kenya Marine Board was the beginning of my national responsibilities. As we were busy strengthening the department, I got a call from the minister's secretary, Ministry of Environment and National Resources asking me to go and see the minister. I drove there to see him.

Mr Andrew John Omanga, the minister, welcomed me into his office and commended me for the good work I was doing at the university. I thanked him even as I wondered why he had called me. He did not waste time in long stories, but gave me a letter appointing me as a Board Member of the Kenya Marine and Fisheries Institute (KEMFRI) which had its headquarters in Mombasa. All he said was that our coastal area was being degraded fast and marine life was threatened; so I was expected to go and protect the environment. He added that our natural resources like fish, mangroves; coral reefs needed laws to regulate their utilization. The same was equally important in our clean water environment such as lakes and rivers.

The appointment was for four years and renewable. Hon. Omanga was a graduate from Makerere University. He was a respected Minister for Environment and Natural Resources during President Moi's regime. He had a passion for education and scholars. He was my Member of Parliament and knew my credentials. The Hon. Minister appointed me to the Board because of my expertise in Environment and he respected meritocracy. The board was composed of deserving men of strong calibre.

I was happy to take on my new appointment, having been involved in environmental protection meetings, discussing the same issues. Interestingly, my chair of the board was none other than my mentor, Professor Reuben Olembo. Other board members included: Professors David Wasawo, John Kokwaro, George Kinoti, and Ken Mavuti. There were also representatives of some ministries on the board. The professors were all from the University of Nairobi, Faculty of Agriculture and Science.

The four preferred to be addressed by their first names: Reuben, David, John and George. Ken and I were budding scientists and we marvelled at the manner in which they casually but seriously conducted board meetings! These senior colleagues in several aspects were my mentors in academics and administration. Mentorship continues even unto old age. We could not dare call them by their first names; we did not measure up to their ranks. This was a powerful team, including ministry representatives. It is during these meetings that I really learnt the etiquette of conducting board meetings and collegiate interactions.

This was a ministerial appointment and I took it with a clear mandate on the protection of our coastal resources. The team was powerful and was composed of my mentors and seasoned researchers. Luckily, we used to meet four to six times a year. The appointment was a testimony that I could contribute my expertise in some areas which needed action. The appointment was gazetted and I wrote to the minister to thank him. As a young scientist, I learnt a lot from my senior board members who were also University of Nairobi professors from various departments. I had practical exposure on the importance of management and conserving our coastal natural resources and upland water bodies. I interacted a lot with the institute staff. This was an eye-opener for future leadership engagements.

My First Book: Weeds of East Africa (Magugu ya Afrika Mashariki), 1987

My first book was co-authored with a colleague from weed science, Mr John Terry, in 1987. The publication of the book was sponsored by Food and Agricultural Organization (FAO). Mr Les Matthews, who was responsible for the project at the FAO Headquarters in Rome, Italy, was very impressed with my work and included me in the panel of Weed Experts from the developing countries.

We were a team of about twelve individuals covering all the geographical regions of the world. Our role was to advise the FAO Director on critical areas of Weed Research and what action must be taken. Les visited me in Kabete two times and supported some research projects. It is through these visits that I mooted publishing a weed book to cover the taxonomy of East African weeds. My request was granted and he provided the funds. Mr John Terry and I made a number of field visits in order to collect weed specimens for the publication. It was a nice experience which reminded me of my post-graduate work at Rutgers University and later in IITA on weed research and identification.

We translated the book into Kiswahili and it became a good reference for weed identification for the students in the East African region. We took very elaborate pictures and used the then latest technique of mounting them for ease of identification. I was able to fund one Master's student, one Saha, under the project. I still possess a few copies.

Reflections: As virtues, humility and integrity have a lot in common. My appointment as a chair of a department by my vice-chancellor had a lot to do with my conduct and interaction with academic and administrative staff. He must have seen some management potential in me. I worked for the department with dedication and did not let down my boss, the Vice-Chancellor. Staff, both in KEMFRI and the department were co-operative. Academic and research programmes improved and I demonstrated my ability as a team player. I made a mark in the department. I later considered this early appointment as a sharpening tool for future responsibilities.

My appointment to the Board of KEMFRI by the Minister demonstrated his confidence in me. As a young scholar I was excited to join the board which was highly dominated by a learned group of professors. I considered the appointment an honour to me. It enriched my managerial skills and, at the same time, exposed me to Kenya's coastal natural resources management. I worked well with my colleagues and did not let down the appointing authority. Instead, I added my expertise to the board and brought in an environmental concern to bear. I would later use their expertise to manage NEMA as the Director-General.

This was also the beginning of my leadership management. What I noticed in both appointments, chair of a department and board member, was that there were no briefs, no training courses, no induction materials and no clear terms of reference. One uses common sense and is expected to perform to his or her best in all circumstances. Learning on the job and high-level use of common sense were imperative. This is the truth in many public appointments in Kenya.

9

A Surprise Appointment to the Office of Deputy College Principal

My busy schedule was the order of the day but I did not feel constrained in meeting my deadlines. As a chairperson, I had to plan the day's work well, attend meetings, deliver my lectures and follow up on my research progress. I also ensured that I supervised my PhD and Master's students during their research projects. In fact I was so used to my daily routine that I had extra time to attend to my personal needs.

A more surprising and unexpected appointment was my secondment to a University College as a Deputy Principal in charge of Academic Affairs of the Jomo Kenyatta University College of Agriculture and Technology (JKUCAT) which was a constituent college of Kenyatta University (KU). I would later be appointed as Chairman of Kenyatta University.

At about 1.30 pm on 15 November 1989, a colleague of mine came in and told me that he had heard my name being mentioned during the one o'clock news. I had just completed my morning lectures, had had lunch and was getting ready to go and inspect my field trials on minimum tillage. As we were talking, a call came in from another colleague to confirm my appointment. I considered this appointment unique since I had barely served a term as chair of department. Under normal administrative progression, one had to go through both academic and administrative ranks before being appointed to the position of deputy principal. In my case, I had the academic credentials but little administrative experience.

When I confirmed the appointment, I called my loving wife, Esther, and told her that I had been appointed to JKUCAT as a Deputy Principal. She had also confirmed the same through a friend. Both of us tried to understand the implications.

I honestly did not know that I would leave the University of Nairobi for any other posting. I had come there basically to work and stay. I always thought that appointments were for the mighty and well-connected individuals. I confirmed the new assignment after calling my Vice-Chancellor, Prof. Philip Mbithi. I knew protocol. I called my dean and College Principal to enquire whether, indeed, I had been requested to leave Kabete. They said that presidential appointments were never questioned. I had to oblige and move.

Later that day, I drove to the office of my boss, the Vice-Chancellor, who had also earlier appointed me as the departmental chair. When he saw me in his office, he of course knew why I had driven all the way to see him. He laughed heartily and handed me an appointment letter.

This time he did not take time to explain to me what the letter was all about. He simply said that His Excellency, the Head of State, had appointed me to JKUCAT as a Deputy Principal to Prof. George Eshiwani. He further informed me that new colleges were coming up and they needed to train young scholars to eventually take over university management. I was therefore supposed to report to my new station as soon as possible.

That was the end of my brief discussion with Prof. P. Mbithi. He remarked that I should go and work hard to lift the status of the college and to learn on the job. I did not have time to enquire about my terms of departure from the University of Nairobi and what I was expected to do in Juja Town, fifty kilometres away where JKUCAT is located. I left his office confused.

I drove to Consolata Primary School where my kids were learning and took them home as was the routine. My wife had already arrived home and we started discussing my appointment. We both internalized the matter but did not know where to start from. This was in November 1989, just a month before Christmas break.

My Immediate Concerns

After a day of considering my next move, I made a list of the most demanding issues that had to be sorted out. The first problem was how to drop my wife and children to their respective destinations and back. We used to live in Kikuyu Township close to my place of work. My wife taught at Rungiri Secondary School about two kilometres from our house and the children attended Consolata Primary School on Waiyaki Way about 5 kilometres from my office. It was therefore convenient for me to drop them early in the morning and pick all of them in the afternoons as I left my office. I needed an immediate solution to this matter. I always wanted to drive my family personally to their schools and did not see the need for a driver.

The second immediate headache was how to handle my third-year students of Crop Protection who were just about to do their final examinations in a month's time. I needed to cover the syllabus, give and mark the examinations. It is instructive that in Kenya, state appointments are taken up at once, lest someone else takes over your seat or you are replaced. I did not therefore have enough time to cover my syllabus.

The third agonizing consideration was to plan on my research protocol. I had conducted my elaborate research programme for over three years, set up field trials and acquired a number of equipment including two vehicles. I was the official contracted researcher while the university was the custodian of finances and general overseer. That is how the protocol was observed. Otherwise, my research assistants would not be able carry on with the technical matters of the research undertakings. Other lecturers had their programmes to run.

The new chair was not a weed scientist, and had little interest in my area as he also had his research undertakings. Research proposals are very personal, and unless one has a group of continuing graduate assistants, the work suffers. I knew I would not complete the work as planned. Luckily, I had published some papers as part of work done. My students continued collecting data and writing their theses.

I approached the donor, International Development Research Centre (IDRC), and we reached an amicable solution of discontinuing the trials but retained the project double-cabin pick up for the department.

A new 504 Peugeot Station Wagon, which I had procured for the project, generated debate as to whether to retain it in Kabete Campus or take it to Juja. I preferred it retained in the department since I acquired it while there. That would have been the fairest decision. But the disagreement between the two institutions made IDRC take back the vehicle. My research work fizzled out. Some funds were taken back while overheads expenses were retained. This appointment disrupted my research at an early age of 38 and I had to plan my next new assignment.

After three days of planning, I decided to drive to the new campus and see the Principal. I had been to JKUCAT previously and so it was not a new place. The total distance from my house to Juja was 56 kilometres. I could not possibly commute that distance. I returned home and told my wife that the place was too far to commute daily from Kikuyu. We had to find an alternative. The easiest was to move closer to town and reduce the distance at the same time maintain our children in the same primary school.

This was done and we settled in Kariba Estate, South B, in February 1990. One concern was solved. As for my third-year students, I promised that I would complete teaching as I reported to my new place. I completed the syllabus, gave a final examination, marked the scripts and provided marks.

My first official reporting day to JKUCAT was on a Monday in December 1989. I drove to the campus only to meet the Principal, Prof. George Eshiwani, driving out. We actually met at the gate. He recognized me, stopped and requested if we could meet later in the week. That was a good decision which gave me extra days to teach and mark the papers. I, however, drove in and spoke to the Academic Registrar, Mr Joel Mberia, whom I knew and he gave me some good tips about the new upcoming university.

The Office of Deputy Principal

I settled in JKUCAT in December 1989. It was originally a technical training college. It trained students at certificate and diploma levels. The college stands on land donated by the Founding President of Kenya, Mzee Jomo Kenyatta. The college was producing middle-level technicians who manned our agricultural and engineering industry. It was supported by Japanese government through the Japanese International Co-operation Agency (JICA). The founding president signed a memorandum of understanding between the governments of Japan and Kenya for technical assistance.

My role as the deputy principal in charge of academic affairs was specific: to be in charge of students' affairs, come up with new academic programmes, handle any academic staff matters relevant to the college's academic progression. A gazette notice was issued by the government to elevate the Technical College to become a Constituent College of Kenyatta University and all academic programmes had to go through Kenyatta University Senate for vetting. I studied the notice in conjunction with the Kenyatta University Act and Statutes. It was clear what my role would be in our college and as far as academic programmes were concerned.

We were three Deputy Principals: one for administration, one Dr Josephat Yego; the other one for research, Prof. Rosalind Mutua; and I for academic affairs. We all reported to the principal, Prof. George Eshiwani, who was seconded here from Kenyatta University. The team was complete and we all had various tasks of nurturing the young college into a university. We actually had an uphill task. Starting a new institution has its advantages and disadvantages. The three divisions: Academic, Administration and Research needed to be set up and be functional just like other universities in Kenya. I was aware of the academic programmes and what the staff and community expected of a new college.

As Deputy Principal of academic affairs, I used surveys, workshops and conferences to develop the programmes. Stakeholders were involved in all the processes during the development of new curricula. Having participated in the University of Nairobi Senate meetings, the kind of resources I needed to develop academic programmes were clear to me. The basic procedures for initiating new programmes were demand. I involved stakeholders from the agriculture and manufacturing industries. This helped the graduates in job placement and created a positive image of the new university.

The college inherited staff who were employed under the Ministry of Education and Teachers' Service Commission. Both teaching and support staff jobs were threatened. I had to deal with credentials of academic staff, while my other colleagues had to handle those of the administrative and support group. It was prudent for me to concentrate on academic requirements because the demand for higher education is ever growing.

The University College was meant to grow into a full-fledged fifth University of Kenya and admit students to pursue degree courses. The long journey ahead of us depicted the setting up and nurturing of a college through some torturous route. I was in charge of all students' affairs, examinations and academic programmes. Many times, Prof. Eshiwani would leave me acting whenever he travelled out of the country or when on leave. He had confidence in me and I took the work seriously. I, however, encountered some problems which had to be solved without delay. My fellow deputy principals and deans were always ready to give some advice. This was good training for the future tasks.

The Principal

After four years of planning, the University College became the fifth public university through an Act of Parliament in 1994. It took five years to put everything in place before it was given the autonomy. Academic programmes were well-thought-out and the Senate had approved marketable undergraduate courses.

The fifth Kenyan public university was set up to be a science and technology-based institution of higher learning. My predecessor, Prof. George Eshiwani, was appointed Vice-Chancellor of Kenyatta University and I took over from him as the Principal who oversaw the transition from college to fully-fledged institution.

There were hurdles along the way as we planned our programmes. The vetting of staff was a problem as we needed to retain qualified ones and release others to their respective employers. Conversion of colleges to universities is a major headache when it comes to staff engagement.

Having been appointed a Deputy Principal and consequently a Principal, I was fully responsible for the upward growth and development of Jomo Kenyatta University of Agriculture and Technology.

10

Elevation of the University College into a University

Once I was elevated to the position of a principal, my immediate focus was to develop and prepare the college to be granted full university status. For this to be actualized, a lot of negotiations took place. This is because no matter what we did at the college to deserve the elevation to full university status, the final decision was always political. This, indeed, was the case with all other universities in Kenya.

The Head of state had to declare the full status of a university. In our case, and despite the reputation we had gained as a college of technology, many technocrats and colleagues from other universities; and most surprisingly, Kenyatta University the parent institution, lobbied against the establishment and granting of full university status to us. I established networks and used numerous operatives to push the agenda. Close allies of the minister for education, politicians and even international colleagues were requested to pass a good word for the college to be elevated into a full-fledged university. Due academic and political processes were followed.

I personally had to get involved. I recall one evening when I was summoned to Nakuru to meet the president, Daniel Arap Moi to justify our quest. My staff prepared briefs and about twenty senior members of staff accompanied me to support our cause. Several were from his ethnic community.

I was bluntly asked to explain why we wanted to disassociate from Kenyatta University. The president was in the company of many other politicians. We slowly explained the developments we had achieved and why this needed to be consolidated with the granting of full university autonomy as the fifth public university of the Republic of Kenya. The programmes we conducted were completely different from those in Kenyatta University. We wanted to keep that character. The following week, the president made the announcement declaring the college a full-fledged university.

Planning which character JKUAT (now a fully chartered University) would take was my fundamental priority. It was now the fifth Kenyan public university. I felt that it was my privilege and turn to make a mark in Kenya's public universities. It would not be business as usual. I had a duty and a brand name to nurture.

JKUAT was declared a full-fledged university in April 1994 with an automatic choice of a Chancellor who was the Head of state. Preparations for the inauguration and receiving of the Act had to be made. The appointment of Council and the Vice-Chancellor were also an imperative requirement to be performed. It is important to walk through the actual preparation and the culmination of the award of the Act. More politics again resurfaced during the appointment of the Council and VC.

Academic Programmes

Transformation of a complex technological college into a university had its problems. The Academic Board wrote all the programmes with a specific mandate that JKUAT's character was based on Science and Technology. There were very few social science courses planned.

The initial courses offered were Engineering and Agriculture. The JICA support was specifically meant for the two programmes. The Kenya government supported the same and when I took over the leadership, I religiously adopted the same. We only added science-based courses to strengthen Agriculture and Engineering. The board, and later the Senate, was also fully aware of the mandate and supported the establishment of a first-class Science and Technology-based modern university. With that understanding in mind, I embarked on the preparations for the grandiose launch of the JKUAT of the twenty-first century.

There were several instruments of authority and personnel which were necessary before we were awarded full university status. The Council was appointed through some consultation with its interim Chairperson, Dr Stephen Mulinge, a plant pathologist (now deceased). Several Council members were named including Uhuru Kenyatta, the current President of Kenya. It was a lean and active Council. In our Act, I was the first Principal to include the Secretary of the Commission for Higher Education, now Commission for University Education, as a member of Council.

It happened to be Prof. Joseph Mungai, my mentor, who had retired as a VC from the University of Nairobi. He was my first employer at the University of Nairobi, and also my close advisor. He was a great asset in the development of JKUAT academic courses.

My appointment as the founding Vice-Chancellor was marred by back door solicitations since many professors had wanted the post. They had seen a great university coming up and wanted to reap where they had not sown. It was a dirty game being played; and I later realized that even other colleagues who had been supporting my efforts then turned against me and started to bring in nepotism. But the Chancellor had the appointment powers and his word was final.

I had excellent rapport with JICA and we were working well together to ensure that all the support agreements were in place and implemented on time. Technically, I had no problem with the implementation of the programmes. I had the competence and drive to enable the young university meet its dream.

After a few weeks of closed-door soliciting, the Chancellor ignored other candidates and officially appointed me as the first Vice-Chancellor of JKUAT in April 1994. Again the appointment was aired in the news and I got down to work. I had seen the growth of the institution having been the Deputy Principal, Principal and now the Vice-Chancellor. This upward mobility gave me inspiration to steer the university to greater heights. I promised myself to develop my own brand of "Leave a Legacy".

Preparation for the Instruments of Power

The university had to be officially inaugurated once the officials were in place. We already had students on campus. A reasonable number of lecturers had already been employed. Every step in planning for the launch had to be expedited to allow for the first-year intake through the Joint Admissions Board (JAB).

There were four vital instruments of power which had to be in place before the inauguration: 1. The University Mace; 2. The University Logo; 3. The University Act; and 4. The University Seal. These were the instruments of power that bestowed legitimacy to an academic process. In addition to these requirements, we had to robe the Chancellor.

All these preparations required my personal input with the blessings of the Council and Senate. I knew what was required, having witnessed similar procedures for Moi and Egerton universities. I just had to ensure that the instruments were made and delivered to us. State functionaries were involved and I kept them posted regarding every development towards the inauguration day.

It was not easy to manufacture the mace. It took me time to get the place maces were moulded. The Asian shop I was referred to was the only place in the industrial area where they could mould one. They also made seals. I asked the owner to manufacture the mace which had been designed by our engineering and architecture department. He gave me a quotation, which I took to the Council for approval and it was granted.

The Asian's foundry was always busy with demands from customers. I literally visited the factory daily to ensure that the mace indeed complied with the University mission and character. I remember taking my son, Michieka Okioga, to the factory to see what was going on and why I kept revisiting it.

It was my sole responsibility to ensure that all academic instruments were in place and ready for the launch. I would not delegate this particular role. I delegated other responsibilities to other senior staff members. Over time, and as part of my administrative leanings, I would delegate roles to my deputies and other academic staff, depending on the nature and importance of the task.

The same factory agreed to manufacture the seal at the same time but at a less cost. Moulding it was not as demanding as the mace. The logo was printed at the university and registered with the relevant office in the Attorney-General's Chambers in Nairobi.

Luckily, as we were preparing the instruments of authority, Parliament was busy debating our Act, spearheaded by the Minister for Education, Hon. Joseph J. Kamotho. I remember attending a parliamentary session to respond to any questions which could arise; as I could not debate openly, I gave technical backup in writing to the minister. The normal procedures were followed and the Act was passed.

I remember one question which was asked was why the university could not be called Jaramogi Oginga University. The minister replied that the founding president provided the land. We had done a thorough preparation during the drafting of the bill and had no major issues. The bill sailed through as I was busy following the instruments of authority.

It was such a headache getting a tailor who would accept to make the Chancellor's gown. After a lengthy search and consultations with State House staff, we finally found one on Mama Ngina Street, Nairobi. Again, I had to get a designer from Kenyatta University to make a blue print and pattern. We got a lecturer from the Design Department who assisted in the designing. The finished product was then taken to the State House for approval, and indeed it was.

All systems were operational for the university's inauguration on 25 April 1994. President Daniel Arap Moi requested for a draft speech which I had to prepare in consultation with my Deputy Vice-Chancellors who had also been appointed. They were academics from other universities and included Dr Josephat Yego, Prof. Henry Thairu and Dr Rosalind Mutua. Prof. Thairu took my earlier post as the one in charge of academic affairs. The new university setup was complete, with the Council as the highest administrative organ, then the Vice-Chancellor, Senate and Faculty staff.

Our University Act was the newest at that time and I took advantage of comparing it with the old universities' Acts. I edited clauses which I knew had problems in other Public universities. There was one clause in the statutes which stipulated who teaches at the university, when, how, where and by what means. The full responsibility of students' affairs was at the hands of the Senate. It was explicit in the statutes that the admission of students into JKUAT was the Senate's affair. The JKUAT Act of 1994 which was later repealed in 2013 changed the operations and compositions of public universities.

My discussion in this book reflects on the then Act and roles as at 1994 to 2013. I had a clear vision for the new university on expansion. This will be covered in a different chapter of this book.

The day for the inauguration came and we had everything in place: The Mace, the Logo and the Seal. I had delivered the president's gown at the State House and requested his details to ensure that they came with it on 25 April 1994. We had done rehearsals with the graduating students and members of staff. They were all excited to know that we were now a full-fledged university, delinked from Kenyatta University.

We graduated a small number of 400 students who had done first-degree courses. The courses had been approved by the KU Senate but taught by the new university staff according to the KU statutes. The relationship between parent or main universities and constituent colleges is often constrained. The Senate's undue conditions to constituent colleges hamper quick development and maturity.

The Senate considers itself the ultimate decision-maker and usually frustrates efforts to nurture and develop colleges under their guidance. Also, University Acts or Statutes did not stipulate how long the constituent college would take before being granted full university status. In some cases, Vice-Chancellors and Principals have to take personal approaches in decision-making.

Many African universities have tended to subdue the fast growth of upcoming colleges. JKUAT was not an exception. We faced challenges, sometimes ridicule, and many times postponement of our agenda in Senate. Frustrations, therefore, were common and obvious resentment by some individuals was a bottleneck. These factors did not, however, deter our own push to be an independent University. The academic programmes for JKUAT were mainly Science and Engineering-based, which the mother university, KU, did not have. The parent university was mainly an education-based institution which trained Kenyan secondary teachers. That was its niche.

It was therefore difficult for us to relate amicably in academic matters. Senate relied on decisions made by our Academic Board. I still do not understand the logic of linking JKUAT to Kenyatta University. KU was an education-based institution whereas JKUAT was science-based. There was no logic whatsoever in this decision. The young University College was, however, now mature to be the independent fifth public university of Kenya and form its own academic character.

A Full-fledged University

On 25 April 1994, we were declared an autonomous university. During our preparations for the big day, my office had invited all types of people from within and outside Kenya. We had invited public and private universities, members of Parliament, university associations, JICA, companies, industries, parents of the graduating students and a few secondary schools. This was a big day for us and we had to showcase it. We had gained popularity as a fine technology college which produced practically-trained graduates. We therefore extended many invitations to industries we attached our students to.

I was not aligned to any political party in my academic career. But the ruling party KANU then had unprecedented presence in all spheres of life in Kenya and even outside. It was therefore implied that I should not extend any invitation to the members of the opposition. I found this odd and decided to defy the quiet rule.

I took it upon myself to invite the leaders of the opposition parties, and most specifically the late Jaramogi Oginga Odinga. I considered him and his group as Kenyan dignitaries and they deserved recognition in our functions. I convinced myself that I should not be afraid of inviting Mzee Jaramogi Odinga. I was anxious to have the task finished and I dared anybody who could challenge our decision since it was not a mistake to invite the group. The worst that could happen to me was to be fired from being the Vice-Chancellor.

I felt that political parties change, and I noticed an opportunity in disguise and utilized it. This is one matter on which I did not bother to consult the State House; the Council, however, was fully aware, and so was the Academic Board. The university was inviting guests and had the full authority and privilege to pick our invitees. I took full responsibility on any subsequent consequences thereafter.

All the invited guests started to arrive prior to the president's arrival. Jaramogi Oginga Odinga arrived early with a group of politicians and I ushered them to the podium. The staff and students cheered him as he raised his white flywhisk high in the air to them. I heartily received him and took him straight to the designated seats for VIPs where most invited guests took their places.

President Daniel Arap Moi arrived a few minutes after Mzeee Odinga. I received him with his entourage and ushered him to his office as the would-be Chancellor of the JKUAT. I surely knew that he had been briefed of the arrival of several dignitaries, including his long-term arch-rival Odinga. As we received him outside the main administration block, I quickly whispered to the Education Minister, Mr Joseph J. Kamotho that Oginga was already comfortably seated at the graduation square. He looked straight into my eyes and said that was okay.

I was relieved of the possible future consequences. My fellow Vice-Chancellors joined the president to have a cup of tea as we all started exchanging pleasantries. One thing I recall was how tense I was as I was entertaining all these people from all over the country and beyond. The Japanese dignitaries who had been supporting the college were present in large numbers and were comfortable amongst the Kenyans. Their role is discussed in another chapter.

After tea and signing of the Visitors' Book, the president was ready to march to the Graduation Square. The chairman of Council and I briefed him on the day's programme and what we expected of him. I had already been appointed as the Vice-Chancellor, the Council was in place and the Act was ready.

All the instruments of power were in place and the day's proceedings were on course. I had prepared well. The past several months of my life had been dedicated to an inauguration that was to last only four hours. I remember the lengthy periods of time which led up to this great event. I had prepared well and left nothing to chance. My wife, Esther, used to advise me on several matters, especially during the drawing up of the list of invited guests across the political parties. She had also assisted in the drafting of the official speech and in the rehearsals.

The presidential procession marched to the dais in this order: Two students, staff, deans, Council members, vice-chancellors, chairman of Council and I. Staff wore their ceremonial gowns but the new Chancellor did not have one. The president was to be honoured first and robed at this inauguration ceremony.

The ceremony started at ten o'clock with the national anthem and the rest of the programme followed thereafter. Senate had approved the award of the first *Honoris causa* to H.E. the President who was going to become the Chancellor of JKUAT.

I declared the first congregation of JKUAT as per the 1994 Act. I read the relevant sections and requested the president to come forward to be robed. I read his citation which had been drafted by me with support from the State House. It was not strong in academics but in politics. It took me about ten minutes to go through the modest curriculum vitae of Daniel T. Arap Moi.

After the citation he walked forward, bowed to the congregation and staff and put on his robe. I raised the hood over his head and dropped it behind his back. I then capped him, declared him the first Chancellor of the JKUAT. The fanfare music was played as the crowd cheered him. I was a proud Vice-Chancellor who then awarded the President the Honorary degree, *Honoris causa*. He was the first official "student" I would be awarding a degree. A similar event would be repeated to his successor, Mwai Kibaki, who took over as Kenya's third president in January 2003. I now knew why Prof Joseph Mungai had explained to me that Vice-Chancellors were supposed to be addressed as "Your Magnificence". He told me that it was the right salutation and one day it would be adopted. The Chancellor was now empowered to hand me the instrument of authority as stipulated by the 1994 Act.

The Academic Registrar, Mr Joel Mberia, handed over the Mace, the Logo, and the Seal to the Chancellor to hand them over to me. The Chancellor gave me the Mace with clear instructions that this was the symbol of power for the VC to conduct the academic affairs of the university. He did the same with regards to the Logo and the Seal. Every activity was exciting and the crowd cheered, accompanied by Police band fanfare music. I do not know whether the current wave of new Vice-Chancellors across Africa get to be fully briefed concerning the meaning and importance of these instruments of power.

Several speeches of goodwill messages followed starting with my fellow Vice-Chancellors: Professors Francis J. Gichaga (University of Nairobi), Justin Irina (Moi University), George Eshiwani (Kenyatta University), Japheth Kiptoon (Egerton University), Freda Brown (United States International University). Other messages were delivered from the Association of African universities, Association of Commonwealth universities, Japanese universities, and JICA Resident Representative, to name but a few. Individuals and private organizations that we had linkages with also sent their goodwill messages.

The messages were brief and we embarked on the other formalities of the day. Our first graduation was held then. Some other speeches were delivered. I delivered mine in about 15 minutes, pointing out that the institution would maintain its original character of being a Science-based technological university with practical-oriented graduates who would meet Kenya's industrial challenges.

I challenged the Ministry of Education for not providing adequate funding for research and staff remuneration. In his speech, the Chancellor congratulated me for steering JKUAT to its status and also producing graduates who were needed in the country. He further stated that since we had started making history, we should continue excelling. Those words were encouraging to us all in the Senate. He also praised the role of the Japanese people for their generous technical assistance.

As the events were unfolding, Mzee Jaramogi Oginga Odinga was listening attentively. He was sitting in the front row, next to a Member of Parliament. My wife, Esther, was sitting right behind them taking her notes as speaker after speaker took the lectern. We would later share the day's events. Nobody had raised any concern about the invited guests. I had been confident in my official speech and had used my past public speaking experiences at various fora to deliver it.

Quoting the relevant section of the Act, I declared the congregation dissolved. We then marched back to the Vice-Chancellor's office in a reverse order with a dean carrying the 25 kg silver decorated Mace. Guests were invited for lunch and everybody who attended was happy. They showered my Senate with praises and the name of JKUAT started rising. This was the beginning of mentoring my staff for future leadership positions.

Mzee Jaramogi Oginga Odinga attended the luncheon which he enjoyed very much. He was heartily chatting with the persons who sat next to him. As he was leaving, long after the President's entourage had left, he called me aside and remarked, "*Rapemo*, (another name for Ratemo as referred to in the Luo language) *thank you very much for inviting me to this great occasion. Your university is a promising one, keep up the good work.*" I thanked him and he left with the students cheering him. He did not address them, which was an excellent gesture on his part. From then on my independence and mode of doing things impressed the community. My bravery paid off and I learned one lesson: Humility and respect pay.

Building an Institution

The inauguration period came to pass. We started to ask ourselves what the future held for JKUAT. I was now the Founding Vice-Chancellor, the Council was in place, and my deputies had been appointed; what next? Most Kenyans were happy that there was a new public university which would admit extra number of students in the next intake.

I matured with JKUAT. There were several tasks which I had to steer forward if we did not want to regress. We had to slowly phase out diploma courses and

replace them with degree courses. That was one way of being relevant to Kenyan needs. Specifically, I was challenged to address the following areas in the transition from college to university:
- Staff recruitment and remuneration
- Staff training
- Development of academic programmes
- Writing the statutes
- Admitting the new students into new programmes
- Infrastructure, lecture theatres, halls, offices
- Positive staff inducement
- Maintaining quality in academics
- Outreach programmes, campuses
- Financial sourcing and fundraising for the institution
- Research proposals, protocols and publications
- ICT as an independent outfit
- Land demarcation and boundaries establishment for the campus
- International linkages
- Equity in gender
- Arid and semi-arid student enrolment, positive discrimination
- Water and sewerage treatment
- Industrial attachment
- Establishment of students' union (JKUSO)
- trategic planning documents, the SWOT analysis
- Environmental concerns – tree planting, a green university and Thika Highway
- Outreach services to the community among the several managerial and administrative responsibilities.

The list was long and we had to start somewhere. I knew that this was a mammoth undertaking, but I was prepared for the task. The tasks enumerated above are not complete, but it should be noted that the transformation of any institution is a complex matter. I neither had clear criteria on how to apply our strengths, weaknesses, opportunities and threats (SWOT) at this point in time, nor SMART objectives. I had a clear vision, however, of what JKUAT could be in twenty or more years.

It was not practically possible to tackle each one of the tasks enumerated above. But I shall endeavour to select a few crucial ones which made JKUAT the university it is today. I shall also specify areas which had stagnated the development of Kenya due to inappropriate attention to critical areas.

From the outset, I convinced myself that institutions reflect the leadership, and leadership reflects them. I also knew that certain qualities like integrity, respect, honour, humility, honesty and earnest consultations mattered a lot in institutional and capacity building. Above all, I wanted to be open-minded and generous to my workers no matter the ranks, because they all counted in achieving the tasks. I embarked on creating a university with a specific image.

The new Council was constituted according to the University Act of 1994. The Initial Council Members were: Dr Stephen Mulinge, Chair – a Plant Pathologist; Dr David Koech, a Microbiologist; Tom Owour of FKE; Uhuru Kenyatta, Prof Joseph Maina Mungai – Medical Doctor and Former VC and then serving as Secretary for CHE (now CUE); Engineer Sharagwe; and Amb. Ali Chirau Mwakwere; Mr Sugiyama (Representative of JICA). This was a Council constituting of highly-respected academics and professionals in Kenya.

They assisted me a lot in laying a firm foundation for the growth of the new university and their support was unwavering. For example, most of the staff the new university inherited had only first and second degrees. My immediate concern was to build qualified staff capacity by ensuring that most members of the academic staff had PhDs. With a lot of resistance from the non-PhD holders, but with the Council's support, a letter was sent to all affected members of staff to pursue their PhD studies with full scholarships provided by the Council. Those who benefited from this directive are still appreciative of what we did. The scholarships were tenable in the USA, Canada, UK, Japan, South Africa and Australia.

The University College was known for training in Agriculture, Engineering and Science. I had to steer the Senate alongside the same disciplines with additional unique academic fields. I had written no strategic plan but I had a vision; a master plan and the kind of graduates to be churned out. The first task I undertook was to ensure that most of my students had gone through practical attachments and could be employed upon graduation. I led the Senate into strengthening the existing programmes and creating new ones which were customer-driven. After extensive consultations with stakeholders, new courses were created as I simultaneously sought out new staff and technicians.

New courses which were market-driven included Architecture, Mechatronics, Geomatics, Environmental Sciences, Horticulture, Landscape Architecture, Intromid, Mechanical Engineering, Electrical and Electronic Engineering, Agricultural Engineering, Actuarial Sciences and several basic science courses. There were also compulsory courses which were more in social sciences than biological or physical sciences.

I strengthened the public relations office which was under my office and made sure that adequate information was given to the public. I knew the importance of timely communication of important events. Knowledge is power.

No programme was passed by the Senate until we were fully convinced that it would add value to the existing ones and graduates would be readily absorbed. How was this notion built up? My staff in the required disciplines would conduct a questionnaire. The public, industry, former students and the country's national plan formed the basis for decision-making. We also cross-checked and interrogated courses offered by other public and private universities. We did not fear competition but aimed at good quality graduates. We got generous assistance from the Japanese government which aimed at strengthening science and technology.

On some occasions, industry, firms and manufacturing companies would place orders to employ our graduates due to the close collaboration with industry in curricular development. The university had very good linkages with industry and we used to have annual University-industry conferences, with funds approved by Council every year. This was an annual event.

The Role of the Japanese International Co-operation Agency (JICA)

Jomo Kenyatta College of Agriculture and Technology had long technical support from JICA. The first agreement was signed in 1977 and the news coverage in one of the local dailies, Nation of 10 December 1977, read thus: *"Japan has agreed to build a Kes. 200 million agricultural and technical college at Gatundu in honour of President Kenyatta. The chairperson of the project, Mr. Ngengi Muigai said the college would be built on a Harambee basis and be of the same status as the Kenya Polytechnic and Egerton College. It will be known as Jomo Kenyatta Agricultural and Technical College."*

The co-operation had been ongoing for over 10 years when I joined the college. It was, however, limited to a few areas of concern. The co-operation aimed at supporting training, supply of equipment and development of structures on the land that had been donated by the founding President of the Republic of Kenya, the late Jomo Kenyatta. It was set up as a certificate and diploma-awarding institution. The ministries of Technical Training and Education deployed staff here.

I joined JKUAT about nine years after the agreement had been signed. At that time several technicians were being trained in Japanese institutes. They stayed for a few months in Japan and returned with a certificate in a technical course. The curriculum was overseen by the Ministries of Technical Training and Education.

My first visit to the college during agricultural teachers' conference gave me a dull impression of a quiet, dusty, low lying environment. Some areas were bushy with overgrown imperata grass typical of an ecological zone five of Kenya's land classification. It is a typical clay cotton soil site. Some areas here were wet and overgrown with the water grass. Some sections of the land were cracky, typical of clay cotton soil conditions.

The low-lying ceilings, walls and hallways attracted my attention as I walked to the assembly hall where we held the Kenya Agricultural Teachers' Association (KATA) conference. It was a two-day meeting and I never bothered knowing more than that. The college was no match to the massive and voluminous University of Nairobi. Little did I know that this was going to be my domicile for a good 13 years of engagement.

I decided to influence JICA, now that we had become the fifth public university. I quickly got to know the team leader, Mr Sugiyama, who was the day-to-day contact person of the JICA project in JKUAT. Our first meeting was formal with a few exchanges of words. I had been given a good briefing by the Academic Registrar, Mr Joel Mberia, on how to handle the Japanese nationals. Mr Mberia was very useful in our negotiations later and he became my right-hand administrator. I learnt their behaviour and respected their culture and humility. In fact, I emulated their characteristics of being calm and tolerant whenever we had an interaction.

Prior to my appointment to JKUAT, JICA supported projects that were already being conducted there. The Kenyan and Japanese governments had signed an agreement through which JICA assisted in technical training, capital development and support for academic development. Our government was meeting its normal obligations of staff remuneration and some recurrent expenditure, whereas JICA was involved in technical co-operation. This was, I think, the best external aid that a university of our calibre received.

I steered clear of any controversy and nurtured the support for all the 13 years I was at the helm of the university. I signed many contracts between the two governments, and I made sure that each contract had a large proportion of academic staff training. Earlier contracts were mainly short-term technical training rather than degree-awarding ones. In fact, I discontinued the training of technicians and opted for Master's and PhD training. This arrangement boosted human resource capacity availability.

I had trained over 50 PhD members of staff before I left JKUAT. My staff retention percentage was high as I did not lose any after training. I guaranteed them posts whenever they trained; and allowed for importation of generous quantities of their personal effects, and expedited processing of duty-free cars. I did everything possible within my powers to ensure comfortable arrival and settlement of my staff from any university worldwide. Many of them brought teaching and research equipment from their universities. Part of our quality assurance was due to modern specialized laboratory equipment.

JICA supported physical structures and laboratory equipment. The Kenya government would not match its obligations at times. I learnt the art of persuasion and techniques of public relationships. JICA, under the able leadership of Prof. Nakagawa, was very responsive to me and the requests I made. The Ministries of Education and Finance were supportive of all the projects we negotiated and I kept them informed. Project negotiations usually took several months and involved a team of experts from both governments. JKUAT negotiated for Kenya Government. The creation of JKUAT character was visible when we graduated students who were productive and practical in their employment stations. I used to get calls for graduates to be absorbed in industries.

My other task once we became a university was to absorb, train, re-train or send away former college staff. The university changed in status and it was my duty to smoothly release under-qualified staff and replace them with those who had the required university credentials. I had to be careful to avoid bad blood between those departing and the ones staying.

As the head of the institution and in conjunction with the Ministries of Technical Training and Education, we carefully crafted the requirements of staff release or retention. My deputy vice-chancellors in charge of Administration and Academic Affairs had to come up with clear criteria. Despite the very transparent requirements for staff training or recruitment, we still had complaints which I personally had to attend to. My handling of these unique cases tested my leadership prowess.

I decided to follow the criteria, keep those qualified to teach or perform other duties and release the unqualified ones as quickly as possible to avoid litigations. There was no directive that the new university would absorb the college staff we found there. We wanted to set up a modern, quality university with qualified and dedicated staff. It was a smooth transition despite a few difficulty cases which needed the intervention of the Council. We reported every action to the Council with well-written minutes. The Council in turn approved and amended, where necessary, our decisions accordingly. I was always very concerned about the Council minutes since they were the final authority to allow us implement decisions.

We had an understanding group of Council members who knew their roles and abided by them. They entrusted all matters administrative and academic to the chief executive and the Senate. We in turn kept the chairman abreast of any developments. The university became a centre of attraction due to the unique and practical programmes we offered and the JICA support we received.

We renewed projects after every five years. There were several midterm reviews before a final big review was done to approve the project. Normally, several experts could come in advance teams before their leader to collect preliminary information. The advance team used to interview staff for specific information in preparation for the signing of the agreement. This was important in order to prove my transparency, prudent utilization of the Japanese taxpayers' money and maintenance of academic quality.

We signed most of the projects after I joined the college. They covered agricultural and mechanical technical co-operation, staff training, and equipment provision, short and long-term training courses for senior management. One unique condition of the project was that a Japanese expert had to be seconded for every programme to work with a Kenyan counterpart. At times, non-qualified Japanese staff was seconded.

I changed the practice of accepting anybody without scrutinizing their academic and technical expertise. I therefore demanded full curriculum vitae of anybody who joined us as a counterpart expert and requested for a PhD graduate in case of teaching and research, or a Master's degree if the person was going to be doing purely technical work. I also demanded proficiency in the English language. I knew that communication could be a problem to my staff and students alike. The team leader was an understanding man and he complied.

Let me take one area of academic programme – Engineering – and demonstrate its broad objective. During the curriculum development, we knew that engineering was the application of all related science and technology to provide solutions to problems of mankind. We further underpinned the fact that the courses in the area were known to spur economic development in Kenya. Rapid industrialization had to be achieved through heavy investment in engineering education, science and technology. The curriculum we therefore drew borrowed a lot from what other universities had done, especially the Japanese ones.

I knew several engineers in Kenya then. They were less than 2,000. The Asian Tigers had over 25,000 with a population of about 35 million. We aimed at producing problem-solving graduates with a wide range of industrial attachment experience. The engineering graduates had to acquire appropriate competencies and skills especially in design. We emphasized problem-solving in all our technical courses throughout the curriculum development. One had to have the ability to identify a problem and solve it; and also be able to provide answers for problems. One had to be an effective communicator.

We were all aware that Kenyan accreditation bodies would need to place great emphasis on problem-solving through Engineering, Science and Technology. We wanted our graduates to attract employers. Indeed, they turned out to be more marketable than those from other universities. The same rigorous curriculum writing was done by staff for all faculties.

The same procedure in writing new courses was followed. Agriculture, Sciences and Architecture were similarly developed. We maintained our character and mission as we debated on the development of various courses. The Senate had a vision and mission clearly pronounced as we set up the new institution.

I had to plead with the Senate and convince them not to allow an excess intake of students in the humanities and social sciences. Our Act was explicit regarding the institution as being agricultural, technological and innovative. Those were our driving forces. Things changed later on and the university was flooded with all types of disciplines.

Training and Visits to Japanese Institutions

All signed agreements had components of short and long-term training. I had also included: Chairman of Council, senior management training for the Deans, Deputy Vice-Chancellors, and the Vice-Chancellor. Many of the senior staff visited several places in Japan to gain first-class experiences in industry and institutions which trained our PhD and MSc students.

The exposure was meant to be an eye-opener for Kenyans to implement part of what could be gathered during the tour. I also had several senior management courses there. For example, a visit to Toyota city, where thousands of vehicles are manufactured, was an exciting experience. I wondered whether our African countries could establish such an outfit. The automation and production of vehicles is so advanced that thousands of vehicles are produced each week. I am sure my colleagues must have also wondered when Kenya would attain that level of vehicle manufacturing.

Every person who went on a training tour was accompanied by a Japanese expert in the relevant field of study. The expert was responsible for all travel logistics, protocol and language translation. This feature was important to ensure value for the trip in order to avoid culture shock particularly for students. A good number of staff learnt the Japanese language and wrote their theses in the language with English translation.

During one of my tours in Japan, I was able to present a paper to scientists from a number of Asian countries. The conference whose theme was, 'Culture Crossroads, where Culture Meets Races' exposed me to various types of sub-races which make the greater Asian block. Through several papers presented, I was able to phenotypically distinguish between residents from Hokkaido and those in Kyoto or Okinawa. There are typical distinguishing characteristics which are unique to the individuals. The significance of these meetings was meant to inculcate some kind of Japanese culture into our institution.

I am not sure how much we gained in terms of attitude change to work. But my staff gained degrees and returned home with duty-free state-of-the-art cars. I am not sure how much technology they gained to transfer to our country.

Staff on long-term training was allowed to bring into the country one vehicle duty-free per individual for personal use. The privilege was awarded to all academic staff of public universities. My staff took advantage of this waiver and enjoyed the benefit to the fullest. This was one of my staff inducements for their retention.

We were able to hire quality teaching assistants with a primary aim of training them to PhD levels and consequently retaining them. The idea worked and I am proud of my training record while I was at the helm of JKUAT. We further attracted staff from the other established universities because of some rare privileges.

I believed in firm, reliable and hardworking staff. I had my own managerial and leadership skills which assisted me in the winning confidence of my staff. As a team leader, I had to perform and be seen to walk the talk.

I knew one thing for sure: that every team needs team work and team spirit; and that the members of a team relies on each other for performance. I encouraged their working together towards our goals. I also encouraged them to seek the best solutions for any problem which could derail our young university. Personal leadership in a business enterprise plays a role in the way workers render their services.

A positive attitude, whatever the circumstances of our working place, was a source of health and happiness. I developed a generous and cheerful attitude. I believed in a corporate business principle, by continuously communicating from top down and bottom up on time. My duty as a chief executive was to integrate every one into a harmonious whole. I knew that leadership rests on this wise power to translate our mission and vision of inspiring leadership which could create a firm academic institution for our country and beyond. My philosophy was to act and not react.

11

Tenure as Vice-Chancellor

Anyone who has ever succeeded in any human endeavour will tell you stories of failures and what he/she learned from those failures. A writer and poet Samuel Beckett, wrote the following words, "Ever *tried; ever failed. No matter; try again; fail better.*" Life is ultimately about these failures. They usher in new experiences, exposures and excellences. My intention was to learn from any mistakes I made and make corrections.

In my earlier chapters, I narrated how I became a departmental chair without an iota of experience. I trained on the job and went along past traditions of managing a department. This learned experience became useful during my tenure as a Vice-Chancellor.

My primary and secondary school days exposed me to some administrative intrigues which hardened me at an early age. As a class monitor and prefect in a primary school, I dealt with both teachers and pupils. I gained more experience as a prefect in Kisii High School; I handled rude and undisciplined students who needed favours. I also had to deal with and appease teachers as well as commodity suppliers, some of whom unsuccessfully attempted to induce me into doing those favours. I made my firm decision then to be a young boy of impeccable integrity. I survived those days and excelled in my academic work.

At Rutgers University, USA, I had to contend with students and lecturers from diverse races. I was discriminated against as a black African student from Kenya who may not compete favourably with the white students. Luckily, I excelled and completed my studies despite those odds. My personal interaction taught me the power of humility and integrity.

My time in Nigeria further exposed me to some unjust racial favouritism amongst international organizations. I learnt to survive under Nigerian military rule as we interacted with the local communities during my field research work. I also avoided "dashing" (bribe) the Nigerian army men and women who manned roadblocks on the major Ibadan-Onitsha highway. I survived the culture of giving in to anything which could compromise my moral conscience.

In many African universities, Principals of colleges and Vice-Chancellors were appointed by either the Heads of State or ministers. These were common from 1980 to 2005. The Kenyan situation started to change with the new government in 2004. The appointments were made with no specific terms of reference, training or refresher courses for the new appointees. I can confirm that my former University of Nairobi Vice-Chancellors were not trained after their appointments. Professors J. M. Mungai, P. Mbithi, F. Gichaga, C. Kiamba and G. Magoha were all appointed and told to report to work immediately.

Lately, there has been a change in the system. The Vice-Chancellors and their Deputies apply for the posts. They are then interviewed and, if found fit, given letters of appointment.

As a Vice-Chancellor, I had to administer the university with a few top managers and get all academic and administrative systems in place. I convinced myself that we must work hard as a team and show the way. I also borrowed a leaf from E.M Forster – *"one person with passion is better than forty people merely interested."* I loved my academic work where I could decide on which project to undertake. But with the new appointment as Vice-Chancellor I had no choice other than to delegate and supervise activities. There was nobody to turn to other than the Chancellor in case I had a problem. There were flaws and advantages of having a Head of State of state as your boss.

It was an advantage that I loved my new position. I worked with zeal and solved problems as they came. There were several administrative concerns that I had no inkling at all on how to manage, one of them being the finances. I learnt on the job how to balance my books and do budgets. I knew my weakness, not being an accountant, but had taken courses in economics and basic financial management. My resolve, however, was to get the best financial manager with a number of good accountants and a tough internal auditor.

I organized seminars and invited experts to be resource persons alongside my own staff from the finance department. I learnt a lot and they also learnt a lot within a short time. We met outside the campus and I allowed for ample networking time. I got to know the government's acceptable procedures of procurement, tendering, expenditures, incomes and balancing books. In my appointment briefs, I knew I was the ultimate respondent to any financial misappropriation.

My Deputy Vice-Chancellor (administration and finance) was also not trained; we were both raw but the many short seminars assisted us in utilizing university finances well. We followed the laid down accounting procedures and never had any audit query for the period I was at the helm of the young university. I did not want to appear before the Public Accounts Committee of Parliament on financial impropriety. Even though I had no formal training in accounts, the staff was aware that I knew a little bit of balancing the university budget. I kept my Council abreast of any shortfalls or savings in the university books.

Another area on which I needed a quick fix was staff recruitment, motivation and retention. In the 1990s the distance from Nairobi to Juja Town where the university is located seemed very far. We were all used to the universities of Nairobi and Kenyatta. Nobody wanted to commute 56 kilometres one way. The cost of fuel, wear and tear of the car and time consumed in travelling was becoming prohibitive. I sold the idea of increasing travelling allowance to the top management.

I decided to hire great scholars from other universities, private organizations and the Diaspora. The incentives were good. I had to convince the Council before I floated the ideas. I sought several PhD scholarships from universities outside. I also encouraged many young lecturers to get admission into Japanese universities with a firm commitment that their posts would be available on return and each was allowed to import a duty-free vehicle. I convinced Council to approve a higher commuting allowance than other town campuses.

We formed a land-buying welfare society to acquire plots for staff to construct controlled houses next to the campus. The Council allowed for a check-off system which was generous in amounts payable and duration. The Council had to approve the borrowing of money towards the purchase. This was the best scheme next to the campus. We approached the Ministries of Education and Finance to allow for a higher house allowance. This was passed, but later almost bundled me to court when it was again being reversed.

The allowances were almost one and a half times those of other public universities. The most attractive package which I introduced was the 80 per cent tuition waiver on qualified staff's children to study for BSc or MSc courses in JKUAT. What this essentially meant was that any child of staff who had performed well and admitted by the Joint Admissions Board (JAB) would join JKUAT and learn almost free. All permanent staff benefited from this arrangement regardless of the cadre. This move became my immediate legacy and I became a friend to many!

This was the greatest incentive which made staff dedicate their services to the university. I had done my homework, however, and knew that a very small number would qualify. Our courses were highly technical, but I encouraged staff to work on their offspring and take advantage of the package.

I also allowed transfers from other universities to JKUAT as long as the cut-off points were met. Two great advantages were eminent: a direct additional financial gain by parents on educating their children and a controlled group of students in case of riots. A critical number of parents in such gains would not wish their children to riot, boycott classes or get into mischief lest they lose the benefit.

I had indirectly turned workers into managers of students. University riots were the order of the day from the 1980s to 2005. I had planned to reciprocate student exchange with other universities, but left before I could implement the idea. For sure, Moi University was willing to discuss the plan. More recently, many Kenyan universities have been giving specified tuition remissions to their workers.

I mooted this idea earlier to curb staff migration and create a university where all workers and students had a say and a sense of belonging. It turned out to be a truly cohesive community. All these activities were funded by the savings we made in the expenditure and government resources. Some research funds that generated interest also boosted the various budgets.

There were areas which I found easy to handle. They included getting out to the public to sell the university. My public relations office was so well equipped that I did not have to worry about our image. Junior staff members were easier to convince on new ideas than top management.

I never discriminated against the cleaners or security guards when it came to meetings or workshops. I organized theirs to fit their terms of service. End-of-year parties and mid-year bonding were my most productive interactions. There was no time as satisfying to a junior worker than when he/she sat on the same table with a dean or professor for a meal or drink. They felt appreciated and respected. They talked freely, and even made suggestions on how they should run their events. I used to tell them that it was their turn and day to tell it like it was. They felt free to interact and express themselves. This freedom was an eye-opener for me in management techniques.

The meetings we held gave me a chance to evaluate the impact of decisions. In such circumstances, I opened up to all staff and became more responsive. The books I used to read on corporate governance advocated staff bonding and allowed for free thinking space. I wanted to achieve many things and this could not be possible without teamwork.

This was when I knew that leadership rested on the wise and sensible use of power. Staff would always trust in you if you identified with them and at times put yourself in their shoes at their places of work. The attrition turn-over of my staff was very small. The incentives were so binding that a good number of them opted to stay at JKUAT and retire from there.

Administration and management of many African universities were many a time through crises. The Vice-Chancellors who had no prior training in administration found themselves making decisions which were not in tandem with people's expectations. My main role was to ensure that academic programmes were adhered to and administration issues followed as well. Senate was my right-hand body. I had a number of senior lecturers and professors who were dedicated to building a modern technical university.

Whenever I had decisions to make, Senate would be convened and I would discuss them freely. I learnt my work on the job, trained on the job, grew on the job and resolved difficult cases on the job. No advance management training had been provided. It was a trial-and-error philosophy. Handling of students was varied. I had a liking for good performance, but we had some notorious students who needed counselling and continuous advice.

They were also to be managed and I played the role of counselling them. No one could predict the events of any given day.

Despite the few incentives and equitable distribution of university privileges, Kenyan public university staff always agitated for higher salaries and better pay packages. There was no single year from 1980 to date when staff in all grades had not asked for some increase in salaries. Several strikes paralysed learning in the universities. The strikes usually involved all universities. In Kenya, we have a powerful committee of Vice-Chancellors who normally meet to compare notes and share common management problems.

The University Academic Staff Union (UASU) is the registered union which agitates for better terms and conditions of service. I recall one strike in 1994 which paralysed university learning for a long period of time. We met with staff on many occasions to resolve the impasse, although my own institution had not been affected. We met as Vice-Chancellors to chart a way forward with the Ministry of Education but no resolution was arrived at.

The Vice-Chancellors had to sort out the problem individually. The Ministries of Education and Treasury gave us a blackout. They bluntly told us that there was no money to increase salaries and other allowances. That was bad news for public universities.

In the very early 1980s university lecturers were the highest paid group. Civil servants were several grades lower than them. The argument was straightforward. To teach at any university worth its salt, a lecturer had to have an earned a PhD degree from a recognized university. This is the basic prerequisite. The lecturer was expected to teach and conduct research and then publish.

To be a civil servant, one did not need to have a PhD. It was not mandatory. It was therefore easier for a lecturer to be appointed into civil service and perform routine strait-jacket duties than a civil servant to come and start lecturing and conducting research. University teaching then was attractive and well-paying. Unfortunately, the pay and benefits have been reversed over the years. This has resulted into continuous academic erosion in terms of brain drain.

Each Vice-Chancellor had to sort out the strike. Several staff members went through disciplinary procedures and their services were terminated by their respective Councils. I fully sympathized with the staff but advised them to continue teaching since their demands had been made. There were a few who were very adamant and could have been sacked, but I did not convene Council's sub-committee on disciplinary matters. I used my own discretion and handled the matter sensibly. No one was sacked or implicated for disciplinary action. The strike fizzled out.

I used to read speeches during my graduation ceremonies. During one of the occasions, I never minced my words. I told the Chancellor about the urgent need to review the university salaries. The request was construed to be rude but in actual fact staff deserved better remuneration packages. Subsequent strikes followed. And even as I write this book, there are pending industrial strike threats. Yet, personally, I have always believed in a contented group of workers who are then motivated to quality work.

My reflections on the role of a Vice-Chancellor are based on the quality of the individuals and their academic progression. Management skills are necessary for negotiations when it comes to university issues. The roles of Vice-Chancellors are known and cut across all universities. Personnel in public and private sectors are better placed to conduct relations, financial management, resources, utilization and dispute handling. Academic matters should be handled by the chief executives and Councils. Learning on the job is cumbersome, demanding and time-wasting.

University lecturers have a lot of respect for their own leader who ably guides them. The Vice-Chancellor must have an impeccable record, morally and academically. It is the only way to win and command respect. The idea of appointing a chief executive of the university without proper consultations is misguided.

A few of the randomly chosen Vice-Chancellors worked well with staff and students. But the majority of them ran into perpetual problems. The current method of advertising for the posts leaves a lot to be desired. The advertisements are placed by those acting in the same positions and the requirements are clearly skewed to fit them. International competitiveness lacks in the whole process and so the status quo usually remains.

The Double In-take, the Creation of Colleges and Campuses

Public universities admit students annually but reporting dates vary from campus to campus. There were usually big backlogs of students due to frequent closures and wasted time. High school leavers had to wait for up to 18 months before they reported to their respective universities. This was common in the 1980s and the public was not amused. Their children had nothing to do for the many months that they stayed home after high school completion.

Public universities still admit students through the Joint Admissions Board (JAB). This is a reputable non-corruptible board of university leaders/heads who meet to determine cut-off points for admission to various programmes in the public universities. It includes the Vice-Chancellors, their deputies, deans and registrars of all public universities. The exercise is carried out in the most transparent manner which is made public through all channels of communication. The new University Act of 2012 stipulates to replace JAB with a Central university Admissions Board for all Kenyan universities. It will be responsible for apportioning students to both public and private universities which are legally recognized by law. This central admissions board has not been constituted by time of writing this book.

The Chancellor of all public universities, who was then the president, once in his off-the cuff roadside speeches in 1992 directed that all students who where qualified should be admitted at once to public universities and report to their respective campuses. Any backlogs were to report to constituent colleges and Vice-Chancellors had no alternative but to sort out accommodation, laboratories, lecture halls and lecturers. The Vice-Chancellors and principals had no alternative but to implement such an impromptu presidential decree! All systems were on motion.

The creation of the current universities has a history which started in 1989. The then chancellor of all public universities directed that all qualified students be admitted to the universities at once. His word was final. A committee comprising members from JAB and the Ministry of Education was formed to scout for suitable colleges and/or institutions to accommodate over 3,000 qualified students. I was appointed the chairman of the task force and we had to physically visit all possible suitable institutions.

The terms of reference were clear: (a) to inspect institutions which would accommodate over 3,000 students; (b) to check for adequate learning facilities, lecture theatres and laboratories; (c) to ensure functional infrastructure, dining and residential halls; (d) to check on any other available facility like vehicles and machinery; (e) specific number to be accommodated by each of the targeted institutions.

The committee under my chairmanship was given seven days to report back the findings. The Treasury was to receive the report and the then vice-president, Prof. George Saitoti, was tasked to source funds for the extra facilities and students. We were assured that our expenses would be catered for expeditiously. This was a directive from the highest office and our duty was to perform.

I assembled the team from the other four public universities to scout for appropriate colleges. I planned the trip and proposed to my team that we start from Western Kenya. All I had to do was ask the Ministry of Education representative to call the heads of the institutes we were to visit. Mr Joshua Terer obliged and alerted them all. The beginning of the creation of campuses started in earnest when we hit the road.

We visited the following institutions: Narok Campus, Kericho, Kisii, Maseno, Eregi, Chepkoilel, Laikipia, Kagumo, Kenya Science, Voi and made telephone calls to Mombasa Polytechnic. We had run out of time and had to compile a report within 24 hours after returning from each visit. I appointed one person per institute visited to compile a draft report immediately after visitation.

We designed a draft guideline following the terms of reference we had received. The writers were to come up with a complete draft the following morning before we continued to our next stop. I further asked one other member, Prof. Everett Standa, to collect all the reports, collate them and start compiling them into one document. The use of current ICT facilities was not available then and so we had to write our reports manually to be typed later. We received hostile reception at some colleges but I warned that stern action would be taken against those who would not co-operate.

After one week, we converged at Kenyatta University Boardroom on a Saturday morning to finalize the compilation of the report to be delivered to the vice-president and minister for finance on a Monday morning for action. Students had to report for lectures to the new campuses in September and systems had to be put in place by then. Yet we had been doing the visits in July 1992 under extremely rigid deadlines! The basic underlying factor was the availability of funds to cater for the staff and basic facilities as needed. The ball was thrown back to the government.

Our mission was simple and specific: to get places for double-intake students. The exercise was not smooth as expected. The heads of various colleges were adamant and did not want us to inspect their premises. Some of them moved away their buses and hid them so that they could not be entered in our inventory column. At one college, the principal instructed security guards not to allow us in because we were intruders and did not have written express permission from the ministry headquarters. We were not bothered. Little did they know that we had a Ministry of Education representative who was their boss. He insisted on seeing such principal himself on identification. I also kept in touch with the chairman of Vice-Chancellors' committee, Prof. Phillip Githinji, who encouraged us. I later learnt that the principals who had resisted our inspection were either sacked or retired.

Acquisition of colleges meant a lot to the staff and property of the affected institutions. There could be a complete overhaul of the proposed new university college in terms of support personnel, lecturers and courses. Staff vetting would be carried out in terms of qualifications required for one to teach at the university level. A good number of tutors then did not have the qualifications to enable them be appointed or absorbed as lecturers. Some of the principals themselves were not suitable and yet they still wanted to continue running the colleges! Support staff, administrators and other cadre were to be interviewed and, if found fit, would be absorbed. We had to take personnel's details as a basis for budgeting.

The exercise ended smoothly and we compiled the report for implementation by the relevant ministries. Our engagement was over after the delivery of the report. A few colleges were selected for the first phase after we had visited ten of them. Those selected at that time included: Maseno Campus, Chepkoilel, Laikipia and Narok College. These met the basic requirements for immediate occupancy after some basic renovations of facilities. They had to be ready to receive students, lecturers, administrators and a principal. The hard work of structuring and steering the new campuses, with no notion of a university setup, had started.

The Scramble for Colleges

My Vision for JKUAT

After my appointment as a Vice-Chancellor, I had my own vision and a specific mission to accomplish: that JKUAT should be a reputable transforming university which could attract students from all over the world. I also knew that for years, our students who had qualified to study locally were locked out of admission due to lack of admission slots and infrastructure. Hundreds of Kenyan students had to leave the country to study abroad. I knew this because the number admitted was small and those left out far outweighed those admitted by public universities. I had also been involved in fundraising for hundreds of students who were admitted into universities abroad and needed financial support.

The travel and tuition expenses in foreign countries were paid in foreign currency. Living costs abroad were nowhere comparable to ours here. What bothered me most was the fact that the courses sought outside were at first-degree level. The country spent colossal amounts of foreign exchange for this purpose. Private universities were few and could not absorb many students. I scrutinized the University Act of 1994 and the statutes. We had to evoke the relevant clauses.

Our statutes stipulated that the University Senate decided who to teach, what to teach, how to teach, where to teach, and when to teach. This conditionality was to adhere to the laid-down procedures on upholding the quality of our programme. I read and understood what was required before new programmes were begun out of the campus. In other words, how legal was it to teach students in appropriately chosen sites for the purpose of taking education to the other parts of the country?

In 1998, I decided to evoke the clause "where to teach". I also wanted to reduce the foreign currency flow out of Kenya. I informed my top management that I wanted to start out-of-campus programmes. I further implored them that there were colleges which would assist in this process. The idea was adopted and I prepared a paper. I also checked what the procedure was in other developed countries. These would be satellite campuses of JKUAT. It was a practice in many universities elsewhere.

The Jomo Kenyatta University of Agriculture and Technology became the first public institution to initiate linkages and the creation of affiliate campuses in Kenya. The senate approved the idea in principle and I led the team to explore the possibility. I knew for a fact that space, staff accommodation and laboratories would be a challenge. I had thought of several options.

The terminology we used then was "*alternative degree programmes*". I recall a heated debate arising from the choice of the name. We eventually settled on it and later called those students admitted under this programme "*parallel degree students*". The University of Nairobi domesticated the initiative and excelled in it.

I had two important missions: to make education available to qualified Kenyans and reduce students' exodus to foreign universities. The amount payable locally was less than that paid externally. Parents, sponsors and finally the government would gain from the arrangement. Private universities only admitted very few students then.

The Senate and the Council bought the idea and we started scouting for appropriate campuses, willing heads of institutions and linkages with the Diamond Systems, Mombasa Polytechnic, Nairobi Institute of Technology, Kenya Teachers College, Kisii Campus, Lamu College, Rosemary College, The Nairobi Campus, KCA, Strathmore, Nakuru, among others. The Senate had the control of the type of campus we needed and the courses we had to offer. Most of them were technical. We selected a few of the chosen ones to teach at Bachelor's levels. Science, ICT courses, Electrical and Mechanical Engineering were approved to be mounted. We also introduced diploma courses in specific areas.

The concern which we had to solve was staffing and manageable numbers of students. Before approving a programme, we had to identify the lecturers to teach, the space and laboratories available where applicable. We made one condition clear: Not to enrol an excess number of students over and above those sponsored by the government.

Nurturing Staff as Future Managers, the 'crème de la crème'

It has always been my belief that I must train staff with the ultimate aim for them to take over from me in various capacities of serving the nation. I believe in a smooth and competent succession. I have even trained my family members to appreciate that we are not immortal.

People come and go; leaders come and go; but institutions stay. I started to train my Deputy Vice-Chancellors indirectly to be fully responsible in their duties and whenever they acted for me, they did so with full knowledge of performing well. I did not fear being thrown over or outshone. I knew my capability, the depth and stature of my energy. Whenever I delegated my duties, I gave clear instructions and limits of the assignments. During my tenure, all my deputies ended up being heads of various higher learning institutions.

My deans, directors and chairmen also became heads of institutions. I knew well that future Kenyan universities would need principals and Vice-Chancellors. At the time of compiling this book, the following ladies and gentlemen were heads of institutions: my former Deputy Vice-Chancellors, the late Prof. Frederick N. Onyango, became the founding Vice-Chancellor of Maseno University, Rosalind Mutua became the founding Vice-Chancellor, Keriri Women's University, Prof. Henry Thairu became the founding Vice-Chancellor of Inoorero University, Prof. Wilson Kipng'eno became the founding Vice-Chancellor of the University of Kabianga. Several deans and directors also took over the helm of many Kenyan universities and they included Prof. Kioni Ndirangu, (Kimathi University of Science and Technology), Prof. Teresa Akenga, The University of Eldoret, Prof. Mabel Imbuga (The Jomo Kenyatta University of Agriculture and Technology), Prof. Josephat Mwatelah (Founding Vice-Chancellor, Mombasa Technical University), Prof. Joseph Magambo (Meru Technical university), Prof. Stephen Gaya Agong (Founding Vice-Chancellor, Oginga Odinga University), Prof. Oyawa (Ag. Vice-Chancellor, Multi-Media University), and Prof. Festus Kaberia (Vice-Chancellor, Multi-Media).

There are several principals who head the colleges which will soon be elevated to fully-fledged universities. They include Prof. John Ochora (Kisii College Campus), Prof. Hamada Boga (Taita-Taveta University College), Prof. Victoria Ngumi (Karen Campus Nairobi). Others are Professors Esther Kahangi, Romanos Odhiambo, Makhanu, Marangi Mbogo, Isaac Inoti, Linus Gitonga, and Francis Mathooko, who serve as Deputy Vice-Chancellors. Others include Prof. Florence Lenga, Mr. Joel Mberia, and Prof. Christine Onyango, who also serve as deputies in other institutions.

The genesis of the said positions was mooted during the creation of affiliate colleges. My vision and mission in the 1990s bore several fully-fledged universities headed by the staff that I had interacted with during my tenure as the Principal and later Founding Vice-Chancellor of JKUAT. I used to hint to them that the support they gave me would bear fruit and that they too would demand similar support in future. It is important to note that the quality and dedication of the named persons elevated them to the positions. There was no outside influence or canvassing.

In Kenya these days, positions are advertised and purported to be filled competitively without any undue influence. Sadly, however, this is not the case in many circumstances. The 2010 Kenyan Constitution was supposed to bring major changes in the way Kenyans run their affairs. This was particularly important as far as basic human rights principles are concerned. There are obvious glaring misgivings in certain quarters, like nepotism and cronyism.

There is evidence of people who merit positions or promotions but certain forces with selfish interests and preferences have sidelined them. We still have a long way to go in order to appreciate meritocracy and its benefits. I have only selected staff that worked closely with me and supported the mission and vision of the university. There are others who joined other positions in the government and the list may be too long. I chose to highlight from the creation of our affiliate colleges which were later converted into fully-fledged universities.

I further believed in training staff to the highest possible degree to replenish the staff attrition or mobility. Hence, I linked up with international organizations and universities to acquire scholarships.

One cardinal rule I believed in was to explain my personal experiences in the job of a Vice-Chancellor. I advised my mentees that there was nothing more difficult to tackle, more perilous to conduct or more uncertain in its success than to take the lead in the introduction of a new order of things, a new university or university college.

Universities in Kenya are created by annexing existing training colleges and turning them into universities. The colleges are affiliated to an older university which ensures growth and development in all areas. The parent university allows the affiliate colleges to be chartered after going through very rigorous induction procedures. It is not possible to enumerate the academic and personnel steps needed to accredit a university; but suffice it say that in the 1980s and 1990s, parent universities had to induct affiliated colleges and ensure that they were complete in staffing, infrastructure, library and any other requirements necessary for accreditation. Most Kenyan universities underwent this process which was, in some case, viewed by the small colleges as punitive.

With the repeal of individual universities Acts, the Commission for University Education will alleviate the need for excessive vetting. Colleges are still linked to a parent university. The scramble for space, market places, towns, shopping centres, high-rises, and technical schools is so intense that communities are confused. There are so many university learning places that competition for students is viewed as affecting academic quality. The creation of many campuses has necessitated hiring of lecturers who may not be qualified in a particular area of expertise. This topic will be dealt with in later chapters under quality assurance in our universities. It is a major concern and needs urgent attention.

Reflections

One of my greatest lessons of leadership is the knowledge that no task was beneath me. I had plans for a university, infrastructure and staff development. I knew what it took to convince my Senate and Council to move the agenda forward. I decided to mentor my potential followers by spending time working with them. I displayed thoroughness in my daily work and had a focused plan. My dedication to work, daily commitment to quality, excellence and persistence, made a difference in my staff and students.

They learnt from me and many gained experience in the process. I will always remember one example given by my colleague at a luncheon. He drew a fish on a chart, labelled its parts and asked us a simple question: when a fish rots, where does the rot begin? We gave various answers. A good number wrote down the word head. I wrote down the brain. He laughed at my answer and gave it an extra mark.

Fish, indeed, start to rot from the head, more specifically the brain. He was comparing governments and institutions in Kenya. They start rotting from the heads of the institutions. If the head is brainless so is the institution.

During the process of leadership, there are key challenges and performance indicators which create a cohesive value system. Team leaders strive to create mutual value systems which are beneficial to all. The leader creates a conducive working environment where expectations are mutual.

I decided to create a shared value system which defined ethical standards, integrity, innovation, and tolerance of all types of diversity. In my building of such shared academic value systems, I came up with stark reality of the diversity of human nature. I could not have one approach in problem-solving or overcoming the challenges. I learnt to understand variations in staff mentoring. I was able to re-mould some staff that later became heads of institutions. They learned how I solved problems facing staff and students and were able to apply the same approach in their future careers. It was on-the-job learning and solving problems in-situ.

My mentoring of future leaders did not proceed without challenges. I went through turbulent times and occasionally wished that I was not Vice-Chancellor of a university. Both staff and students would make demands which would be beyond my comprehension. I, however, used the same people to solve some of the problems and we eventually came up with solutions. I will highlight a few examples in later chapters.

In principle, I had a few words of wisdom to pass on to my colleagues. I always reminded them to obey and respect authority, observe the employees' working hours, perform duties timely, be responsible and accountable, provide quality services and remember that employees were dispensable.

Moi and Mori Meet

Normally, any Japanese delegation which came to Kenya had to visit the famous JKUAT. It was the ultimate visiting place after the Maasai Mara or coastal region of Kenya. I used to receive numerous courtesy calls and hosted all types of dignitaries. They included politicians, emperors and empresses, technocrats, tourists, professors and business magnets. The University was a household name in both Japan and Kenya. The Japanese team apparently wanted to come to ensure that their taxes were being put into profitable use and not squandered on unnecessary projects.

There are two outstanding visits which gave me sleepless nights: the courtesy call by the Princess and Prince, and their Excellencies the President of Kenya and the Prime Minister of Japan. During the Princess's tour, the Japanese Ambassador and JICA were to be fully involved to ensure that everything was right for their visit to JKUAT.

On my part, I had to prepare the gifts, touring of the campus and snacks. My wife, Esther, and I were to pick appropriate native wears for the Princess and a souvenir for the Prince. I recall protocol being concerned that we should not pick certain colours which are detestable in Japanese' culture. My wife was able to pick an outfit which was highly appreciated by both parties. Their visit was one of the most covered in the media both locally and internationally. After the tour, they praised the work done and even boosted JICA's support to increase technical assistance. They spent about two hours and each passing minute was a blessing to me as I wanted them to leave for other state tours. The pressure of the protocol involved had been rather too much on us!

The other visit was by the two Heads of State, Moi and Mori. We got information that the Prime Minister of Japan, Mr Mori, was to visit Kenya and targeted JKUAT as one of the sites he would visit. The Ambassador of Japan, JICA, Kenya Foreign Affairs and the university had to prepare accordingly. This was apparently the first time the Japanese Premier was visiting Kenya. We were also told that he could be accompanied by the President of Kenya, Daniel Arap Moi. This was the first time for me to receive two dignitaries at the same time on my campus. I had interacted with President Moi at least several times and knew how to handle him: but for Mori, perhaps it was going to be a nightmare.

All systems were in operation. I was given the itinerary by the protocol officials, advised on the routing within the campus and preparation for high tea. We had to rehearse and rehearse on how to receive the two guests without causing scenes. Vehicle arrangement was a concern to the Japanese team, in terms of who came first or last. Anyway, after several runs we agreed on the basic movement pattern.

Both men could not directly communicate because of language barrier, but Foreign Affairs officers took care of this issue. When the Excellences arrived, official security details disorganized us so much that whatever rehearsals we had gone through for weeks came to naught.

I found myself next to President Moi instead of being closer to the Prime Minister of Japan for any explanations. We were thrown into disarray and everybody seemed to move about as if they were enjoying themselves, not bothered with our earlier arrangement. In fact I now found myself more comfortable with President Moi as I explained to him what we do in field research. The Prime Minister Mori would pass by me with a wide smile and I had to explain to him slowly the work we do through a translator.

He was a jovial Prime Minister who later in the afternoon enjoyed a Taekwondo show which was put up by my students. In fact, this was the best official visit I had ever hosted. The two Head of States of state had spent the whole afternoon in my campus inspecting the projects. Again their memorable visit boosted the name of the university and even led to increased JICA technical co-operation. I had an excellent public relations team who prepared for modest gifts for the two dignitaries. I only wished they knew what I had gone through the previous two weeks in preparation for the visits. A sample of speeches I used to make during the annual meetings is hereby reproduced. A typical message that I delivered to the former Japan-trained colleagues is hereby reproduced.

Message from the Patron, Prof. Ratemo Waya Michieka – Annual Meeting 2013

I have been a JEPAK Patron for over now 10 years. As a group of people who have benefited from JICA sponsorship under different programmes, we will forever be grateful to the government of Japan through JICA for this very good relationship. As compared to other bilateral donors, the government of Japan has always stood with the people of Kenya even during very difficult times globally. The challenge for us now is how much of the knowledge we acquired in Japan has been put into practice on return to Kenya: How much impact have we made in Kenya, now that we have received the training in Japan? In other words, how much can we show after the training?

To some extent as the Patron, I have personally witnessed some impacts especially during the Annual National Conferences whose themes touch on various issues of concern for this nation. The other such activity is the medical camp that shows our social responsiveness to the people of Kenya and especially those economically and socially disadvantaged among us. Indeed this is the only alumni in Kenya that engages itself in such activities as a way of paying back to the nation to which gave the participants this great opportunity.

I, however, challenge you members to come up with more activities where you have the potential. I look forward to that time.'

The JEPAK annual calendar was effectively implemented as planned with the members going to Olkaria geothermal plant in Naivasha for the educational tour, medical camp at the City Cotton slums, and the end-of-year orientation party for the former and new participants. Of great impact was the Annual National Conference whose theme was 'The Future of the Boy Child'. In the words of the Chief Guest, Dr Naomi Shaban, who was then Minister for Gender, Children and Social Development, 'the theme was a clear indication that JEPAK has noticed the alarming trends in the general care and protection of the boy child who seem to have been forgotten'.

Traditionally, in the African culture, the boy child was more adored than the girl child. This, however, changed when the many crusaders for the welfare of the girl child, affirmative action and policies came into being and seemed to place the boy child to the periphery.

In appreciating the theme of this conference I was hopeful that we could develop a solid paper and the genesis of a working committee to articulate the challenges of the boy child and especially what is hurting him. I mentioned the contributing factors to this problem as ranging from moral degradation, absence of a father figure/authority, alcoholism, drug abuse and neglect among others.

The harsh reality of growing up in sub-Saharan Africa does not make things any easy for the boy child. In some situations, going to school is a luxury that has to be put aside for more important things. The upshot of all this is a boy child who ends up in the dumps or in a dangerous state of Nihilism. He becomes an enemy of society who has to be paid back in equally harsh measures that is how we lose potential future leaders.

The points to ponder over this situation include: Is the alienation of the boy child deliberate and calculated? What are the current trends and consequences? How do we identify the myths from facts regarding cultural beliefs, equality before the law among others? One other serious question would be: Where is the future of the advantaged girl if the boy child is not there? The country has to reflect and seek a way out for depicting the true meaning of gender equality and discrimination.

In addition to the above points, specific recommendations that the conference felt would be pointers to the way forward included; reviewing of the Sexual Offences Act which is too severe, need for more attention for the boy child in disability status, programmes that would bring boys together with men as a way of mentoring, commensurate rewards in education should be well defined and implemented among others. I actually strongly recommend for quality education especially in science and technology.

As an organization with a rich pool of human resource power, I would encourage you to continue tapping this wealth in order to shape the destiny of this country. I urge you to proactively contribute to the economic development of Kenya in meeting its aspirations of Vision 2030 which calls for the country to be globally competitive and prosperous nation with quality life for all citizens.

Finally, I wish to appreciate the continued support from JICA, without which none of the above activities would have been accomplished. I particularly register my profound gratitude to Mr. Eguchi, JICA Kenya Chief Representative, for his participation in almost all JEPAK activities. *"Thank you all and long live our friendship"*.

This is the kind of speech I usually make whenever we have the meeting. I cover contemporary topical issues which need attention. They are usually widely covered in the press and I get to be contacted to give ideas or solutions on the same. I am still the patron of the organization and it is useful to meet and compare notes.

Challenges and Mischiefs from Students

It would be totally misleading to depict JKUAT students as trouble-free. No university in Kenya can boast of having a perfect group of students. Just like all public university Vice-Chancellors in Kenya, I also had my fair share of headaches from students. The difference was in the management of the same.

I would like to demonstrate this by citing a few examples. I usually had an Occurrence Book kept by my Chief Security Officer who recorded all the events within the 24-hour period. I read every entry and made remarks against each one.

If I felt that some were more serious than others, then I would ask for more details on the same or make appropriate recommendations for the relevant department to take action. By and large, many of the reported cases were trivial and would need action by the deans or deputy principals. Most students' problems ended up being handled by the Deputy Vice-Chancellor (academic affairs).

My very first appointment to JKUAT was to be a deputy principal, academic affairs. All academic programmes and examinations were overseen by me. I therefore had a very clear understanding of what it meant to be in charge of students' affairs. As a university Vice-Chancellor, Prof. Henry Thairu served as my first DVC (AA). I knew what he had inherited from me and the kind of life he would lead thereafter.

Most occurrences in the registry involved students' behaviour. Besides recorded cases, I used to receive calls whenever there was any major problem. My landline then was available 24 hours, seven days a week and had to be operational throughout. My wife, Esther, and kids knew what to say and how to handle any university calls.

My previous interaction with students was an asset in handling difficult cases. Sample the following incidences. The university is located on a major highway, Thika-Nairobi, where motorists speed with no regard to pedestrians. There is a flyover that the students can use without danger. But others decide to cross the highway without using the flyover.

I got a call one evening that a student had been hit by a motorist who did not stop for fear of being killed by the public. The students who witnessed the accident mobilized and started pelting innocent motorists with stones. They damaged several vehicles of unsuspecting motorists. The police took some time to arrive at the scene. They, however, dispersed the demonstrating students, apprehended a few and booked them in.

I received another call from OCS office at 2 am that some of my students had been locked in at the Juja Police Station for causing disturbance in a public place and they needed advice urgently. I got up and called the Deputy Vice-Chancellor in charge of Administration to get in touch with the relevant authorities at the Police Station and decide on the action to be taken. He delayed in following up the matter and within two hours, at about 4 am the whole campus descended on the Police Station demanding the release of their comrades.

I was again called by the OCS of Juja and so I decided to personally go there to sort out the issue. To avoid too much hullaballoo, I requested for the release of the four students on bond to be charged later in a court of law. The boys were released in the morning and life went on. I negotiated for the release of the students to avoid unnecessary confrontation which would have resulted in further damage of property. I had an excellent rapport with the officer in charge of the station.

Another incident was when one lecturer erred in his teaching materials in Chemistry. He had a good course outline which he circulated to students as required. He had to give continuous assessments as required by the Senate. I think by mistake or sheer carelessness, he set an examination based on topics he had not covered.

The paper was difficult and the whole class walked away on him, boycotting the examination. As if that was not enough, the students demanded the lecturer's immediate sack. They further went ahead and frog-matched him past my office and then out of the gate. I heard some whistles as they were escorting the lecturer out. On enquiring, I was told that I had no business in internal class problems and inconsiderate lecturers. I got the story later and shelved it as the DVC (AA) had already handled it.

Another unfortunate event which I had to personally get involved in was when a fifth-year architecture student was tossed out of a 4th floor window. One evening at around 11 pm, students were doing their projects in the studio. They normally require space and a quiet working place just like a library.

One student came in with a radio and kept on playing some music as others embarked on their projects. He was requested several times to turn the radio low or use earphones. The young fellow ignored the pleas from his comrades. The students were preparing the projects for examinations. In a matter of minutes, a group of them walked to where he was sitting, lifted him and literally tossed him out of the window.

He landed on a concrete floor and ended up with broken limbs. The students locked up the studio from inside and continued with their projects. They indeed took the law into their own hands. I was called at midnight and I advised the medical officer on duty to handle the matter and ensure safe custody and care for the injured student.

The following morning, things were back to normal but I ordered for a speedy investigation in order to bring to book the students who had been involved in the heinous act. As the injured student was recovering in the University Hospital he was able to name the culprits. The fractured leg was put on a caste and the student would support himself to classes and other places. He had a friend who assisted him whenever necessary. He confessed that he had been playing loud music against the advice and pleas from his fellow classmates. I advised all of them that in certain circumstances, common sense must prevail.

There was always very high tension among the students whenever examinations were approaching. The student who eventually got injured should have known better and ought to have been considerate. I ordered those who tossed him out of the window to pay for his treatment and a penalty fee instead of pressing for justice outside the campus. Luckily, the young man recovered well and the culprits apologized in writing to the student and to his parents. The disciplinary committee was satisfied with the decision and let things rest there. I later ordered for visible postings on the walls that no music or noise is allowed in various reading rooms and libraries.

When I was an undergraduate student in the 1970s, I remember being locked in the library while reading past 2 am. I did not observe the closure time because I was so engrossed in reading a Chemistry subject in which I had an examination the following morning. Time passed unnoticed. I recalled my experience as I was advising the young men. Students' minds are completely occupied; and slightest provocation would result into an unprecedented reaction.

Reflecting on my whole tenure as a university leader and my relationship with students, I would say that I managed to contain any tensions among the student community and minimize any violent conflicts which in most cases are a feature of public university life in Kenya. The only violent riots that took place leading to the closure of all public universities in Kenya were those sparked by the implementation of World Bank-supported cost-sharing policies in 1991. Then, I had barely started my tenure as principal (academic affairs), and the riots, in a sense, welcomed me to the upper echelons of university administration. A brief description of the context within which this happened and how I handled it at the institutional level is necessary here.

Forced by the World Bank to implement cost-sharing measures in higher education in 1991, the government did mandate public university Vice-Chancellors to meet and recommend how best such measures would be implemented with little disruption.

The Vice-Chancellors held a meeting on 27 June 1991 at the University of Nairobi where they discussed issues regarding the University students' loan scheme and how best it would be managed to create an effective revolving fund. Besides addressing strategies of sustaining the loan scheme, the Vice-Chancellors went ahead and restructured the manner of tuition fee payment which, if implemented, involved direct token payments by all students towards tuition fees. Bursaries were also introduced for students from poor families based on criteria of the level of need to be determined by the universities.

As expected, there was a general negative reaction to this announcement, with various newspapers, to some degree, misrepresenting the actual situation as announced by the Vice-Chancellors. *The Daily Nation* of 30 June 1991, for example, ran a headline, *"Boom times may soon be history."* 'Boom' was the name given to the allowances that students used to receive from the government. *The Kenya Times*, another daily on 29 June 1991 had a headline, "New Varsities Fee Structure". It is needless to say that the total effect of the media stories contributed to violent student riots in the universities, which led to the closure of the institutions for the better part of the second half of 1991 and the loss of some lives.

As deputy principal (academic) and reacting to the newspaper stories and the tension they had sparked off among the student community, I prepared a memo on 29 June 1991 to various section heads of the college. I informed that I had personally gone around the college and noted the tension among most of the students. Given the tension that was evident among the students, I requested the various section heads to observe the following to calm the situation:

> That each of us be involved in explaining to the students on the exact figures involved and correct the distortions from the media that all their allowances had been scrapped.
>
> Each of us makes rounds within the campus, to talk to any group of students who may require explanations and remove all or any notices or posters on the boards from students that may appear to be of inciting nature.
>
> Assure the staff of my availability in the compound on the subsequent days until midnight to sort out any crisis.

Later on that same day, at around 5.15 pm, I wrote another memo to all students on behalf of the college principal. Noting of the various meetings I had held with students and staff earlier on that day, I indicated that all the misleading distortions that had been passed on to the student community regarding their allowances had been explained. Consequently, I informed the students that a meeting of the college academic board had decided the following in nature:

> That normal college activity should continue undisturbed. Any interference with fellow students would not be tolerated.
>
> There could not be any more unauthorized meetings/processions held by students in the University college premises.

Attempts by any students to hold meetings/processions outside the University College would be dealt with appropriately.

In the event of any of the above being contravened, serious and prompt action would be taken against the individual or group of students concerned.

My Reflections

How was I coping, attending to national duties and to my nucleus family? I was used to multi-tasking and delegation. I delegated a lot and knew the strengths and weaknesses of my management staff. I knew how unbelievably hardworking and competent many of them were. That confidence I had in them built them up to greater heights. I was also committed to seeing that my family did not miss me on account of excessive national duties. I finished all my day's work in the office and very seldom did I bring any work home.

I knew that 90 per cent of my success was showing up early at work and being present unless I was called elsewhere outside to attend to other meetings. I always made sure that I was the first one in the office and the last one to leave. I was not necessarily at my desk always. No. I walked around in the lecture theatres, laboratories, dining halls, farms, libraries and even halls of residence.

I got a clear first-hand picture of the happenings on the ground. I would at times make some decisions on the spot and even then follow up with confirmation notes.

My memory of the names of staff and students was and is still sharp. I still remember many by their names. That in itself was an advantage to me as we could freely interact anywhere with respect. Members of staff felt proud and appreciated me for identifying with them.

On the family side, we were able to travel together on holidays. My family visited major African cities like Dar-es-Salaam, Zanzibar, Cape Town, Cairo and Alexandra. For each trip we made, my children appreciated the different cultures and lifestyles of the locals and compared them with the Kenyan ones. We still recall the amazing tourist sceneries like the Table Mountains and Roben Island of Cape Town, the Cairo Museums of Egypt, the oldest Library in Alexandria, the Cloves Industry in Zanzibar and many other attractions. The visits gave them some impetus to work hard for their future.

12

Vice-Chancellorship and Networking for University Development

The thirteen years I spent as the Vice-Chancellor of JKUAT was the most rewarding period in my career development. One of the things I learnt during my tenure as VC is the multiplicity of roles that a VC performs. A VC literally is expected to perform duties that range from administration, fundraising, public relations, to academic work. The VC is the national and international public face of the institution. These roles can sometimes overwhelm and one can easily overlook paying adequate attention to some things.

At a personal level, this was the most active period in my life when I made impact locally and internationally. Despite my tight and demanding schedule, I was able to multi-task and perform other roles which I now consider my legacies. When I look back at the many meetings, travels, assignments and accomplishments, I wonder whether I was efficient or just a routine administrator.

The following are remarkable undertakings which occupied part of my time. All these assignments were accomplished between 1994 and 2003. I was the chairman of Kenya Agricultural Research Institute (KARI), a trustee and chairman of Kenya Education Network (KENET), a chairman who revitalized the Inter University Council for East Africa (IUCEA), manager of the African Institute for Capacity Development (AICAD), a founding member of University Science, Humanities and Engineering Partnerships in Africa (USHEPIA), a World Bank Committee member, chairman of Joint Admissions Board (JAB), chairman of Ibacho Secondary School, and finally a board member of Nairobi School.

Each one of these positions had specific terms of reference and timelines were set to achieve specific goals. I accomplished milestones in several goals as set.

Chairman of Kenya Agricultural Research Institute (KARI)

Board members of many parastatals are normally appointed by the Head of state, with advice from relevant parent ministries. These are non-executive positions but important for the smooth running of an organization. KARI is a large agricultural parastatal which oversees all research programmes related to crops and livestock. It has several research stations which are charged with specific mandates. There were over 40 such stations which covered all Kenyan ecological zones from the high-potential to arid and semi-arid zones. Research on specific crops and animals was conducted in each station and findings presented in annual fora.

My role as a chairman was to give managerial, scientific and policy guidelines to the directors of stations. I had a powerful board comprising reputable scientists in either crops and/or animals, including culture. I am a crop and environmental scientist. I therefore had knowledge of what had to be done in order to make KARI an outstanding parastatal amongst the many that existed.

My six-year tenure as the chairman of KARI saw many changes. I combined my administrative skills and academic achievements to improve the working conditions of staff, in particular sciences. It is during my tenure that we made clear distinctions amongst staff of higher qualifications from those who had lower academic credentials. Before I was appointed the chairman, all first-degree holders were paid the same salary and other remunerations equal to second-degree holders.

There was no distinction between BSc and MSc holders. I asked the Appointment and Promotions Committee (APC) to come up with clear terms and conditions of service showing specific progression of any member after attainment of a higher academic degree. This was done within the first six months and we were able to place scientists in their respective cadres. It was shocking to find qualified scientists working as laboratory attendants instead of being in the field doing research. Staff morale went up, papers were published and external donors increased research funds.

We set up a very liberal scholarship scheme for anybody who wanted to go for further training either locally or outside. We allowed academic progression and received hundreds of scholarships from the USAID, World Bank and local universities. We however, set up some training conditions. Candidates had to have worked for a specified period of time and had to return to the station. We bonded whoever wanted to go for further studies. The returning scientists had to put in three years of service after return or pay a penalty equivalent to the period he or she was out.

With the increment in remunerations, several staff returned to KARI after their training and contributed a lot to Kenya's research sciences. I recall visiting several PhD candidates who were stationed in the University of Missouri and encouraging them to complete their degree work and return home. We did not lose any staff through this arrangement.

During my chairmanship, we introduced the KARI Biannual Scientific Conference. It used to take place every November with wide presentations of papers. I also introduced trophies for the best research papers during this time. The conferences were open to all scientists in related areas and edited proceedings were published.

Promotions to next levels depended on papers, and I remember being accused of bringing university standards to research stations. One had to publish or perish! These days it is publish, innovate or perish. I defended my stand by referring to the relevant Act which established research centres in Kenya. It was stipulated clearly that their objectives were to conduct research on their specific mandates and disseminate them for the country's development.

My director, the late Dr Cyrus Nderitu, was a hardworking individual who had published many papers in his area of expertise. He confided in me and told me that the former board had been so incompetent that he had been running the institute single-handedly. He appreciated my innovative ideas which put KARI on the world map of reputable national research centres. We continued attracting research funds for specific projects and came up with programmes for livestock and patented them.

I visited all the stations and encouraged young scientists. The parent ministry, then Science and Technology, was very supportive as we opened new centres which were put up by the World Bank funds. Modern research laboratories were established for areas which were unique for specific commodities. The accurate accounting of funds from donors was my priority. I knew the importance of using donor funds transparently for future attraction. The director had to be alert to ensure that proper books of accounts were kept when timely contract renewal was to be sought. I however, did not run the institute as a chief executive but chaired board meetings. All accounting and prudent management of resources and assets were the responsibility of the Board.

My chairmanship was not smooth all the way. As a board, we had to tackle very sensitive matters which touched on land. KARI had the largest land allocation, second to Kenya Wildlife Service. It was the responsibility of the board to ensure that all KARI assets were in inventory and guarded safely. These included moveable and immovable assets. The board was liable for all the research land in the country.

I was appointed the chairman of KARI at a time when land grabbing and transfers were at their climax. The highest degree of land corruption was the hallmark of the 1980s and 1990s. The director, Dr Cyrus Nderitu, told us that he had not acquired any custody of the KARI parcels of land. No title had been given to him by the relevant authority. The previous board had not bothered to account for the parcels of research land which KARI was entrusted with.

This was a mammoth task and a dangerous undertaking because virtually all the land titles which were missing were owned by the most powerful government officers in Kenya. It was the who-is-who in Kenya who had access to the title deeds. In fact my research was being carried out on people's lands. KARI was viewed as a squatter. I am talking about thousands and thousands of hectares. Several pieces of prime land for breeding research were sub-divided into unmanageable portions for breeding purposes. Yet some disciplines in agriculture needed thousands of isolated hectares to carry out research. My board was, hence, charged with the responsibility of reclaiming and ascertaining KARI land titles.

The director furnished us with the available document of ownership. I used my university experience and advised the board to hire a competent lawyer and contract another competent firm of lawyers to pursue the KARI title deeds. Our internal lawyer was to provide all details and/or transfers and furnish the law firm we hired with relevant information to be used in court. As a chair of the appointments committee, I did not mince my words to the lady we hired. She was told that her first and immediate duty was to acquire all KARI title deeds and redeem any which could have been sub-divided.

We knew big names would surface but decided to carry on. The scientists, many of whom were my former students at the University of Nairobi, assisted in detailing the lawyers how KARI lands had been axed or sub-divided. They welcomed the idea and supported the Board to revert the research lands into their original purpose. The affected parcels were those for livestock holding and high-value crop research sites in high-potential areas of Kenya.

I recall specific former ministers in the Ministry of Science and Technology who had acquired large tracts of land. One of them summoned the board to a meeting in Eldoret's Sirikwa Hotel. I had a hint that he was going to arm-twist us to sign some documents and sub-divide two parcels of land. The centres were in Molo Potato and Pyrethrum Research Stations and Grassland Centre, both in Nakuru County. They were large tracts of land which were specific for highland research for animals (sheep) and potatoes.

The minister chaired the meeting and beat about the bush on very irrelevant subjects. He later on, during the discussions, asked me straight to my face what I thought about the many KARI research lands (indeed government land) which were not put into use. He proposed that they be sub-divided and, as a minister, he wanted some hectares allotted to him. I promised the members that I would take him on as long as I had the minutes ready. I had prepared myself and told him straight to his face that the board had passed minutes that all KARI land must be protected. They were meant for research and future generations.

I further explained that it was not prudent to sub-divide large-scale farms as this would affect mechanization of the farm. I further told him that we had engaged a powerful law firm to acquire all the titles. The minister had a few titles in his custody which he was not entitled to keep. By law, they were the assets of the board. One board member supported me and said that allocation of any portion of research land to an individual would jeopardize future research protocols. Board members kept quiet; the director who was sitting next to me would not believe that I had turned down a minister's request for a piece of several hectares of land.

He surrendered the two title deeds he had brought to me to endorse for sub-division and left for other functions within the city. I honestly did not care what the consequences would be thereafter. I cared more for just actions than end results. I did not budge on this request. I am an accomplished researcher and had been fully briefed on the future consequences of research if the land was reduced to small parcels.

Research and food security go hand-in-hand. The more advanced a country is on innovations and new discoveries, the better for its citizens. My insistence and firmness regarding KARI parcels of land saved several thousands of hectares. I was able to register and acquire 100 clean title deeds for several parcels of land. The board was supportive of my efforts and we all left KARI satisfied that our research mandate was on course. Indeed, we lost some hectares in Kitale, Tigoni and Molo Research Centres. The cases were so complex that our lawyers could not unravel several fake documents which had been purportedly prepared by the lands office in Nairobi.

Kenya Education Network (KENET)

The Kenya Education Network Trust (KENET) was formed in 1999 as a membership organization to serve higher education and research institutions in Kenya. KENET is recognized by the Government of Kenya as the National Research and Education Network (NREN).

KENET is constituted and registered as a trust under the Perpetual Succession Act and its beneficiaries include students, faculty, and staff in the member institutions. It is governed by a Board of Trustees. The founder trustees and the institutions they represented were as follows: Dr Freida Brown VC, USIU, Mr Augustine Cheserem, Managing Director – Telkom Kenya, Mr Samuel Chepkong'a, Director-General, CCK, Prof Crispus Kiamba VC, UoN, Prof Raphael Munavu, VC, Moi University, Prof. Stephen Talitwala, VC, Daystar University, and I as the founding Chairman and Trustee. The Trustees were assisted by a Management Board representing founder universities and research institutions in Kenya. The day-to-day operations were run by an executive director, a group of 11 talented and skilled young engineers and technicians, three interns, two young accountants, a management consultant and a research projects administrator.

The KENET Trust had six objectives, namely:
- To provide a sustainable and high-speed internet connectivity to Educational Institutions;

- To facilitate electronic communication among beneficiaries in educational Institutions;
- To support the sharing of teaching and learning resources among educational institutions;
- To support teaching and learning over the internet for beneficiaries in educational institutions both in Kenya and outside;
- To collaborate in the development of relevant content of syllabi in educational institutions
- To collaborate in research in educational institutions.

The Kenya Educational Network (KENET) was a brain-child of a few Vice-Chancellors who came together and decided to form communication channels amongst universities and schools. The project was aimed at using high frequency systems to link up the learning institutes through various nodes. We formed a committee of trustees and I was nominated the chairman of the trustees. Each founding university paid Kes750, 000 (US $15,000) towards the funding of the project. Professors Raphael Munavu, Freida Brown, Francis Gichaga, Henry Thairu and I were the founders. The project now covers several learning institutions in Kenya and has a full complementary staff. As a chairman, I initiated the project and saw it grow to involve many universities, colleges and schools on the same frequency routes (nodes).

Partnerships with African Universities

One project which assisted in training a number of PhD students in some African universities was the partnerships we formed. University Science, Humanities and Engineering Partnerships in Africa (USHEPiA) was launched by eight universities to boost PhD training in selected science areas. Soon after the South African apartheid regime was dismantled, the Deputy Vice-Chancellor of University of Cape Town (UCT), Prof. Martin West, came to visit me in JKUAT. He had two specific objectives: to find out how Kenya fared in terms of racial co-existence; and, to forge a linkage.

He was explicit in his first objective and I could tell from the questions his team asked me. The primary objective of his mission, however, was to forge linkage amongst eight African universities. It was a UCT initiative to promote collaboration amongst established African researchers in the generation and dissemination of knowledge and to build institutional and human capacity in cash-strapped African universities. The goal was to develop a network of African researchers to address development requirements in sub-Saharan Africa.

I welcomed the initiative. I wanted to have my staff trained in UCT and to share research findings through collaboration and staff/student exchange. I was very concerned about my staff acquiring quality training so that they could take care of JKUAT. The university was young and needed all assistance possible in stabilizing its ambitious programmes.

After our initial meeting, Prof. West visited Makerere University, University of Dar-es-Salaam, University of Nairobi, University of Zambia, University of Botswana and University of Zimbabwe. There were eight universities collaborating with UCT and a memorandum of understanding was signed between UCT and 21 Vice-Chancellors and deans of Science and Engineering. The agreement was that UCT would be the host and training university and the eight would be sending their trainees there. We got initial scholarships from Carnegie Corporation, The Rockefeller Foundation and Mellon.

The linkage was boosted by appointing four of us to the International Steering Committee for fundraising and selection. We sourced substantial money from friendly donors and were able to train several PhD students who helped boost faculty members in the collaborating universities. I gained a lot in having over 10 PhD trainees under this arrangement. In the process, UCT built a magnificent African House with the funds we raised. The house accommodated students and had special rooms for visiting professors plus conference halls. This was a landmark for UCT.

The University of Dar-es-Salaam also set a telecommunication institute through this arrangement. As a committee member, I bargained for more science training slots and I was able to attend the first PhD graduation ceremony at UCT. I was very happy when I saw my staff walk in the procession. I recall that Prof. Luhanga, Prof. Wole Soyinka and I were the special guests during the first graduation ceremony. I had met my goal of human capacity building for JKUAT.

It was during this time that we met the UCT Vice-Chancellors, Profs Sunderland, Ramphele Mamphele and Dr Njabulo Ndebele who also gave financial assistance to USHEPIA. The programme still runs with more universities but less funding. Personal commitment was the primary prerequisite for the success of the partnership. This was an excellent undertaking by a few selected vice-chancellors and I considered the training beneficial to our African universities which always relied on sending trainees outside the continent for staff development. Although the partnership did not stop further training in developed countries of the north and west, the few we trained strengthened our human capacity in the science and engineering disciplines. We still maintained the other linkages on course.

The Bilateral Agreement Between the Governments of Japan and Kenya

The Japanese government accorded substantial financial, technical and personnel assistance to JKUAT. Through the Japanese International Co-operation Agency (JICA), JKUAT benefited from this generous technical assistance for over two decades. In my earlier chapters, I wrote on the collaboration which started in 1977 to develop a mid-level technical college but which ended up creating a first-class fifth Kenyan public university. I was at the centre of its development and therefore accountable for JICA's technical assistance.

JKUAT gained in many ways: Staff training, technical assistance, capital infrastructure, provision of equipment, middle-level training, supply of project vehicles and assorted ICT components. The Japanese evaluation teams which visited the University used to be so impressed by the impact created by their support and the prudent use of their taxpayers' money that they kept on renewing programmes for over two decades. The teams recognized authority and funded projects on the strength of trust and proven individuals' integrity. The teams worked with individuals who were the key architects of change within the institutions.

I remember several occasions when the team leader of the programme, Prof. H. Nakagawa, would explain the aims of the project. He stressed technical assistance, exchange of ideas between Kenya and Japan and transfer of technology. We did not fully succeed in technology transfer. Even after training over 100 higher-degree personnel, we still lacked commitment to translate and own that technology.

Japanese people usually go out of their way to give the best of training. But the trainee or learner must be able to translate his/expertise into tangible results. The results are time-bound. Their training is industrial-based, and staff alternate from lecture halls to industry and vice-versa for a specified period of time. Practical-oriented courses have made Japan what it is now.

The support that JKUAT received through grants in aid, in addition to the government's subsidy, transformed the young university. We were training first-class industrially-based graduates who fitted in the job market soon after graduation. However, the many staff that finished their PhDs came back and were caught up in a myriad of local problems and even forgot the quality training that they had received. A few still display their talents but a large number have drifted into other activities which have diverted their concentration. The blame is not fully on them but on the large number of students to teach with absolutely no support facilities. The staff degenerate into mere lecturing with no practical activities. Upholding quality and high standards in academia is not only through lectures.

The current situation does allow for critical evaluation of what is taught and how it is taught. Lecturers have no time to spend in the preparation of their lecture materials. They do not engage tutorial fellows to follow up on what was taught earlier. Even when they are engaged, they get their pay after several months. I know of cases where they were never paid and quit for other places.

Professors who should be engaged in research and postgraduate guidance end up being loaded with several contact hours that they have no time to do quality academic work. They do not have consultation times. Worse still they cannot cope with the high number of students. Students at times are overcrowded and learn under very horrible conditions. This issue is discussed at length in another chapter.

For the thirteen years I worked in JKUAT, the interaction with my staff and those from Japan was so enriching that we used to have bonding outings yearly to learn the different cultures. The number of students and staff was small and we could afford generous outings. I made several travels to Japan under the senior management training programmes. These were the only formal induction courses I attended during my tenure as a Vice-Chancellor. I met other senior managers from other countries, especially universities where JICA had collaborations, and share experiences. I learnt a lot in terms of management, financial appropriations and accounting, public relations and the management of crises. During my extensive visits, I was able to appreciate different cultures and food habits of far eastern countries.

I attended seminars. One conference exposed me to diverse Japanese phenotypes and I was able to spot out a native from Hokkaido Prefecture and another one from Okinawa. The conference was themed "The Crossing of Culture in the Eastern Pacific". One belief which has been a hindrance to our own development is the misconception that time is elastic. Kenyans and, to a large extent in Africa, as a whole, people have no sense of time and that the importance of keeping deadlines. We have proudly coined our phrases and happily say 'African time' when one is late for more than two hours to a function. We have never appreciated the fact that late performance of an assignment contributes to underdevelopment. Time matters and time is all that makes a difference in one's well-being.

In Japan, you are either on time or you get left behind. I learnt the importance of time management and continue to respect that virtue up to now. In fact, when I negotiated for the JKUAT project, I used time to gain mileage during the discussion. It was simply having the Japanese negotiating party come late and apologize. Once I accepted the apology, I knew in my mind that they were already in a positive but disturbed mind and I could make subtle references on the effect of lateness and that I would have been in another meeting. We would then expedite the negotiation matters.

I knew the Nairobi traffic jam could always delay meetings. African countries have yet to appreciate the power and implications of keeping time. Time wastage in traffic snarls and service counters is retrospective. I met many Kenyans in the management meetings and they proudly talked about the timely bullet trains from Tokyo to Hiroshima or Kyoto. I have yet to see timely movement of buses from Kampala to Nairobi or Dar-es-Salaam. We generally do not put into practice what we learn.

I am the patron of the Japanese Ex-Participants of Kenya (JEPAK). We always have conferences to discuss topical issues. We have been meeting for over 20 years and the total registered number is over 5,000. There are Kenyan people who have been trained in Japan and who have visited the country severally. I am a life member of JEPAK and have attended all meetings since inception. I listen to the informative deliberations on Japanese experiences.

The most eloquent speakers in Japanese language attend. They speak in that language and move the crowd. There are also fluent English speakers who express themselves in the most figurative manner as to how Japan shaped, taught and nurtured them.

I listen pensively and wonder when our country will be transformed even in one service only like healthcare so that ordinary people benefit. There are many obstacles to achieving what was learnt in a foreign land which cannot allow the ex- participants to practice the knowledge and skills acquired. This was true even for those who studied in the north and west. It is, however, noteworthy to mention that a good number of those who attended training for a long duration do practice certain skills for the good of the country.

Time-keeping is now catching up in Kenya; open office spaces and open door policy are now a norm.

Project completion period on a JICA-funded programme was always ahead of schedule. All structures which were put up by the Japanese assistance in JKUAT were always two to three months within schedule. They would be complete in time and await handing over. There was never any occasion of varying the contracts upwards or asking for extra days to complete the buildings. In Kenya, and indeed in many African countries, if a project is not re-valued upwards, and request for extension in completion period is made, then that project is 'aborted'. Nearly every government capital structure must beg for an extension and be valued upward. This was not the case in all the structures put up under the JICA technical co-operation.

As the University was becoming consolidated in academic programmes and building structures, I had an idea of forming an Eastern African organization. I needed a Centre for East Africa and eventually Africa as a continent. We created the African Institute for Capacity Development (AICAD).

The Offshoot of an Institute

The African Institute for Capacity Development (AICAD) is an international organization established in the year 2000, with its headquarters located at the Jomo Kenyatta University of Agriculture and Technology (JKUAT). We had to have some continuity before JICA finally disengaged with the University. The main goal of the institute is poverty reduction in the African region through human capacity development. The institute has three country offices at Egerton University (Kenya), Makerere University (Uganda) and Sokoine University of Agriculture (Tanzania).

AICAD was initiated by the East African countries with support from the Government of Japan through the Japan International Cooperation Agency (JICA). I was the first chairman, having initiated the project. Its vision is: 'To be the leading African institution in building human capacity for poverty reduction', while its mission is: 'To link knowledge to application within communities in partner countries in Africa in order to reduce poverty'.

AICAD aims at achieving poverty reduction and socio-economic development by facilitating the indigenous people to solve their problems through:

- Utilizing existing knowledge and technology;
- Creating new technologies suitable for local conditions;
- Developing and utilizing the potential capacity of local expertise;
- Building a bridge between institutions creating technologies and the communities using them;
- Exchanging information, experiences and practices;
- Sharing human resources and information in the region and beyond.

AICAD collaborates with both governmental and private sector organizations such as universities, R and D institutions, industries, NGOs, community-based organizations, and government agencies/departments; and utilizes human resources and facilities in such institutions to advance its mandate.

AICAD pursues its mandate through its three mutually-interactive functions of Research and Development (R&D), Training and Extension (T&E) and Information, Networking and Documentation (IN&D). This interaction is facilitated through an administrative function, Administration and Finance (A&F) which provides logistical support and management of its human and physical resources. Its programmes range from in-country training targeting various community groups with problems, residential in nature and conducted at its headquarters and country offices, to grassroots training targeting community groups with specific development needs, non-residential and conducted at identified venues within the community. There is also knowledge and technology dissemination (KTDP) by which its research results are disseminated to target communities through various media including a training and community empowerment which embraces an integrated approach in which various identified community issues are addressed at the same time.

The main guiding philosophy in AICAD interventions is *"development for the people by the people"* using a bottom-up approach where identified community participants are accorded an opportunity to identify and prioritize areas in which they need assistance, with AICAD playing the facilitator role. This approach has been responsible for successful project identification, formulation, implementation, and sustainability of AICAD's development initiatives.

My registrar, Mr Joel Mberia, and I realized that, at some point, the support from JICA would come to an end. I asked him to write a concept paper for presentation to the Resident Representative, Mr Eiji Hashimoto. We were very good friends and shared the same birthday. He would always remind me about it whenever we met. Hashimoto had all along supported our efforts to see JKUAT the best in the region. Mr Mberia and I made an appointment to see him since he was always jovial and welcoming whenever he saw us. I had hinted to him that we had some idea on what JICA could do to support JKUAT become a regional giant within Africa. We met him and went through the concept paper we had written and he bought the idea in principle.

That was the beginning of establishing AICAD. From there on, I embarked on perfecting my public relations and sold the idea to both the Senate and council. I had some strange resistance from some of members of Senate who wanted to know what the university's take would be. This was a strange notion since I had kept everybody in the know.

The Council was very understanding and President Uhuru Kenyatta, who was then a council Member, endorsed the move. I also held several meetings with the Japanese Team Leader who was also supportive. The word of the team leader meant a lot to the JICA headquarters as he was the most trusted professional in ensuring that the Japanese taxpayers' money was used well. AICAD's main objective was to be a training centre for East African researchers and scholars and over the years cater for the continent. We had five years' projections and JICA was willing to support the initiative by contributing to AICAD. I appreciated the support given by my colleague vice-chancellors, and the ministries of Education and Finance from each partner state.

My critical involvement in setting up the institute saw a robust working force of staff from each partner state. As the first manager, I made sure that we got a team which would promote the tenets of AICAD without compromising those of JKUAT. As I doubled up as the vice-chancellor and manager, I was able to second my reliable academic staff to AICAD to push its agenda. We were able to construct modern conference halls with the latest ICT facilities. I literally bulldozed the construction of 42 rooms for hiring to visiting researchers. The centre was detached from the main campus despite its physical location within. It was an institute within an institute which had all the necessary facilities for living.

I had to organize for an independent title deed for AICAD which made it an autonomous institute. I faced challenges from students and some staff when I announced the taking over of the games pitches. Students were on my neck and wanted to know the value and use of AICAD. I, however, convinced JICA to develop another site to accommodate all sporting facilities to replace the ones we had annexed. The university gained by having a well-landscaped games pitch with all the modern changing rooms and the provision of water. AICAD stands tall by the main entrance of the university and serves as a research, cultural and academic centre for the African continent.

The continuity of its functions relies heavily on the financial support of the partner states. I came up with a formula for each country to pay and support AICAD's activities. JICA was to oversee that the operations were in place before pulling out. This is one centre I am very proud of and which I started as a small project. It caters for the continent, but must have a forceful chief executive to design useful projects.

The Inter-University Council for East Africa (IUCEA)

A closely related regional institution like AICAD which I enjoyed working for was the IUCEA. The East African Community which broke up in 1977 had several organs which were very useful and active. In 1980, the Vice-Chancellors of the three universities – Makerere, Nairobi and Dar-es-Salaam – together with the permanent secretaries of the ministries of Education met and drew up a Memorandum of Understanding(MoU) committing the three universities to work together. The MoU led to the transformation of the Inter-University Council (IUC) into the current IUCEA. When the community broke up, some institutions remained intact and IUCEA continued functioning. It was not, however, vigorous due to the weak involvement of some partner states. Academic co-operation, research and staff movement were not affected. The council used to meet seldom but functioned normally.

Initially, three countries – Kenya, Uganda and Tanzania – formed the community and committed funds for running the council. The IUCEA was therefore established by the three East African universities: Makerere, Dar-es- Salaam and Nairobi. The function of the council was to provide avenues for working together and sharing experiences as was prescribed in the protocol which was signed by the three founding vice-chancellors. It was one of the best academic arrangements which existed after the collapse of the community and I had the opportunity to revitalize it.

I was the chairman of the IUCEA from 1996-2000, during which the three countries experienced an unprecedented growth of universities. Despite its continued services, the IUCEA had basically been dormant in its roles and the committees which were supposed to be in place had largely been absent. The then executive secretary, Mr. Erick Kigozi, whom I had the pleasure to work with, ran the council single-handedly. He, however, kept it functional whenever he could. The council had skeleton staff, and the Uganda government met most, if not all the running costs since it was housed in Kampala. During his farewell party, I acknowledged Mr Kigozi for keeping the functions of the council alive even after the 1977 collapse of the East African Community.

When I took over as the new chairman, the Association of Commonwealth universities, through a private consultancy, Commonwealth Higher Education Management System (CHEMS), requested that IUCEA roles be reviewed and revitalized to meet the challenges of the present-day demands in higher education. When I took over the chairmanship from Prof. Geoffrey Mmari, former VC of the University of Dar-es-Salaam, I accepted the challenges of reorganizing the council. Chairmanship rotates from one partner state to another and it was Kenya's turn.

In 1998 my committee was commissioned to carry out a study to establish how IUCEA could be revitalized, and develop strategies for its sustainability. The committee included representatives from the three partner states, with the secretariat in Uganda. I made sure that the two important ministries, Education and Finance, were included. My earlier experience in the establishment of AICAD and the creation of Kenyan colleges during the double intake came handy. I was aware of the imminent tasks ahead noting that I was now dealing with three diverse government setups; leave alone the academic staff from several universities. It was, however, a pleasure to work with a typical academic group of people with one goal, the revitalization of the IUCEA.

A team of about 20 experts embarked on the task. We were given specific terms of reference, deadlines, and the budget was made available through the Kampala office. Our first meeting was held in Kampala where we drew up the plan of action and assigned ourselves roles which were specific to each state. We needed basic statistics and agreed to meet within one month. It was easy for me to co-ordinate the discussions through e-mails which were slowly being adopted.

It was a pleasure to work with people who respected deadlines and answered calls promptly. We had a total of eleven meetings with a final one being the launching of a revitalized Inter-University Council for East Africa in Arusha, Tanzania, the EAC headquarters. I chaired a meeting of all East African University Vice-Chancellors and their top management staff, permanent secretaries for the ministries of Education and Finance and senior staff from the community which culminated in the endorsement of our recommendations. The meeting suggested some changes which were useful for the strengthening of the council. I recall recommending the Republics of Rwanda and Burundi to join the council after the ratification of the community. It was a memorable day for me.

The salient recommendations included:
- The creation of a stronger body and staffing with appropriate academic qualifications;
- The inclusion of Rwanda and Burundi in the community;
- The preparation of a protocol to the East African Community making the council more powerful and independent within the region;
- Enhanced funding by each accredited university pegged on student populations;
- Enhanced funding by the partner states and clearance of outstanding arrears;
- Efforts to be made to fundraise externally among several friendly organizations identifying with IUCEA and which could willingly donate research funds;
- Encourage proposal writing for sourcing of funds and personnel;
- Increase staff and student exchanges across the region;
- Revitalize thematic groups/disciplines immediately;
- Claim the grabbed land of the IUCEA from the Uganda Government and put up a modern office facility with adequate guest and conference halls;

- Harmonize the academic programmes, especially secondary school/ requirements;
- Recruit the Executive Secretary competitively on a rotational basis as prescribed in the document we produced;
- Furnish the current offices with modern ICT equipment;
- Start a newsletter which would circulate widely amongst universities highlighting the academic events in the region;
- Enhance and encourage East Africanness in an attempt to live harmoniously even outside the academic sphere;
- Promote Kiswahili as a common regional language.

We all signed the document, presented it to the then Community Secretary General, Mr Francis Muthaura, who was very glad for the work well done within the stipulated period of three months. Among the very many recommendations made, several of them have already been achieved. As a continuing chairman of the IUCEA, I had to advertise and interview the Executive Secretary for the revitalized council. I had the same committee which included several Vice-Chancellors from sister universities who conducted interviews for the Executive Secretary. The terms of hiring and conditions of work were stipulated in the new document. The first executive secretary after the IUCEA was revitalized was Prof. Chacha-Nyaigotti-Chacha. He secured the position from among other applicants. Prof.. Chacha took over the council and implemented most of the recommendations we made.

I cannot enumerate all of the glaring achievements made so far, but it is worth mentioning a few. The entry of Rwanda and Burundi into the Community was initiated by my committee. We always invited the countries as observers in all our meetings. The representatives from the two countries desired to be included in the academic activities of the region despite the prevailing communication hiccups. We got along very well. In fact their entry raised the profile of the IUCEA and eventually the cooperation and also assisted in financial contributions.

Other obligations which were met included: staff recruitment respecting the country quarters; funds were raised, for example, the ViCRES research project; a protocol was signed which fully integrated the IUCEA into the larger Community; the IUCEA land which had been grabbed was recovered and the headquarters was put up as passed; and, funding was enhanced by partner states. IUCEA is well respected and internationally recognized, and has enhanced funding by each accredited university with current offices being well connected via ICT. A newsletter is published regularly. These are but a few of the accomplishments that IUCEA has achieved. It is one of the strongest and most respected organs of the Community.

The final report was presented to the Vice-Chancellors and deans of partner universities. This was my joy as we got compliments for work well done and accomplished on time. I was later requested to present a summary of the report to the Heads of State during the high-level summit meeting in Arusha, Tanzania. The then Foreign Affairs Ministers, Hon. Stephen Kalonzo of Kenya and Hon. Jakaya Kikwete of Tanzania commended us for those institutions which stood the hard times as the community disintegrated.

The EAC was re-established by a treaty in November 1999 and it was recommended that the Republics of Rwanda and Burundi become members. My committee had made similar recommendations earlier on. The new EAC recognized the revitalized IUCEA as one of the surviving institutions of the community and would support it fully. A protocol was signed in 2002 making it a corporate entity under the EAC. My committee was appreciated for its efforts.

The revitalized IUCEA is now a well-established regional academic powerhouse manned by a team of about 40 talented and skilled personnel. My reflection on this organ of the community is simple. Scholars are known to relate and build bridges along their disciplines. IUCEA demonstrated this and did not care whether the community survived or not. Responsible academicians kept the organ going and ignored the political mess which was created by opportunistic and insensitive politicians. I was proud to be a part of those who saw the rebirth of the current East African Community and a very well-energized IUCEA

My personal involvement made a mark in my academic life as I witnessed my colleagues, Professors Chacha-Nyaigotti-Chacha and Mayunga Nkunya, steer the IUCEA to greater heights. They have both done a sterling job in putting the institution in the world map. Its funding from the partner states is timely and generous. It is the pride of the community as one of those organs which survived the breakup. I am happy to have been associated with it.

Two other international associations which I enjoyed being part of are the Association of African universities (AAU) and the Association of Commonwealth universities (ACU). The two bodies perform similar functions and bring together vice-chancellors to compare notes and discuss topical issues which affect them. The AAU, whose headquarters are in Ghana, organizes meetings on specific themes to be discussed. I recall the interesting stories told by various Vice-Chancellors on how staff and students caused havoc on issues which were common to all of us. The topics I enjoyed most were those dealing with quality education in our continent. The cries for additional funding for research and capital structures were common features by my colleagues. Inadequate government funding, hence the brain-drain, was an obvious topic which surfaced time and again.

I got a chance to meet many dignitaries and share the strategies of university management. In fact when I reflect on my experiences, I consider my management style reasonably better or above average compared to others. I heard of horrible stories amongst staff and students in some universities which I found strange and alien to Kenyan ones. The Association of Commonwealth universities is an elite academic club which brings all former British colonies together to compare academic and political notes. As members, we used to meet in various Commonwealth countries and we would spend a week talking on all types of problems, issues, soliciting for scholarships, or simply networking. We used to invite powerful world leaders to give keynote speeches on topical issues.

I recall one such a meeting in Ottawa, Canada, where Mwalimu Julius Nyerere, the then President of Tanzania (now diseased), gave a moving speech which covered education and poverty alleviation in the African continent. He moved the crowd when he equated the level of a country's development and the academic attainment. Developed nations spend more resources in education and research than those in economic transition. African states never learn or emulate facts which contribute to poverty reduction, he concluded. He further challenged the scholars present to persuade their governments to take up major problems which confront the continent. Nyerere was a great speaker and he inspired us all.

Mwalimu Nyerere, as he was commonly referred to, is known world over as a great leader who promoted socialism in the United Republic of Tanzania. His teachings which succeeded in making Tanzania one great, united country were promoted through the use of Kiswahili as a common language. He taught its citizens to live as brothers and sisters and one cannot easily identify them along tribal lines, which is common in the neighbouring countries. The language is now being promoted and adopted by several African countries, especially those in the Eastern African region.

His popular teachings through the media and public gatherings indeed made the country an envy of many as every Tanzanian you met referred to one another as brothers and sisters. He promoted education, social justice and equal citizenry. He preached humility, love for the country and a culture of respect for one another. Mwalimu Nyerere alluded to these tenets as he delivered his great speech to the scholars in Ottawa, Canada. He stressed that Vice-Chancellors had a big role to play in changing the world to make it a better place to live.

The late president is still remembered as a visionary leader who appealed for a United Africa along the likes of Ghanaian President Kwame Nkrumah. He steered the formation of the East African Community along with the late presidents Jomo Kenyatta and Milton Obote of Kenya and Uganda respectively.

"You teach the youth and have a chance to mould them to be better citizens," he echoed. Other countries which hosted the ACU meetings were the UK, Singapore and Malta. An interesting thing happened in one of the meetings held in Cyprus in 1999. Two of my Vice-Chancellor colleagues had to cut short the meetings and fly back to their respective universities because students were rioting and had taken over the streets in the cities. The VCs were recalled to go and handle the students. It was not a unique occurrence, I had been recalled once and my close friend, Prof. Francis Gichaga, former Vice-Chancellor of University of Nairobi, had been recalled on several occasions over students' riots.

The ACU meetings followed the agenda of the Heads of State meetings. The networking made it possible to secure scholarships, negotiate for staff exchange and sabbatical programmes. During these meetings, I was able to secure several scholarships for staff development. I was also able to fly high the name of my University, JKUAT, amongst the world giants. Commonwealth countries are so many that one could easily live up with great universities and strike academic business deals which would assist the younger colleges.

My own observation was that as a club, the ACU had enough comparative agenda to hold annual meetings. It was simply maintaining the colonial umbilical cord whose purpose I could not decipher. This is one conference that I think needs to be refocused. Yet the head office was in London which made it easy for ACU to continue with the practice of keeping the status quo.

13

University Leadership and the Donor Community

Structural Adjustment Programmes by World Bank

Structural Adjustment Programmes are the policies implemented by the International Monetary Fund (IMF) and the World Bank (the Bretton Woods Institutions) in developing countries. These policy changes are conditions for receiving new loans from the IMF or World Bank or for obtaining lower interest rates on existing loans. Conditions are implemented to ensure that the money lent will be spent in accordance with the overall goals of the loan.

Structural Adjustment Programmes (SAPs) are instituted with the goal of reducing the borrowing country's fiscal imbalances. The bank from which a borrowing country receives its loan depends upon the type of necessity. SAPs are supposed to allow the economies of the developing countries to become more market-oriented. This then forces them to concentrate more on trade and production so that they can boost their economies.

The Structural Adjustment Programmes in Kenya were felt in the areas of education and health.

The Impact of Structural Adjustment on Health in Kenya

It is reported that Structural Adjustment Programmes have had a negative impact on infant and child mortality. There is evidence that non-adjusting countries with low levels of debt in sub-Saharan Africa (SSA) have succeeded in accelerating the rate of improvement of their infant mortality rates during the 1980s; that the rate of progress in severely indebted, non-adjusting countries has remained broadly unchanged; and that progress in severely indebted, intensively-adjusting countries has slowed markedly. Some countries claim evidence of increases in infant and young child mortality in several SSA countries over the past few years.

In addition to the negative impact on women's health associated with the general decline in communicable disease control and health care provision, there is evidence that mobility and mortality associated with pregnancy has also been aggravated. In a number of countries the introduction of user charges for antenatal and maternity care has been associated with an increase in deliveries conducted at home, as well as those occurring in hospitals without previous antenatal care or assessment. The rising costs of transport together with lack of money on the part of poor women have been other contributory factors. Finally, there is evidence, mainly of a qualitative nature, that risk behaviour in relation to HIV transmission has been influenced by deteriorating economic circumstances which have forced an increasing number of women into commercial sex activity.

The UN Coordinating Committee's Sub-Committee on Nutrition has documented deteriorating standards in child nutritional status in sub-Saharan Africa between 1975 and 1990, while nutritional outcomes are as a result of a complexity of factors, including disease, diets, droughts and war. There is substantial evidence from a number of countries, particularly in SSA, that child nutritional status has deteriorated after the introduction of SAPs, including situations where mortality data has stagnated or even continued to increase.

The impact of World Bank policies on Kenyan politics has been to usher in political pluralism, greater democracy, accountability and respect for human rights, just to name a few. However, there are heightened ethnic tensions, polarized communities and increased violent ethnic and tribal clashes. All these have resulted into a fragmented society.

Poverty is a major concern in Kenya. The 1996 Policy Framework Paper estimates that at least ten million Kenyans are living below the poverty line and that despite the progress made in structural adjustment and economic stabilization programmes, poverty remains one of the greatest challenges in Kenya.

This is due to economic depreciation and developmental decline. There is even increased inflation, poverty, unemployment, underemployment, retrenchment and forced retirements. Although the government was committed to the reduction and eventual eradication of poverty in Kenya immediately after independence, this has not been realized. In my view, the major causes of increase in poverty include: poor planning and priority setting for the small and medium business enterprise for the youths; wasted manpower in the youth who may not get a chance to proceed with their education after Form Four; lack of clear-vision job planning; lack of incentives towards the value addition in agricultural products; poor perceptions in certain areas like environment; health care; quality artwork and packaging; quality construction; for job creation. One practical example is the fact that our countries are littered with solid wastes which could be well managed by employed youths. Such an engagement could easily create thousands of jobs in the country and, at the same time, encourage a health nation.

The Impact of Structural Adjustment on Education in Kenya

Higher education was considered very important at the start of Kenya's independence. The government spent public resources on education since it was viewed that in the long term the whole country would benefit. This was inspired by the belief that it would lead to:

(a) Generation of more wealth within the nation;

(b) More equitable distribution of resources and opportunities to access such wealth;

(c) Better socio-economic development;

(d) Increased political socialization and democratization.

The demand for university education increased. In 1963, there was only one University college (Nairobi) with 565 students. By 1973, the number of students had increased ten-fold (Republic of Kenya, 1988). By 1990/91, there were four public universities with 30,000 students; this increase in university intake adversely affected the financing of university education (Ministry of Education, 1988). The government's full responsibility in financing university education ended in 1974. A new policy of cost-sharing was implemented in 1974/75 academic year.

Financing of university education was to be shared between the students and the government, whereby each student was automatically entitled to a loan to meet their accommodation and catering services when they qualified to join university. The government paid tuition fees and released loans to the University Students' Accommodation Board (USAB) to meet the said costs of the services. Loans for book allowance or practical attachments were, however, paid directly to each student. This system of loaning ensured that the needs of students were taken care of adequately. But this form of lending did not last long. The Higher Education Loans Board (HELB) came into being in 1992.

From 1985, Kenya started implementing structural adjustment programmes (SAPs) as part of the reform initiatives driven by the World Bank and bilateral donors. This programme required, among other things, reforms in the education sector, especially the reduction of government subsidies to university education. Since public higher education previously benefited from the state's approach, any reforms towards privatization or liberalization of its provision was to shake the very foundations on which it had always stood. When cost-sharing was implemented, it sparked nationwide riots.

Students and their parents considered these changes quite drastic. Resistance from the students culminated in rioting and subsequent closing of all public universities during 1991/92. From the 1991/92 academic year, the government stopped paying tuition fees and loans were no longer automatically accessible to all university students. Only students who could prove that they were unable to raise full or part of the fees were awarded loans.

Also affected was the cafeteria system where students were to pay directly for their meals. This new system of meeting students' personal catering needs became popularly known as 'pay as you eat' (PAYE). The students were further to pay for their accommodation and stationery needs. These were indeed drastic changes.

Structural Assessment Programmes in Educational Institutions

Structural Adjustment Programmes were linked to increasing rates of non-enrolment, grade repetition and dropout in educational institutions. Cost-sharing made the cost of education unaffordable to students from poor backgrounds. Those who failed to enrol due to lack of funds found it difficult to secure employment, and hence ended up as social misfits.

Students resorted to several ways of raising money. Cases of criminal activities such as stealing of university property or other students' property became common. Students also became increasingly involved in Income Generating Activities (IGAs), which were time-consuming, immoral and anti-social.

The Higher Education Loans Board (HELB) has, however, assisted thousands of would-be dropouts. The money given out to students as a loan is not enough due to high demands. This has affected the quality of education because students have to find means and ways of fending for themselves. Some postpone the academic period, they forgo lectures, and engage themselves in chores which are not related to their education. Poor families cannot afford the cost of university education. In future, some more innovative ways shall be needed to supplement the good work that HELB is providing.

World Bank Conditionalities

Under the Structural Adjustment Programmes (SAPs) which were instituted by the World Bank, African universities were in dire need of assistance. The countries which embraced the notion of SAPs made sure that they implemented the conditionality to the letter. I am not a financial manager, but what I knew was that the World Bank demanded many developing countries to liberalize their economies and allow a free market economy where commodity trading was left to float. Imports and exports of goods were liberalized and all markets were open to all. The most affected commodities were manufactured goods and foods.

The case for universities was unique. When the Vice-Chancellors of public universities got news that the World Bank was ready to finance university programmes, a committee chaired by Prof. Shem Oyoo Wandiga, the then Deputy Vice-Chancellor of the University of Nairobi, was formed. All public universities had two representatives, including finance officers. I represented JKUAT with my Deputy Vice-Chancellor, finance.

We were told that the World Bank:
- Was ready to give a loan to Kenya government to finance its education programmes;
- Would lend the money on a long-term payment basis;
- Would invest heavily in IT and provide the latest equipment to enhance LAN (computers, etc.);
- Would assist in the refurbishing of all science laboratories and also provide modern equipment;
- Would support various training programmes up to MSc., MA, PhD levels at a University of one's choice;
- Would facilitate and enhance staff exchange programmes and sabbaticals;
- Would facilitate country seminars and conferences on the latest technologies, especially in sciences;
- Would set aside reasonable funds for the purchase and importation of all types of vehicles to assist with the transportation problems that we frequently faced;
- Would frequently dispatch a number of experts to advise, inspect and commission the physical structure renovations as prescribed (travel costs to be borne by the recipient country;
- Was clear that every item ordered from outside the country, including vehicles, computers, would meet all specifications from the universities.

The committee chaired by Prof. Shem Wandiga was explicitly told that the available university grant was US$69 million. I remember this figure vividly as the five public universities started to divide the cash by five. In my mind, I knew I would get at least US$13 million to enable me cross over the university line. It was now a psychological game of the available money to be utilized. We never missed any meetings which were called to discuss the grant. The University Implementation Programme Committee, as the name suggests, was to just implement the requirements!

We opened an elaborate office with capable staff mainly from the University of Nairobi where the work station was located. We were given clear terms of reference and held in very high esteem by the external reviewers from overseas who had been contracted to oversee our planning processes. The consultants would send advance notices detailing arrival and departure dates, with an elaborate travel itinerary within Kenya as they visited the benefitting universities. The high-level consultants were selected by the World Bank from all over the world, mainly from universities. They were nominated according to their disciplines. I honestly thought that we had hit a jackpot.

The World Bank gave explicit conditionality on how to utilize the funds, which included:

- De-link university services from academic matters;
- Charge students on all items they received at cost, including food;
- Give students loans to be repaid;
- Liberalize education;
- Use university savings to service major building renovations;
- Put up adequate facilities like gas chambers, air conditioners, exhaust pipes in chemistry laboratories and paint halls;
- Provide trucking for connectivity in all buildings which are to receive computers;
- Where possible plaster, tarmac dusty areas for computer dust safety;
- Set up committees in relevant areas to address 'corrective' issues;
- Install grills in all buildings which would receive the World Bank equipment;
- Ensure adequate and safe parking garages for the World Bank vehicles;
- Give elaborate training programmes as necessary in World Bank-related concerns like laboratory equipment;
- Get several literatures on the needs and demands of the World Bank-borrowed money and its repayments;
- Select some senior officials to be trained in management and procurement procedures in other countries.

All the above sets of conditionality were to be met by individual universities at their own cost. It was not clear whether the loan would re-imburse the expenses. Every step we took, every renovation we made and every grill we put up was inspected and certified by a World Bank consultant. Perhaps the notion of thefts, car hijackings and other issues was a driving force for these inspections. They did not want any item bought under the UIP to chance. Safety was the prime mover of the programme. We all complied and effected modifications as demanded.

There is an analogy which my former Vice-Chancellor of the University of Nairobi, Prof. Phillip Mbithi, gave. He said in one of the briefs that World Bank would normally ask the recipient to do several things before they award or loan out any money. That it was like asking someone to remove all the clothes including the underwear, and then tell the same person to stand naked as you start supplying him with the needed clothes one by one since that person would have responded to your request.

You start with the underwear, vest, a trouser, a shirt, socks, shoes and maybe a tie. These items are given to this individual after meeting the laid-down conditions, one by one. I thought that was a classic analogy of a third-world borrower, who must fully comply with the lender's demands before any assistance is given. Some of the demands created problems in the institutions. Structural adjustment programmes have their effects and I cannot delve into them. I will, however, give the highlights of the consequences of the World Bank projects in our universities. We were given deadlines on what to do and the kind of announcements we made to the public and students. Some announcements caused havoc and all public universities went into an unprecedented strike. This was in 1992/93.

A good number of the conditionalities given to the public universities were not met. After several years of negotiations, we were able to get a few computers; defective buses and cars whose specifications were ignored and blatantly rejected; obsolete equipment like laboratory equipment were imported; and the whole procurement process was chaotic. Specifications for several buses, cars, pickups were not adhered to and those of us who returned them never got a replacement. We actually lost by refusing equipment which did not conform to the original approved and signed documents. These had been ignored. I am not sure what may have happened to the rest of the requirements.

One mammoth headache that all universities experienced was the occurrence of "the mothers of all strikes". The UIP office was ordered to announce major financial implications in line with the World Bank aid. Prof. Shem Wandiga, the chair of UIP, issued a press release which in principle cancelled students' allowances (the boom) and itemized cost-sharing measures.

The local dailies wrote a detailed analysis on what the government would undertake and students' responsibilities. I remember the announcement was made on a Thursday morning. This was to be done before the 1996 budget deadline. Everybody hesitated to make the announcement but asked Prof. Shem Wandiga, who was the chair, to do it.

The strikes which followed across the country were gross and devastating. It all started from the mother University of Nairobi and word spread across the other universities like fire. It is not possible to enumerate the reactions from each Vice-Chancellor. It would take another whole book. All I can say here is that Moi University students burned vehicles and a student died. The university was closed for over six months. The University of Nairobi was shut down, followed by Kenyatta University. That marked the beginning of the longest closure of public universities.

I resisted the closure of JKUAT for two days and had a hard time to contain the students. We very much tried to advise them that it was a government decision not the individual university. I was marooned in the campus and my office for two days and two nights with no change of clothing, but supplied with food from the cafeteria. Other Vice-Chancellors were in University of Nairobi holding crisis meetings with Prof. Phillip Mbithi who was then leading the team during the announcement.

We were in touch with the Chancellor, President Daniel Arap Moi. He left the matter to be dealt between us and Ministries of Education and Treasury. The intention was not to close all public universities but have one or two to stay on course. This would not be! They kept on calling me not to close. Closure of all universities could create a defeatist impression.

I tried to do everything possible to persuade the students to continue with their studies. Early Monday morning, after all the ordeal, I called the academic board and ordered the closure of the campus. Luckily for me, not a single structure or car was damaged, except for the looting of groceries and cooked food.

Down all Kenyan universities went, and we had to face the reality of implementing the new order of academic dispensation as per the World Bank's negotiations and recommendations. All systems were paralysed and Kenya went quiet on the academic front. We closed for half a year. This caused the state a colossal sum of money in terms of wages and related expenses, not mentioning the psychological agony of students and parents!

The closure of any university for any reason comes with unprecedented consequences. The process can result into loss of lives, destruction of both public and private property, wasted time and opportunities and numerous socio-economic implications. Everybody, both staff and students, gets disoriented. Timetables become chaotic and general planning of events is thrown into disarray. That is why Vice-Chancellors stay alert 24 hours, seven days a week to avoid such occurrences. It was not a position to envy and yet it had to be done by somebody.

Despite all these problems, I attempted to create strong academic linkages in all areas of study. The many bodies that I headed had direct benefits to JKUAT. I was known as a performer and my interaction with other people and bodies demonstrated the same. I always glorified my university and painted a good image of it to the public. I was able to secure scholarships for training and my graduates were easily absorbed into the job market because of our unique academic programmes. The World Bank assistance was well intended, but the consequences were so negative that they outweighed the inherent good.

Chairman of Universities Joint Admissions Board (JAB)

One of the most demanding responsibilities in academic circles revolves around students' admissions to universities. Kenyan students who qualify to join universities exceed available infrastructure. Most of the students would wish to join the public universities because the government subsidizes costs.

It was only in 2012 that a Commission for University Education (CUE) was created to replace Commission for Higher Education (CHE). All individual University acts were repealed under the new higher education dispensation. The CUE now oversees all matters which affect higher education in both public and private universities. Admissions are centralized for all the Kenyan public universities. Before then, a powerful Joint Admissions Board (JAB) was totally responsible for the exercise. This was an agreement which was not formalized but served the purpose without the slightest condemnation.

The Vice-Chancellors of five public universities chaired JAB on a rotational basis. My university became fully-fledged and I then qualified to chair the meetings when my turn came. Besides admissions, a few other managerial matters were discussed. It was, however, clear that the boards' role was admission of qualified students to all public universities.

We had a central processing office managed by competent workers who were responsible for documentation. I recall that during my time as chair, Prof. Tony Rodriguez and Mr Mungai Gachui were so thorough that the whole process of admitting student took half the time we normally used to complete the exercise. We had a responsible and competent team in the ICT offices.

My tenure as chair was smooth and I recall that only less than 10 per cent of qualified students were admitted to the five public universities. The Kenya School Certificate of Education (KSCE) normally presented a large number of qualified candidates. We had to trim them and pick only the top-crème. The cut-off points and spaces available in each university and programme determined the final entrants. A very large number of university admissible candidates were left out. They joined either private universities or other tertiary and middle-level colleges.

I must state that JAB was never manipulated or corrupted during its admission procedures. During my few years in the parastatal, I came to regard the board as one of the most transparent, competent, trusted and corruption-free body that ever existed without legal instruments. The rules and procedures of the game were so clear that the public knew who was admitted, where and when to report. Individual universities got a central admissions list for all the programmes and students' names.

It was the board which determined the cut-off points for programmes and reduced one or two overall cut-off points for girls and students from the disadvantaged areas. This was seen as a positive move to uplift the girl child and students from disadvantaged areas. The public was made aware of these rules and regulations.

Although JAB was not officially registered, it was a recognized and an acceptable committee composed of the Vice-Chancellors, their deputies, deans and registrars who formed the bulk of individual university senates. The challenges of managing student admissions and undue pressures were not evidenced during my entire 13-year membership.

Undue influence by outsiders was not entertained. The chairman usually briefed the press on all matters to do with admission on behalf of the five public universities. This was done immediately after the admission meetings at a specified centre. All admitted students and the colleges to join were made public for scrutiny.

The only time the Head of State commented on admissions of students was when he hinted that we reconsider a second double intake in 1997. I was the chairman of JAB then and had to politely refuse the directive. The local media picked up the story and accused me of disobeying my boss, the Chancellor of all public universities. The dailies were courting an imminent confrontation between us, but the then Head of State ignored the innuendos. The matter was not pursued any further.

Once students are admitted to their respective universities, it is usually the role of the Vice-Chancellor and his/her Senate to manage other logistics. They are allowed to transfer from one university to another depending on the capacity of the programme, or can also transfer from one faculty to another, again depending on the declared capacities. It is noteworthy that even a member of JAB would not be favoured to have his/her unqualified sibling admitted. I recall many staff members whose children missed by just a point and were advised to seek alternative places for admission. JAB was that strict and transparent. It lived to its obligations and commitments. Nobody wanted to be discussed in Parliament and our board was composed of eminent Kenyan scholars. JAB had the face of Kenya and integrity prevailed.

I must emphasize that the Head of State did not put undue influence on JAB to push for admission of any student who was not qualified. He respected the board's decisions and had very high regard for scholars. This was a commendable characteristic Daniel Arap Moi as president and Chancellor of public universities. His ministers did not even attempt to make any such requests for unqualified students. JAB was sensitive to the gender disparity and students from some arid and semi-arid parts of Kenya. They were given special considerations. JAB was an institution in itself.

The replacement of JAB with the Central Admissions Board to include all public and private universities follows the basic admission procedures which were set up by JAB. Now all universities and colleges declare their programmes and student capacities before the first sitting. This is a demanding exercise as other programmes and capacities are not divulged. They are hidden under parallel degree programmes and they create anomalies which distort officially admitted numbers. The admission task is daunting now that Kenyans expect more equitable admission procedures, which include all on the basis of the enhanced numbers and equal opportunities for higher education.

14

University Leadership and Quality of Academic Programmes

University Appointments, Promotions and Award of Honorary Degrees

Universities worldwide consider teaching staff as the core of their functions. Basically, the role of lecturers is to teach, conduct research and disseminate the findings thereof. They are expected to publish their work for the common good of humankind. Promotions to high cadre are normally pegged on the number and quality of the papers, financial attraction or sourcing and seldom administrative roles. That is why one is always advised to publish and innovate or fade into oblivion. In any university worth its credibility, scholarly work is the only consideration used for promotions.

Peers respect persons of high academic credentials. They compare notes and enjoy reading and critiquing others' work. They love to debate and correct their colleagues without any intimidation. For example, a course in Critical Thinking is designed to teach several strategies for increasing one's abilities to react critically and form opinions after reasonable arguments. Those arguing seek to gain the acceptance of others for their points of view.

The participants will learn and acquire the art of asking the right questions, including self-criticism about one's own thoughts. One will learn the art of reasoning. One will also research for useful data, pinpointing the real issues and offering critical options based on those evaluations.

Critical thinking, therefore, is not just an art but also disposition and a commitment. As compared to formal debating skills, one's informal strategies for advocating and arguing positions will be sharpened since one examines reasoning capacities of others in speeches, conversations, essays and group deliberations. University lecturers and professors are always entangled in debates on issues they consider important. All areas of discipline will always provoke debates and proven research.

Universities in Kenya expect persons who have attained a higher degree to be appointed at a lecturer's level. He or she might be expected to have published some papers. This is the norm in many universities. At the time of writing this book, there are 22 public universities and 17 accredited private universities. Of the 22 public ones, there are over 800 degree programmes spread in all of them. Ideally, lecturers with earned PhDs should teach the courses within the programmes.

I have, in my opinion, given a clear method of awarding a degree and the expectations before its award. What I did not cover is the reasoning behind the award of *honoris causa* (honorary degrees).

Honorary degrees are bestowed or conferred to persons of distinction. They are conferred or rewarded for exemplary services, roles, duties, distinction in service to mankind, university, nation and academia, and unique assistance. Possessing an honorary title or post is not a qualification without necessarily performing services towards the earning of such an honour. One simply receives the reward on the basis of some exemplary undertaking. The recipient does not go in lecture halls to be taught, write examinations and then dissertation.

During my tenure as a Vice-Chancellor, I awarded four *honoris causa* degrees only. I consider honorary degrees very special and should be awarded to the most deserving as examples to the present and future generations. My Senate thought I was mean. But, afterwards, they appreciated my stand since other universities dished out the same to non-deserving individuals. As the current chairman of Kenyatta University Council, I have advised Senate to consider carefully individuals they wish to be conferred with honorary degrees. The name of a university is held high in tandem with the awards it gives. Awards of diplomas and conferment of degrees is such an honour to the recipient that if such a process is abused, the reputation of a university goes down the drain.

The award of *honoris causa* in 1994 to the retired President Moi was done as a routine activity as he was the Chancellor of all public universities. It was therefore an obligation to award him the degree. He could not perform the function of a Chancellor without the said award. It took me some planning and preparations to have him gowned and declared my first graduate soon after JKUAT was elevated into a full university in 1994.

President Moi was my first Chancellor. I had to repeat the same when I again awarded President Mwai Kibaki (now retired). Soon after winning the 2002 presidential elections, Mwai Kibaki performed the first public function at JKUAT. I had to confer him with the honorary degree in his capacity as my second Chancellor before the Acts changed. The two honorary degrees were awarded because they had to be, according to the 1994 Act of Parliament. President Kibaki served as my Chancellor for less than 24 hours. He had posted me to the National Environment Management Authority.

Mr Koichiro Matsuura was the Japanese Director-General, UNESCO. He was influential in education matters in Africa and particularly the JICA support to JKUAT. It was my humble submission that we, as senate, award a re-known Japanese national an honorary degree. This was due to the generous technical support of JKUAT by the Japanese government.

I also had a choice of awarding the JICA president but opted for the UNESCO Director-General. He was also an academic in his own right who was nominated to head the powerful UN office in Paris, France. He later created a chair in our Biotechnology Department. The chair was also an honour to our university as it was the first one by a UNESCO Director-General. The Japanese Embassy to Kenya commended us for the coveted award.

The Senate was convinced that Mr Matsuura deserved the award and Council approved it. The exercise followed the full normal process as prescribed in the university statutes.

Prof. Risley Thomas Odhiambo was a scholar, an innovator and researcher. He was the founder of many academic institutions, including International Centre for Insect Physiology and Ecology (ICIPE). It was my honour to bestow him with an honorary degree. The Kenyan scientist deserved it. That was the very last one that I conferred as a VC at JKUAT. One disturbing trend emerging both within public and private universities is the award of honorary degrees either to politicians because of some patronage networks or financial considerations, which awards do not benefit the academic or financial reputation of the universities.

University appointments and promotions follow laid-down procedures with specific academic requirements. During my tenure at JKUAT, we were specific and clear on the requirements for upward mobility. I was accused of slowing promotions or being too strict. I was certain of one thing: to retain the staff. I wanted motivated members of staff who would be my best assets. But I also knew that career development was every employee's target. They looked at the opportunities ahead for career advancement and considered training as one criterion to enable one to proceed to another level of grade.

When I reflect on how I retained staff, one reality comes into my mind: we had reasonable package allowances like housing and commuter then. There is a major problem currently because many universities in Kenya are experiencing an unusual number of academic staff departures. Maybe, the employees are feeling undervalued by their employers.

I cannot over-emphasize the importance of staff welfare. There are specific reasonable allowances like commuter, students' fee waiver, books, per diem, computer packages, medical care and utility which, if enhanced, would retain staff and reduce excessive attrition/ mobility. Many universities, for example, are unable to reimburse medical expenses. Staffs give up following the refund and develop low working morale. A medical scheme is one area which is so vital, yet so neglected. This does not build the institution.

Competitive travel grants can be introduced for staff to attend national and international meetings. It can be conditional. One has a reputable paper to present or attend a meeting which would benefit and promote the university. Universities need visibility and global image/fame in their academic and educational pursuits. Councils can always oblige to some reasonable package requests. My motto was to always retain good members of staff even if it meant creating extra duties for them to engage in as long as I would justify it.

The University of Nairobi, for example, has among the most demanding criteria in staff promotion. I sit in senior appointment committees and realize that despite those demands, staffs enjoy work; they stay in the same positions until their time comes when they have met the stipulated requirements. Despite those who leave for higher competitive posts like Vice-Chancellors, the rest of staff go through the rigour of interviews in order to qualify for the next positions.

Currently, Kenya has 32 universities, both public and private. There has been unprecedented exodus of staff from one campus to another, as academic staff seeks fast promotion. This is affecting the quality of education negatively. The University of Nairobi has a policy on such an exodus which I consider very well considered.

If a lecturer moves to a sister university on promotion and wishes to return, he or she will revert to the same grade he or she was at before leaving, despite the higher grade received from the other university. This condition has made it difficult for migrating members to return to the University of Nairobi because they fear being downgraded. The reality is that promotions at the University of Nairobi are strictly determined by the laid-down requirements which cannot be circumvented. This is part of upholding the institution's high quality and standards. The university still maintains its staff despite the continuous squabbles from some quarters.

In the early 1970s and 1980s, there was only one University of Nairobi. Things may have been different then. In the effort of attracting and retaining staff, it is most likely that hiring regulations and procedures may have been flouted.

I have heard of exchanges during interviews to the extent that interviewees even stormed out of the boardrooms. Possibly, such panels may not have been competent enough to interview some senior members of staff. The promotions to full professors are demanding and controversies exist even unto this day.

The Chairs of Council those days were businessmen who knew little, if anything, about university promotion requirements. They did not have academic and teaching qualifications to chair interview panels. They wholly relied on the Vice-Chancellor's guidance and recommendations. It was therefore likely that major rifts and favouritism could prevail during such high-position interviews. Requirements could be flouted in favour of particular people, especially if there existed some differences amongst the panel and/or the interviewee.

Currently, in the UoN, the clear guidelines have reduced complaints considerably. Senate passed the regulations and they are hitting hard on them.

I have no clue as to what the young universities do as far as associate and full professorial positions are concerned. Other than the older Kenyan universities which might have laid-down procedures for appointments, the younger ones are yet to come up with appointment requirements comparable to international standards.

It is instructive to note that if the head of an institution does not measure to the position, so shall be the followers. The head of the institution, in this case the Vice-Chancellor, must be a person of proven academic track record. This is the only way that laws and procedures can be followed. African universities have had appointed Vice-Chancellors.

The recent move on competitive hiring of the same has reduced rifts amongst senate members. Sycophancy has reduced and proper criteria have been put into use. Not all Vice-Chancellors fall into this trap of sycophancy. African universities, I must admit, have had some of the best Vice-Chancellors who have made a mark in the academic development in their respective countries. Building institutions to appreciate standards is an uphill task. Building and acquiring competent staff to run such institutions is not easy either. The head must be able to understand the dynamics of politics and make rational decisions accordingly.

In my own case, promotion from lecturer to full professor took a long path. My appointment to lectureship in 1980 called for a PhD and some publications. I had them already on my first appointment. It took nine years to ascend to senior lecturer and associate professor. This was with additional publications and mandatory MSc and PhD supervisions. I got my full professorial position in 1996, again with more published papers, PhD supervision and publication of books. It was not therefore an easy ascension.

It is in order to enumerate some MSc and PhD candidates who have gone through my tutelage as follows: Sibuga Piri Kallunde, Leonard Wamocho, Wariara Kariuki, Safary Ariga, Joseph Oryokot, Boniface Ita, Flora Kiririah, Ongusi Bongo, Ben Wafula Wanjala, Margaret Karembu, Bala. I was also able to write papers with them.

I was supervising the named students while I was doing my other administrative duties. It was a pleasure to lead young scientists and see them graduate. A good track record of a committed scholar matters a lot, as a role model, since future generations would always ask for tangible examples of work done.

All these demonstrate that mine was not an accelerated promotion rather I put in time and sacrifice. It is also important to mention here that greatness is not born but grown and nurtured. That is why I always advise the youth not to idolize those people who attain success overnight on a fast unexplainable track. One needs to go through a process to get to the summit. There have been cases of unfair appointment based on tribal considerations even when the appointee did not qualify. The appointees did not deserve the promotion and ended up frustrating the systems, hence little, if any, development is witnessed. Those who deserved the post got frustrated and did not exert their potential.

This is a very common scenario in Kenya and nasty cases of nepotism have been brought into the public arena and embarrassed those with such tendencies. Even when laws prohibit the practice, many senior officers have been indicted in the courts of law with devastating revelations.

The hasty elevation of several Kenyan colleges into universities, from a mere seven to 22, is a case in point. Virtually every university's top echelon is dominated by one community or tribe. This is against Kenya's constitution; and the appointing authorities has been unable to unravel this misnomer despite the full powers they have. No wonder the nation is crying of falling academic standards! There is little distinction between some universities and *harambee* secondary schools which are run by the board of governors from one community. Members of the academia see the flaws and level hard questions at these institutions.

With the new University Bill 2012, there is order in the appointment of Chancellors, board chairpersons, Vice-Chancellors and top university organs. The powerful Head of States of state used their positions to directly or indirectly influence actions which would otherwise be questioned. The constitution demands for gender inclusion in all committees. The practice is now entrenched accordingly and qualifications are adhered to in the universities and public sector. Public appointments, however, are still tribal-based despite the academic requirements. Many African Head of States of state have yet to appreciate that they run countries which require public good to all. By and large, new Kenyan universities have unfortunately taken tribal dimensions in their setups.

The problem has been made worse by a number of complex issues which include: over-enrolment of students; an extra demand in teaching module two students; lack of adequate learning facilities and materials; de-motivated staff; insensitive management styles; random and occasionally unilateral decision-making which affects harmony and focus negatively in respective institutions; prolonged procedures in staff evaluation and promotion; and, staff favouritism by a clique of managers.

With all these discouraging revelations, what then is the future of quality education in our schools, colleges or universities? It is gratifying that we are aware of them and talking about them. There needs to be immediate action on these anomalies. In the meantime, relevant authorities have decided to bury their heads in the sand. These trends will highly balkanize Kenya a country which for many years has been struggling to be a nation of one people. This issue will be discussed in the next chapters.

Struggling to Uphold University Academic Standards: Are Our Academic Standards Falling?

The population of Kenya stands at approximately 40 million people. The number of students in secondary schools keeps on increasing every year. Those who qualify for university admission exceed available infrastructure. There has therefore been cause to increase institutions of higher learning. The increase has had its consequences.

In recent years, Kenya has had an unprecedented increase in the number of universities. Between 2010 and 2013, fourteen new fully-fledged universities were created across the country. Several of them were converted from middle-level colleges into independent universities. This ambitious move has brought about major and far-reaching implications on all facets of education in Kenya. The planning may have been too abrupt with minimal, if any, consultations.

University colleges are generally nurtured by their parent universities. This is how all the current East African universities came about. The idea of a university college, if well organized, is to run programmes just like the mother university, but under careful guidance. They need to have all the facilities, staff and students just like their parent campuses.

The only difference is that their courses and staff assigned to teach them are monitored by the parent university. In fact, they can be very strong in some disciplines and can even perform better than the mother university. But because of other complementary courses, they are still subservient to their parent campus. In my view, the arrangement is meant to cut down certain costs, ensure maturity of programmes, reduce course duplication, monitor market changes for graduates and have complementary services like teaching, equipment and laboratory utilization universities in other countries practice the collegiate system of higher education.

The contrary may be seen to be true. Such a system allows full autonomy early enough for independent development. These include having a full complementary staff, residential houses, developed programmes, their own budget and capital development. Our country's national budget may not sustain such fast growth.

Many countries tend to have no financial allocations for university development. I am not implying that we should not have many universities, but rather we need to develop them gradually. The current situation is pathetic, to say the least. Kenya needs to be a first world in the near future, but the development plan in place may not take the country far if we do not put efforts in quality university systems.

The following paper which I delivered at a private dinner meeting summarizes the problems we face in higher education. It analyses facts which affect quality in academia. It is reproduced here verbatim.

Compromising University Education: Quality versus Quantity

Our country has witnessed an unprecedented growth of universities and student population. The increase has happened within the last eight to ten years and is commensurate with secondary school growth. The demand for university education is the ultimate academic desire for any average literate person. Kenya has witnessed increased enrolment for all cadres of degrees. What can nationally be done about this abnormal increase?

I have been in lecturing and administering university business for a long time and would like to offer my thoughts on the quality and quantity of our education progression against the backdrop of Kenya's Vision 2030. Teaching is a noble call. It is a profession which deals with human skills, knowledge and development. Quality therefore is affected by the deliverer and recipient of a subject.

What is quality? In simple terms, it is the fitness, the finish, the attraction on the package and wholesomeness of a processed product. In our nurseries, primary, secondary schools and universities, the deliverer is charged with the role of nurturing toddlers, children, pupils, and university men and women. Those responsible for teaching the said groups must themselves be knowledgeable and fully equipped with the processes that lead to finished and functional products.

I will, for the time being, ignore the lower cadres and address university concerns. To qualify for a lecturer's position, one must have completed his/her PhD from a recognized university. This expectation presupposes the rigorous stages that the individual went through in his/her area of expertise.

To declare that someone has been taught, supervised, examined and qualified for a specific award means that the person has had adequate contact with the lecturers and professors during his/her training. What does this mean? It is the duration taken to teach, consult, assign, mark quizzes, assessments, examinations, conduct practicals (where necessary), read theses, be available for academic guidance, write examination papers, write make-ups and be constantly on the laptops responding to open distance learners (ODL/ICT group) – who are usually at large. This is a typical scenario of a dedicated lecturer.

Consider the following counter-productive scenarios:

> for example, due to high cost of living, a lecturer may opt to take up several part-time lectures across Kenya traversing from Western, Eastern, Coast and Central Kenya to the City of Nairobi where universities have campuses and where teaching opportunities exist. This is where the highest number of workers who yearn for knowledge resides. The transit lecturer who is on a permanent payroll elsewhere must be able to do all that appertains to the fulfilment and award of a desired course in another campus several kilometres away. He/she has to be a great time planner considering the state of our trunk roads and airports. This is not, however, a unique situation. I once had a lecturer who would fly from the University of California,

Davis, to Rutgers to teach us for 3 days a week and fly back to his base. It worked out for him because he was a great planner and a renowned world scholar.

'The sum total of quality education calls for every input by a lecturer to a student in terms of content, coverage, feedback, mentoring and role modelling. Meaning? The sum of quantity production is merely to churn out graduates who do not apply their expertise and cannot think critically to create jobs for themselves or for others. They are not to blame. It is just the overall craziness of university degree acquisition.

The proliferation of universities has raised the need for unprecedented number of lecturers and yet these are already in short supply. They must circulate at a cost, do part-time and offer half-baked commitments. Departmental chairs are compelled to follow up on part-timers for results and the consequences are quality degradation.

This practice inevitably compromises on the quality of education. How then will Kenya realize her Vision 2030? Senates decide who teaches, what they teach, when and how they teach. A professor may choose what, when and how to teach. Some disciplines call for very elaborate practicals, which may not be possible to conduct due to unmanageable number of students and the cost involved.

What are the possible solutions? There are several alternatives which I would like to offer. Firstly, we need dedicated trained human power. I suggest we form a human power consortium of more established universities to train young upcoming graduates under staff development programme, pay them well and bond them for not less than six years. The ministry responsible for higher education should meet the cost of training specifically meant for the very newly upgraded universities. They have to be retained in specific areas of their expertise.

Secondly, the newly established universities should work together and identify critical areas to mutually contract a university or universities to train their PhD personnel. Some years back, a few African universities came together and formed an association with the University of South Africa, (UCT) to train our staff. The USHEPiA project assisted JKUAT in training PhDs in critical areas. This was an excellent arrangement for staff development.

The contracted university should give approximate period within which they graduate the trainees. In other words, the training university ought to put in extra time, expedite the supervision and degree award. Extra funds should come from the requesting university.

Thirdly, freeze for a while, any academic programs which are understaffed and then seek outside scholarships in the area. This will allow for the release of space and personnel for a quick jump-start.

Finally, create strong industrial linkages and appoint substantive chairs specifically for human power development. Name buildings, streets, and laboratories after these sponsors and strike lasting assisting deals with enterprising, resourceful and philanthropic individuals.

The continuous discontent with the deteriorating quality of education can be overcome by having enough dedicated university lecturers who are well motivated to arrest the situation. Equipment, supplies and learning facilities are equally critical in maintaining high standards. Other universities elsewhere are known for greater scholarly work because their countries know the importance of retaining trained personnel.

Solutions are available for Kenya's public universities, but lack of concerted efforts and funds have created this dismal scenario. We can uplift the academic standards of our universities. This is possible if only we make our management styles more responsive, consultative and visionary. The managers must take cognizance of their mandates. Universities are not commercial ventures but academic and scholarly in nature. They should desist from being seen as business entities. They must focus more on genuine and productive scholarship.

What I consider important in our efforts to create competent human power is the quality of degrees. There have been a lot of write-ups criticizing current university degrees.

My sincere belief is that any programme can be useful. The graduates have an option of utilizing what they trained in rather than wait for job advertisements. It is true that the speed at which colleges in Kenya have duplicated degree programmes is a worrying trend. I know for a fact that at present we have close to 1,000 degree programmes in the 22 public universities and their seven constituent colleges. New ones are being developed and others are being offered by their partners.

Kenya's university education system is robust in accessibility. Was this the best investment? We can trace the problem back to the year 2000 when the World Bank encouraged commercialization and entrepreneurial practice as the sole means to drive university education. All public universities ignored their original missions and characters and went on a competition spree. Demand-driven courses were introduced. Student involvement in parallel degree programmes was quadrupled and duplication of courses soared. Hundreds of new courses were introduced with almost the same titles across universities.

The worry is that some degrees represented poor value for the resources spent and the parents and guardians remained confused as to whether their money had been put into good use. I have alluded, in my earlier chapters, to the fact that lecturers in many public universities in Kenya are over-burdened and have no time for research. The crowded teaching timetables may as well be transferred to junior colleges to be handled there.

The reality of the matter is that some students never cover their prescribed lectures and practicals in their entirety. Those that opt to go for module two pay a lot more and never realize full potential for their money. They are short-changed, compared to the regular ones, and because of these crush programmes, it is not certain to speak of quality or quantity education.

The former Commission for Higher Education (CHE) which was responsible for quality control has been quite weak in the context of providing effective leadership and vetting the degrees being offered. It is hoped that current constitution will bring some order in the management of higher education. The enactment of the new Education Act, 2012 will hopefully curb the academic degradation.

Currently, there is little control of rapidly growing programmes. Commission for university education has to provide firm leadership and vet the degrees being offered. universities exist for public good and indeed are the precursors to technological development and innovations.

The idea of attracting thousands of students under the scheme of privately-sponsored group to raise funds has been over-stretched. Some public universities seem to short-change the students, hence perpetuating poverty. Most rural folks find it very hard to raise school fees for their children. When a programme charges so exorbitantly, it leaves the sponsors poorer and hating education. CUE should standardize fees charged to all students but differentiate as per the programmes of study.

My personal take is to create a strong tertiary academic cadre which is practical and innovative-oriented. There is need to drastically reduce and or merge course programmes at our universities. Besides, others can be transferred to middle-level colleges so that the national focus in education is on internationally acceptable academic standards. universities are not polytechnics.'

The University Hierarchy and Tenets

Let me not over-emphasize the importance of and respect for the university's hierarchy. As alluded to earlier, a university uses its Acts to produce statutes and regulations for governance. A belief in the guidelines for good corporate governance in any state-owned corporation must be well understood. Although I took a course on corporate governance several years later, I still fully executed its requirements. It is instructive to explain briefly my personal interaction with each organ and why I succeeded in my management of JKUAT even during the most turbulent times in Kenya.

The Chancellor was the President of the Republic of Kenya and served in a titular position. He attended all my graduation ceremonies personally and awarded the diplomas as well as conferring degrees.

As the Chancellor of all public Universities, and Head of State of state, it was not possible to reach him readily but he was always ready to receive the Vice-Chancellors for consultations whenever need arose. I had total respect for him as my boss and as the President of the Republic of Kenya. My relationship with President Daniel Arap Moi was cordial and he had a lot of respect for me. I did not use my position to my personal advantage for the thirteen years I was at the helm of JKUAT. We used to brief him, especially when things got tough. The individual universities Acts specified so. We were hence compliant with the law.

My respect for and continuous briefs to the Council gave me opportunities to handle issues competently. I viewed my Council as the final in policy matters. The formulation and final drafting of both the 1994 University Act and thereafter the regulations gave clear roles for each organ. I always called my chairman of Council, Dr Stephen Mulinge, to brief him on any issue that needed the Council's attention. All matters passed by Senate were carefully analysed, apportioned for action by various bodies and I could only pick on those which needed the Council's action. I did not hold back any information which had direct bearing on the smooth running of the entire university.

The public expected the chief executive of the institution to be answerable to all. I was therefore the main link both internally and externally. I did not have any problems with my Council which was composed of the chair, Dr Stephen Mulinge and other members who included Dr Davy Koech, Mr Uhuru Kenyatta, Eng. Sharawe Abdullahi, Hon. (Amb.) Ali Chirao Mwakwere, Mr Tom Owuor, Prof. Joseph Mungai, two senate representatives, representations from ministries which were relevant to higher education, two student representatives and one alumnus.

Although the Council used to meet quarterly, I used to brief the chairman occasionally in his office and usually sent him what I considered urgent for his attention. He would then advise accordingly. I also had a duty to brief the senate on matters I considered necessary. This kind of communication and trust for free flow of information was very vital to me as a means of easing stress.

The University Senate, in my view, was the most critical organ of the university functionaries. The decisions from the senate had to reach the Council for ratification. My senate consisted of my deputies, full professors, deans who were elected by their respective faculty members, directors, chairmen of departments whom I had appointed as per the statutes, the registrars and two student leaders (who could not attend deliberations on examinations).

The Senate was the most critical body which could determine the smooth running of the university. The composition of my senate included elected and nominated members. I had to develop confidence in them and run the business of the university as a corporate entity. As the Vice-Chancellor, I made sure that I followed the rules to the letter.

The University Act of 1994 and the attendant statutes were my additional benchmarks. I hired a lawyer to sit in all the deliberations and interpret clauses.

The elected leaders, deans, for example, had to appease their constituents at all times, while the chairmen of departments would side with one of them. Also, the two students' representatives had to please their union. The only persons who would be totally independent in thought and would make rational decisions were the full professors and the two alumni.

I made sure that I held my Senate meetings on schedule as per the almanac. They were always brief and this encouraged members to attend, knowing that they could not spend the whole day in meetings. I had constant consultations with the deans and directors on urgent matters which needed quick action and would always brief full senate meetings.

The role of Senate was stipulated in the statutes. It was the ultimate academic organ which recommended graduation procedures, awards of honorary degrees and processing of examinations, among other functions. It was my firm belief that all meetings were to be held at the university compound, especially if we had to close the university prematurely for any reason. One could not hold any such meetings elsewhere but within the campus as a way of keeping a keen eye on the institution.

I used to have several seminars, workshops, symposia and even conferences involving my senate. I would occasionally take them out of town to a more relaxed environment where they would bond, think freely and open up. My outside meetings were usually held at the coastal town of Mombasa or tourist sites like Maasai Mara. They enjoyed the interactions, and I learnt a lot from the bonding.

I shared with them openly that my job was not to be envied. I was only carrying out a national duty which any one of them could hold. In any case, I was a lecturer like them and their interactions with me should be collegiate. Our annual bonding outside the campus created an extremely enjoyable environment. Perhaps it was through these interactions that my university did not experience premature closures and staff unrests.

The processing of examinations can be one tedious and agonizing exercise. This was one task I was overly sensitive to. I set up an isolated, lockable, out of bounds premise strictly for examination processing. I vetted the officers in charge, and the DVC (AA) had to personally ensure total compliance to the requirements. Lecturers had to acquire passes to access the centre.

I did not experience or witness an examination leakage for the whole duration in JKUAT. I had trust in the staff manning the place and they also had confidence in me. I used to give them surprise visits and have tea with them.

The Senate therefore had full confidence in the examination processes and knew those responsible for them. Once one group knew that we worked as a team, it was automatic that all the others would join and share in the success. Two other groups were equally important for the university's successful operations: Faculties/schools/ the departments and student unions.

The university's basic grounding was centred at the departmental level. The chairmen were appointed but the deans were elected. However, this could potentially create several centres of power if the Vice-Chancellor did not have a good grasp of the university's complex system.

JKUAT had four faculties, one institute and one school. For the purpose of this discussion, just assume that all of them were faculties. The statutes specified that the deans were to be elected but directors and chairmen of departments were to be appointed by the VC.

In all my stay in JKUAT, the election of the deans was done every three years without fail. They had to be holders of PhD and be senior lecturers. The incumbent would contest only once and return to class to teach. I considered that if anyone stayed too long, promotions to higher grades would not be easy. Such an individual would not have enough publications through research for upward mobility.

Deans' meetings were often many and could not allow for enough time for research. This was also the case with chairmen of departments. In my case, nobody could contest for deanship if he/she had not attained a PhD degree and was not a senior lecturer. The deans were also expected to command respect, especially when they chaired meetings and appointment committees.

Departments are the foundations of academic, research and publications bases. They have to be well anchored and productive. I considered them the anchor of the institution. The chairmen of department can be powerful as, indeed, they should be. In fact I used to joke knowingly that department chairmen were my direct deputies and my representatives at that level. I had full confidence in them and gave them automatic access to my offices. I attended to them at any time, any place, and under any conditions.

They were the persons on the ground and had continuous contact with students and staff. They were the examinations chief officers and presented the same in the faculty board meetings. They were responsible for ensuring that courses were taught, examinations were administered, marked promptly and results were processed. No university could do without these organs. I trusted those I appointed but also had an occasion of sacking a few for gross misconduct in delaying the examination processes.

Finally, there was the students union, the Jomo Kenyatta University Students Organization (JKUSO). Students' unions in Kenyan public universities were banned because of their riotous behaviour. The ban was later lifted conditionally. When I was appointed the principal of JKUCAT in 1992, I decided to convince the Senate and the Council to allow me initiate the formation of a students' union. I was given the green light.

I appointed a team of three senior professors and two students to start the process. Students had approached me and requested for an organization through which they could air their views. In my mind, I knew that the organization could be both negative and positive in its functions. I was totally right. The dean of Science, Prof. Job Ndombi (deceased), chaired the drafting group and came up with a very workable constitution. We presented it to the senate which endorsed it and eventually the Council ratified it. This was in 1992.

This was the only organization which was officially functional since the other universities had banned theirs because of excessive riots. From 1992 to 2003, I conducted annual elections for the student leaders. I must admit here that there were times during the campaigns when I felt like disbanding the students union. But I put on a brave face and managed to control the would-be students' unrest. I had a few very bad secretary generals and chairmen.

By and large, JKUSO was indeed instrumental in maintaining stability at the university. They kept senior management on the alert and I had fulltime deans of students. Looking back now, JKUSO was the best students' entity I ever had because they made genuine complaints which we could attend to. One thing they knew me for was that whenever I said "yes" to a request, I would deliver; and whenever I said "no", I sincerely meant it and would not change my mind. I was firm in my decision and could not allow for any canvassing because I could not relent.

Consider the request that the VC should provide buses to ferry students from downtown Nairobi to Juja every morning and evening because they had seen a rich private university do so. I told them that this was untenable not even for the staff. Some of the requests were so ridiculous that I would tell them in their faces that even their sponsors would not be amused. I however enjoyed the students' leadership and they did not doubt any realistic promises. At times they were more supportive than some of the lecturers.

If I were to apportion my consultation ranking in terms of percentage and importance of each organ, the simple tabulated summary below reflects my feelings. It highlights their importance, dependability and time of consultations.

Department	Faculty	Senate	Senior Management	Council	Total
25 %	20 %	30 %	5 %	20 %	100 %

If any one of the above setups failed or jammed on a Vice-Chancellor, then there would be chaos. It was my secret management weapon which I used effectively to succeed in decision-making. In fact to ensure my collegiate belonging, I identified myself with one department, Horticulture, as a staff member. I taught and examined a course.

I used to attend departmental meetings as a bona fide lecturer, but not at the faculty. I also supervised two PhD students who were under staff establishment – Dr Leonard Wamocho and Dr Wariara Kariuki.

My interaction and presence with lecturers demystified my office without compromising my executive role. I remember my department chairman proudly telling the Senate that my marks were submitted on time for a course I had taught.

He was indeed my chairman and I respected his memos on examination marking and submission of results. I later humbly re-joined my original University of Nairobi, the Department of Crop Science and Crop Protection. I toed the line with pleasure and enjoyed seeing my young lecturers run the show.

What am I saying? Lecturers are very accommodating once they are accorded opportunities to relate with their bosses. I was easily understood and my occasional outbursts did not offend anybody but built a cohesive, great institution.

15

Scholarship and Academic Community Service

At some point in my career, I had to get back to my rural home and provide some services to the community. In Kenya, many city dwellers are either from the rural settings and have simply come to settle in urban areas or some are born here. I moved from rural Kisii and happily settled in Nairobi, got married and set up another home here. I still have a comfortable home and tree plantation in my rural Nyamasega village. The home has trees on it and has a serene environment.

As a family, my rural house serves as a retreat during holidays. The amenities like power, water and access roads were not there. Whenever we visited home, my wife and I had to prepare for lighting equipment and water provision. I decided to apply for power which I paid for and the whole community of over 1,000 residents benefited from this service. It was expensive in 1985 to pay for power and tap it from over four kilometres from the mains. It was my first rural project which I am still proud of.

I also provided the first telephone line connection to my house which the community used freely in cases of emergencies. It was later replaced by the current cell phone system. But I still paid for the landline alone and maintained it by paying the monthly charges.

As a community leader, I initiated and set up one of the best performing SDA Boarding Primary Schools in the area. The school started with a few students who first used my house for classes and the teachers' house, and then we later built up a boarding school on a public communal land. The school now has over 500 students. I organized fundraisers to build it.

Within its vicinity stands a moderate health clinic with basic medicines and equipment for the area. It took me several years to set it up and I also spent our family resources and time to apply and pay for licences. The area leadership was always negative for such development and yet the community deserved it. I am proud to date that I succeeded in constructing the facility.

Next to the health clinic stands a chief's camp office which I put up again for the community for closer administrative services. The initial temporary building was put up by me as we planned for a more permanent structure. The community has remained grateful for my initiatives since they enjoy using the facilities.

The secondary and primary schools in Nyamagesa Village are government-sponsored but I still provide for essential books. I have managed to put up one classroom alongside the main block, and have provided power to the classrooms. I do supply required books to many schools in the area and beyond as I encourage all students to work hard and pass their examinations commendably.

Other contributions to the Nyamagesa community include fundraising for church buildings, financially assisting needy students who have done well in their examinations. As a family, we always give financial support to young people from the community for their university education. I also give free advisory services whenever I am available.

I have been requested to construct tanks for clean water provision, but have not been able to do so. Plans are underway for the project. This is because of the complex nature of our terrain and the need to involve everybody in the community. Provision of electricity was easy because I first connected my house, lit it up and everybody saw the advantage. Piping for water and setting up common collection tanks calls for all interested parties. Our homes are scattered all over the hills and provision of any infrastructure is very complicated. We are still exploring the possibility.

Board of Governors in Schools

In my earlier chapters, I narrated my early childhood and education at Ibacho Primary School. I had the privilege of serving as a board chair after several years. As a university professor, I had to go back to my roots and give guidance to the young boys and girls on what it takes to become a professor. I wanted to be a mentor and pay back in kind.

When I was appointed the chairman of the school by the Minister of Education, Hon. Stephen Kalonzo, I was surprised by the amount of work that was needed to uplift the academic standards of Ibacho Secondary School, my young alma mater. Many schools in Kisii and Nyamira counties performed badly. Anyone who chaired any school's board of governors was expected to raise the standards of academic performance. Failures at all levels were and are still at an all-time low.

The community was delighted to have one of their alumni head Ibacho Secondary School. My four years as a chairman were educative, revealing and disturbing. The cause of poor performance in examinations was a common denominator in many Gusii schools. I had previously followed annual results and marvelled as to why Starehe Boys or Alliance Girls performed so well that they beat many Gusii secondary schools. My first meeting at Ibacho was to ask the board of governors and teachers to write a paper on how to improve the standards. It was a long explanation why poor results were the norm in Gusii schools.

I learned several things during my first board meeting; the composition of board members was gross. Persons who had never gone to school formed the bulk of members. They could not advise on any academic matters. The teachers were not bothered to teach. Many were untrained, and ran businesses for extra income. Unqualified students were admitted to higher levels without adequate preparation for the next classes. Finally, the head teacher was so weak that he could not control staff with regard to time-keeping and absenteeism.

One of the most disturbing contributions to failure was political interference. The area members of parliament had strong influence on who served as the head of the school and who served as board members. They had to be diehard sycophants of a particular member of parliament regardless of their qualifications and competence in running the high school.

The poor academic performance in Gusii schools necessitated using my family resources to conduct a survey and come up with a publication which I presented and distributed to schools within the region. My strong plea that we should have qualified staff was heeded and there was a slight improvement in performance. My other intervention was the supply of reading materials and the completion of syllabi.

I organized several academic day meetings with the area members of parliament and discussed openly what I had found out in my survey. I was categorical in implicating and condemning the heads of schools that produced poor results. I also highlighted my findings in many leaders' meetings whenever I had an opportunity. I bought books and supplied to some schools which later produced appreciable results. Specific recommendations that I made were implemented and gradual change for better performance was seen. I am proud that my small input in the promotion of education yielded fruits.

I also served as a board member of Kisii High School, my secondary school alma mater, and Nairobi School. The differences in students' performance were so diverse that I made one conclusion: teachers and especially the head teacher made a marked difference. A disciplined group of teachers and pupils, provision of reading materials and timely coverage of the syllabi were a necessity for high academic performance.

My presence and advice to the youth stimulated many and I met them later as my students in the university. The little book I wrote stimulated many and it revealed the rot in our primary and secondary schools. The findings would apply to any schools within the African continent. It is at this level that science and technology should be emphasized to trigger innovativeness in the minds of the youth. I have always stressed the importance of good performance in science-related courses for national development.

My other service to the community involved church construction. I wanted to demonstrate my role to the youth by putting up two campus churches. Kenyatta

University (KU) Senate provided some land for Seventh Day Students to construct a church. KU is where I am the current Council chairman. Until 1995, students used to meet at a family's house.

When I was told about the provision of land, I encouraged the students to plan for the construction of a church. I was then Vice-Chancellor at JKUAT. They invited me for a service in 1995 which I attended happily and spoke to the gathering.

Several years later, in 2013, we had another funds drive to complete a beautiful church which now stands on a one hectare piece of land donated by the university administration. The last funds drive towards the completion of the church was on 13 July 2013. My wife and I pledged to provide some of the funds. It was a privilege to speak to so many pastors and to encourage them on the critical roles that our churches play towards the well-being of the youth.

A Catholic church I was personally interested in was the St Augustine Catholic Church, just outside the JKUAT compound. In 1999, His Grace the Archbishop Mwana wa Nzeki visited my campus to talk to the students. I was invited as the head of the institution to receive His Grace, a request I graciously obliged to. Despite being a Sunday when I would normally take a break, I drove to my office and received His Grace. We had tea together and he requested me to hire a priest for Catholic students who would also teach some courses in the universities. He gave me Father (Dr) Lawrence Njoroge's impressive CV. I advised His Grace that the relevant appointing body would consider Fr Lawrence alongside others who had applied for the post. Fr Lawrence, who was then completing his PhD studies in the US, was in the country and so he attended the interview.

He was found capable of teaching some units in the Institute of Human Resources Development (IHRD) besides his pastoral service. He had to double up as a lecturer and chaplain. I later provided him with a house on the campus.

Later on, His Grace Mwana W' Nzeki introduced a subject which I took up as my own pet: to put up a church on a plot they had acquired many years before. I accepted to assist in its design and provide some basic infrastructure like water and electricity. I had a very active Catholic University Students body. Two of them were fifth-year Architecture students. I challenged them together with their lecturers to come up with a beautiful design of the intended church. In the process, Father Njoroge had reported to work and I assigned him the duty of ensuring that the church was completed.

My students came up with a wonderful drawing of the St Augustine Catholic Church in Juja which is also almost complete. His Grace sat with me several times in his downtown office to organize for fundraising. He liked my efforts and loved to work with me on any important project. What was more exciting was the fact that JKUAT students of architecture took up my challenge to them and were able to provide a free architectural design of the church. The bond between JKUAT and the Catholic Church has continued even after my departure.

Today, under the guidance of Father Lawrence Njoroge, St Augustine stands prominent outside the university boundary next to the AICAD Centre. I regularly visit the Church and the community has remained very grateful to our efforts.

I remained selfless in my service to the community as long as those involved were focused, honest and equally committed.

Perhaps a significant venture was the provision of a modern primary school which I personally undertook in order to assist my staff's children. For a long time, my staff's children were being ferried to several schools in Nairobi. They would leave Juja at about 5 am and return at about 7 pm.

I never fancied this arrangement. It was demanding on young children, cumbersome to the driver and a big concern to my staff who I wanted to dedicate their services to building the young university. I knew the agony of waking up very young kids long before 5 am for them to be transported to nursery and primary schools scattered all over Nairobi, 46 km away. Even driving to Nairobi and returning was in itself a real challenge.

At times I would be called that a van had been involved in an accident or broken down and my transport officer would be forced either to send another vehicle or hire one from town. I thought that this was a ridiculous and costly arrangement. Yet staff argued that Nairobi had better quality schools, hence their preference.

I had a solution: to put up a nursery and a primary school. I called my management staff and told them that I was concerned about the safety and health of the children being ferried to Nairobi and back every weekday. In fact nobody had thought of my idea that we could also start an excellent primary school to rival those in Nairobi.

Management gave me an okay and they left it to me to give guidance. I immediately called all parents whose children were in city schools. I gave them a talk on the negative effects of waking up nursery and primary school kids to go to school at the ungodly hours of 5.00 am. I told them that I could provide a plot for construction of a school which could first and foremost serve my staff's interests and then those of the immediate community.

The parents endorsed the idea, supported me fully and I requested for an interim school planning committee. They chose an excellent committee from amongst themselves which included my finance officer and registrar. I served on the committee as an advisor and patron. I personally attended several meetings during the planning stage. Soon the project was on its feet. Within one year, we put up a block of classrooms through parents' contributions and I terminated further ferrying of young children to Nairobi in search of quality education.

The JKUAT Academy, the name we adopted, became number one in the division. I was again personally involved in the selection of the teachers to the school. I needed to know their past performances. The staffing officer in Kiambu was my friend, and I promised him that parents would indeed top up the teachers' salaries as an encouragement.

I could not believe that after a mere three years, the academy was among the best in Central Province and we were limiting new entrants. There were several schools in the vicinity, but they performed KCPE poorly. Many of the teachers were my lecturers' spouses. And if there is any set-up I am still proud of, it is the JKUAT Academy. It stabilized my teaching and support staff and the community around the university.

I learnt several lessons in the process of building the primary school. I had solved the anxiety of parents and children from early wake-ups and late arrivals, and also saved them from the highway uncertainty. Good quality education could be attained in a short period of time if we earnestly tried. And most importantly, perhaps, I stabilized my staff and made them more committed to serving the university by providing them with the most important facility for their children. They could now not transfer or quit the workplace for lack of quality primary education for their children.

The community was impressed by my visionary approach to education. I was in both the university and primary school set-ups. The end result of this noble venture was that I created a conducive working environment for my staff. They were now able to concentrate on their work and produce quality work for me in turn. The constant worry about their children's safety was no longer there. They could now teach, conduct research and publish – the key indicators of productive work. Value and quality for my time was achieved and my staff had more time to themselves.

When I reflect back on this specific venture, I conclude that one can hit a target of set values within a short period of time. It is not possible to hit a target you never set. You only need an initiator who remains focused on set values. We could have easily taken up an existing primary school and boosted it up in terms of books, staff, water and classrooms. But we would never have become true owners of such a school. Perhaps we would even have been seen as intruders!

Apathy is the biggest destructive and retrogressive element which is hard to change in our society. That is why I enjoy taking risks and trying them. But I always weigh my options. Human beings always need an initiator and an achiever. Then the rest falls in place.

One similar development I would like to share in this book is the establishment of a university staff residential village. This project was one of those meant to induce staff and have Juja town as their permanent residence.

Real estate in Kenya is the only thing that an elite person can afford invest in. Land and houses are an ever-increasing demand which many financially able citizens are nowadays engaged in. I knew for a fact that many members of staff wished to own property near their areas of work.

There was our neighbour, Ms Mary Mburu (now diseased), who had a flower farm next to our campus. She had an excellent rose farm and a school on her land. I knew her well as an excellent neighbour who always invited me whenever there was a function in her residence. She also allowed students' attachment in the farm. Her brother, Mr. George Muhoho, a former chief executive of Kenya Airports Authority, and a brother to Mama Ngina Keyatta (the mother of Uhuru Kenyatta, Kenya's fourth president) was also a good friend of mine. He was a chairman of Mang'u High School's board of governors when my sons were students there.

One day, in 1994, Mary Mburu offered to sell part of her land to us. My predecessor, Prof. George Eshiwani, initiated the process in 1992 but shortly after he left to become Vice-Chancellor of Kenyatta University. Mary and I held some discussions on the same and I was convinced that we needed to pursue the offer to a logical conclusion. I had to move fast and set the negotiations in place.

I called my chairman of Council and briefed him as this project would involve substantial financial outlay. He gave me an okay to proceed with further exploratory surveys in the area. I set up a committee comprising the DVC (administration and finance), the finance officer, and a few other members. I was a member of the committee but the DVC took charge. I sounded senior management and called the Senate to inform them about the same.

The land was about 25 acres to be purchased by interested staff members and subdivided it into an eighth, a quarter and a half of an acre. The price was agreed on by the land owner and my committee. There was a lot of resistance from some members of staff.

The final price was eventually decided on and the Council approved the purchase of the land using the salaries of those staff members who showed interest. The purchase price was to be deducted from the salary for those who showed keen interest. Several meetings were held to iron out modes of payments which included legal fees, surveyors' fees and stamp duty. I then drew up a final list of those who were indeed interested in buying the plots. Payment schedules were drawn up and agreed upon.

We got a surveyor who subdivided the parcel of land into eighth, quarter and half-acre plots. Every university member who wanted whatever portion qualified as long as their check-offs could sustain repayments. This was a project which I had been focusing on as realizable several years ahead. We made bye-laws and specified the kind of residential units to put up. We also left adequate space for future university development, some space for a school and playground.

Again the staff residential site stands unique as dream houses have been erected by members of staff. Overall, we all benefited from the initiative. I got my half- acre plot.

Setting up a staff housing scheme was a fulfilled dream to several members of staff who could afford to buy the plots. My initial idea of staff recruitment and retention was now being realized in earnest. Although not all staff could afford to buy, the few who succeeded remember me with deep appreciation.

Many have built their dream houses and have permanently settled there as they continue to perform their noble duty of serving the institution and country. Their children attend JKUAT academy, a few minutes away, while they lecture or carry out other duties on campus. This was our promising environment which, in my view, was perfect for a young nation like ours where competition for property had become the norm. I certainly succeeded in this capital investment venture which remains prominent on the Thika-Nairobi Super Highway. The current market value of the property now is unbelievable.

The Juja business community had confidence in my work. They respected me and we engaged in some capital development advisory services. I told them in the several consultative meetings we held that, one day, JKUAT would turn around the business targets of the area. For example, the staff working here would be many and student numbers would soar. They would therefore need living houses and hostels, so Juja would eventually become a university town. I urged them to consider developing it.

As a service to the community, I offered them free of charge architectural designs of universally recommended students' hostels. I knew that no matter what we did, the university would not invest in putting up more hostels. We could not match the high demand. The number of students would be overwhelming.

I told the business community to put up standard residential houses for students. I even detailed in the design, the number of students per flat, how they would share common facilities like the kitchen, sitting rooms and a small number of shared bathrooms. I drew comparisons from the western countries where students' dormitories are very simple in design but neat and easy to manage.

The idea was taken up by financially able investors and I understand that thousands of students now reside in Juja town, and have reduced the accommodation pressure on the universities. The town was initially isolated, but with the new highway, we now have a fast-growing metropolis connecting the city of Nairobi and Thika.

My reflections on what I did to retain staff can be summarized as follows:

> "Human beings must be appreciated in order to serve an institution with dedication. A good working environment is a fundamental prerequisite for quality service."

The family needs the staff and staff welfare in general were of critical interest to me. But patience pays because the benefits of a well-planned venture are witnessed several years later.

My sincere advice on housing development by the financially able investors is now bearing fruit. The staff housing scheme in the university's proximity is now an area considered the up-market of Juja and the Nairobi environs. Property value keeps on escalating and the investment made by risk-taking colleagues is bearing fruits. My vision and positive selfless actions are slowly unfolding for the general good of the entire Juja community.

Academic Survey

As part of the community service, I was requested to chair a team of scholars to find out why children in Gusii schools performed poorly in national examinations. As a young man who had grown up and studied in Kisii, I agreed to undertake a survey and come up with the causes of poor performance. I was happy with the task.

The genesis of the survey was a meeting in 1994 of the area academics and members of parliament. They noted that mass failures were the norm for hundreds of children in Gusiiland, now Nyamira and Kisii County. In 1994, I assembled my team; we drew up the terms of reference and embarked on the survey. We met once but realized that there were no resources for the project. I offered to conduct the field survey using the family resources. I knew that the team would not complete the exercise if there were no funds to use.

I designed a questionnaire which I was going to administer to 70 public schools at primary and secondary levels. I personally visited the schools and issued the questionnaire which included, among other things, some indication of time management by teachers – especially the head teachers. It was a detailed survey which covered many aspects of poor performance. I was able to compare results of several schools in the community.

I came up with a monograph/booklet which I shared with the community. It was entitled, *Improving Academic Performance in Gusii Schools*. It was widely circulated to the stakeholders. It revealed virtually every factor which was contributory to poor performance.

In Chapter Six of my booklet, I proposed 17 interventions and recommendations which would reverse the trend. To mention but a few: poor school management; rampant staff and student absenteeism; inadequate coverage of the syllabi; poor child foundation; lack of reading and teaching materials. The 36-page booklet makes recommendations which have been revisited by other schools nationwide. The problems are not confined to Gusii schools only, but also afflict other parts of Kenya.

My 1994 study corroborates with the recent survey of World Bank, as relayed in *The Standard* on Saturday, 2 July 2013, almost 20 years after I did my survey. The World Bank Survey of Service Delivery Indicators (SDI) for Kenya states that children in Kenyan primary public schools are taught for 2 hours and 40 minutes a day instead of 5 hours 40 minutes. The study which was very elaborate condemned the teachers, including the head teachers, for absconding classes. It said that seniority and years of training did not correlate with higher levels of knowledge. The survey had a larger sample size than the one I did. It focused on 300 private and public schools while I surveyed 70 public schools only. The very damning report concluded that, "while teachers at public and private schools were likely to show up for work, public school teachers were 50% less likely to be in class."

My findings had the same conclusions as those of the World Bank almost 20 years later. The comparison between private and public schools was interesting because the former pay well for work done, hence the commitment. I was shocked at the findings and the fact that nothing much had been done to arrest the situation several years later. I have continued to publicize the report and share it with opinion leaders.

My observation on this important subject is that there is no substitute to working hard. Success comes with sacrifice. Parents, teachers and children must know that education brings about innovations. The country needs leaders and educators who make a difference in national development. Also, early foundations to the nursery and primary school-going children can determine their ultimate level success. Time is not static, and cannot be rewound. Hence, teachers have to account for their services. The time allowed to cover syllabi can determine the quality and the pass-rate of the children.

16

Exiting University Administration

My tenure as a Vice-Chancellor was eventful, rich and fulfilling. The ascent from a deputy principal to the rank of a university leader had its consequences. My interaction with all types of individuals within and without the university precincts taught me various lessons and hardened my resolve and approach to issues.

The sudden news of my appointment to JKUAT, which is discussed elsewhere, took the same mode on departure. I was taught one thing by my father when I was growing up. He always said that as one grows up, one expects the very best in life. The key word here is "expects". He would then remark, *"Young boy, life ahead is full of challenges and big responsibilities."* I literally witnessed and experienced the same.

The 2002 Kenyan General Election

Kenyan politics are unpredictable. I grew up hearing of the Founding Father of Kenya, Mzee Jomo Kenyatta. He was the first Chancellor of the University of Nairobi. His two Vice-Chancellors were Dr Josephat Karanja, who became the vice-president of Daniel Arap Moi and Prof. Joseph Maina Mungai, who served under both presidents. Graduation ceremonies at the university of Nairobi were pompous, colourful and unifying. It was the only university then and it was mandatory that the Chancellor personally presides over the ceremony.

I worked under President Moi for virtually all the period I was a Vice-Chancellor. Vice-Chancellors in Kenya's five public universities were appointees of the president. These positions were therefore political in nature. The individual university acts specified so. My VC colleagues were all KANU followers.

That was the only party for a long period of time, from independence in 1963 to 2002. In fact no one survived an appointment without being a staunch KANU follower by action and/or expression. I happened to be a silent follower, which almost cost me my second-term appointment as a VC. I suspected I was only reappointed because there was no shortcoming regarding my work to stop Moi from renewing my tenure for a second term.

Politics and manipulation played a role in the appointment of Vice-Chancellors and heads of other parastatals, as well as permanent secretaries. I considered the sycophancy as the ugliest culture in our Kenyan society.

Elections were conducted in 2002 and President Kibaki won, becoming the third president of Kenya and, by law, my new Chancellor. The status quo remained for a while and life continued normally. We all celebrated President Kabaki's victory and had big plans for pushing for many pledges as politicians had promised during the campaigns.

My university always held its graduation celebrations between February and May each year, several months before the other sister campuses. As earlier stated, we did not experience strikes and closures as was the case with the other universities. JKUAT had a predictable timetable, a working almanac and students knew that they would indeed graduate within their stipulated study schedules. That was my pride and celebration for the 13 years I was at the helm of JKUAT.

The Shortest-Serving Chancellor

My graduation ceremony date was set for 24 March 2003. Council and Senate approved the date and expected I take charge of the rest. The date was soon after the 2002 elections and President Kibaki took office on 30 December.

We settled for the date with state functionaries and it was my duty to arrange for the ceremony. It is instructive to point out here that the remaining four public universities' graduation ceremonies would be presided over by President Kibaki in his official capacity as the new Chancellor. But their turns would come much later after ours. This was the understanding as per the individual university Acts. It is also worth noting that President Kibaki came into power on a Democratic Party ticket and KANU was no longer in power. It had been thoroughly humiliated at the 2002 general elections and many of its staunch supporters did not register any wins for their favoured candidates.

During the 2002 presidential campaigns, Mwai Kibaki had a road accident and one of his legs suffered fractures which had far-reaching effects even after recovery. I have had to recall this because we had to walk some distance to Graduation Square and also climb up the stairs for lunch. I had also spoken with him during one of my visits at the State House and confirmed that all would be okay despite the fractured leg.

Elaborate arrangements were therefore necessary to convince his aides and security that His Excellency could access both venues with ease. We succeeded in this assurance. And I had direct access to Mwai Kibaki just as I had had to the former Chancellor, Daniel Arap Moi. But this was for a brief period.

Elaborate preparations were in place for the 10th graduation ceremony at JKUAT on 24 March 2003. I saw the then State House comptroller, Mr Kereri Matere, and secretary to the cabinet, Mr Francis Muthaura, who were supportive of the function. This was going to be President Kibaki's first public function after winning the general election. It was, therefore, an opportunity His Excellency' to share his views as he made his inaugural speech. The Council and the Senate were happy that we were allowed to go ahead with the 24 March graduation ceremony.

Robing the Second Chancellor

In 1994 when JKUAT was declared a fully-fledged university, I had had the privilege of robing the president and awarding him an honorary degree of Doctor of Science (*Honoris causa*). It was therefore an easy task for me. The Senate had met under my chairmanship and had agreed to award President Kibaki the said degree which was also endorsed by the full Council chaired by the acting chairperson, Dr Davy Koech, since Dr Stephen Mulinge had passed on. The other details like the citation and making of the gown were left to me and my top management to deal with. Luckily, I am a fast mover and time-conscious when it comes to accomplishing challenging tasks.

I made an appointment to visit the State House to take the measurement of President Mwai Kibaki for the ceremonial gown and also to request for his CV for the citation. I was granted permission to take his measurements, assisted by a university tailor. I was also given his draft CV with all details on his bio data. I set up a committee chaired by the Deputy Vice-Chancellor (AA) to polish the speech and citation appropriately.

When I visited the State House, President Kibaki was very receptive and responsive. He called me into the room where he was recuperating after the road accident. We had very good discussions and we even recalled when we used to meet in the past.

We particularly referred to the time when he came home to Kisii for the funeral of my late uncle, former minister James Nyamweya, and how I was host to him and Mr Joseph Munyao. He recalled the heavy rainfall we experienced on the burial day. All along, I had kept him and his company until the funeral service and burial ended. During this State House visit, the first lady, Madam Lucy Kibaki, was around and so she joined us for tea.

The university tailor and I took the measurements for the presidential gown. I collected a copy of his CV from the secretary and off we went. Mr Kireri Materi and Mr Francis Muthaura were in attendance all through, cracking jokes. This was a month before the graduation time. The gown had to be ready in good time.

On 25 March 2003, just one day before the big event, I took the gown for the occasion to the president. I again met Mr Kireri and Mr F. Muthaura there. They informed me that they were on their regular briefs to the Head of State of state. My mission to the State House this time was to brief the Chancellor-to-be. I had to explain to him the role of Chancellor and how I would robe him. I feared any embarrassment during the prestigious occasion.

Mark you; this was the first time for him to appear to preside at such a function. Moreover, he was still limping since the accident. He had not fully recovered, and so I had to be very careful. I actually had to perform a brief rehearsal in State House as we recalled the past meetings we had had.

At the same time, we talked about the meeting I chaired in Green Hills Hotel, Nyeri, where he attended with all MPs from Central Kenya to discuss the province's development agenda and politics. I chaired it because I was a Vice-Chancellor of a university located in Central Province.

In fact during this meeting, I slipped off a stair and had a fracture on my right foot. I had to use crutches for six months. I recall travelling to Nakuru from Nyeri and my foot kept on swelling without my realization that I had suffered a fracture on metatarsal five.

During that meeting at the State House, Mr Kereri Matere was talking a lot and he dominated the conversation. That was perhaps his nature. I remember him asking about high government positions for his people. I did not know how to respond to the demands since Mr Francis Muthaura, who was the head of the Civil Service was in a better position to assist. However, we shared with him details about JKUAT and its programmes.

I kept my informal talk brief because my aim was to get back to the university and finalize preparations for the next day. I knew that if there was any hitch, I would be the ultimate victim. Good planning of events reflects seriousness of the respective management. I left State House at about midday and went back to JKUAT, very anxious to witness the installation of my second Chancellor the following day. This was going to be another memorable moment and I really looked forward to it.

A Terse News Announcement

On the eve of our tenth graduation ceremony, I got a phone call from a colleague asking me if I had listened to the 4 pm news. I had not. I wondered why he asked me the question.

I decided to continue with my preparations and wait for the 6 pm news brief. I did not imagine what kind of announcement would be aired at 4 pm other than that of the graduation ceremony. Normally, when the Head of State of state is about to visit or tour a place, his itinerary is announced to the public. I thought that it was my function that my colleague was perhaps referring to. It turned out not to be the one! Instead, it was a bombshell: that I had been removed from JKUAT as a Vice-Chancellor and appointed to head the little-known parastatal called the National Environment Management Authority (NEMA) as a Director-General.

Confusion Reigns at the Eve of Graduation

My appointment to JKUAT as a Deputy Vice-Chancellor (academic affairs) in December 1989 was a sudden announcement. It was a pleasant 1 pm news bulletin on a national radio station, Kenya Broadcasting Corporation, and several other stations. A colleague of mine relayed the message to me as I was winding up a lecture to my third-year of a crop protection course, weed science. The announcement for me to quit JKUAT for NEMA on 25 March 2003 was, again, relayed at 4 pm on the local radio station, KBC. Again, a colleague passed on the message to me. This was a huge surprise.

First, I was not sure whether the message was meant for me or some other Vice-Chancellor since I had a graduation ceremony the following day, 26 March 2003. Decency dictated that I should have been informed several months earlier to prepare for a respective departure. The news went around the country like a bush fire and I got calls from everywhere soon after the 7 pm news. I was still at the campus putting the final touches for the big day.

I called my wife, Esther, immediately and told her of the interesting but ridiculous timing of the change of leadership in JKUAT. She advised me not to quit but to continue with the preparations. I had also decided to go on and ignore the announcement for the time being. It was an announcement after all and arrangements had been made under my general guidance. I had to move on. It was much too late to alter anything!

I called the acting chairman of Council, Dr Davy Koech, and passed on the message. He had already received the news and was not sure which way to go. In fact he suggested that we do further consultations and I should not proceed with any further preparations. This was the only time when I totally rejected his proposal and affirmed my stand, to go ahead as planned and only take note of the announcement.

First, it was too late to change the programme and reverse the graduation date; parents had already arrived and camped in the compound. Second, we had also announced to the whole nation that there would be a graduation ceremony the following day. Third, all the graduands and their parents had already travelled to Nairobi and the environs of JKUAT for the celebrations. Most importantly, I had put in my time and energy for the ceremony. The Council and the Senate were set to receive the president.

All along, I kept asking myself what this last minute announcement meant. Was the abrupt change meant to embarrass me? Ridicule me? Test my tolerance or simply throw me off balance? None of these was evident. I vowed to perfect the ceremony notwithstanding the changes! It was now seven in the evening and I was still at the campus going round the very fine details like receiving the printed citations, ensuring that high table chairs were correctly labelled with names of distinguished guests. Most importantly, I had to assure the staff and students who may have heard the shocking news that all was okay and our graduation ceremony would go on as scheduled. I was focused on this matter.

I left the office for my house in Spring Valley in Nairobi. The driver dropped me and we agreed that he would come for me the following morning by seven. I found my wife, Esther, and our children having supper but evidently worried about the unfolding events. I was at ease with myself as I got into the house and repeatedly told them that we would continue with our ceremony despite the announcement.

My wife served me dinner and she may have noticed some concern on my face. But I still assured her and the children that there was nothing to worry about regarding the events of the following day. I had not received any communication from either the Minister of Education, George Saitoti (now deceased) or from State House operatives. In my case, my staff and students were all set for the event. As usual, my wife had to attend the graduation ceremony.

Despite all these personal assurances, several phone calls were coming in as people enquired whether there was still a graduation ceremony. I firmly said yes, please attend. My invitees doubted the legality of the function. But I still insisted that a letter and gazette notice were the final authority to adhere to.

I did not imply that I would not leave JKUAT. I was simply saying that my plans to have a successful graduation ceremony ought to go on! Period! After which I would gladly move on to NEMA.

Back in my house, I rehearsed my speech as usual, which I thought was very good, touching all facets of national development and industrialization. I also implored the new government to look into the plight of university staff in terms of poor salaries and continuous brain drain. I amended my speech accordingly, toned down on some aspects of my future plans and assured the graduands of a bright future.

I made the speech shorter than I had originally planned. I called the acting chair of Council and all Council members to advise them to attend the ceremony. I made reference to the announcement and further assured them that it would be futile for us to cause confusion on the eve of the great day. I then retired for the night.

The following day my driver, Mr Joshua Maina, arrived early and off we drove some 46 kilometres away. I left my wife at home, expecting her to join me later at the graduation square. The morning of 26 March 2003 was more uncertain. I found several students outside my office asking what was going on.

I addressed them briefly and told them that my term as their Vice-Chancellor had come to an end, and I had to move to NEMA after the graduation ceremony. They wanted to demonstrate against the move, but I advised them that changes were normal although they called for decency in effecting them. I was addressing students who still had more time to stay in school.

I told them that the graduands deserved peace and enjoyment as this was their ceremony, together with their families. I pleaded with the students to let me complete my day's programme successfully and thereafter proceed to NEMA. It was not a big deal but simply a change in roles, I asserted. The students then dispersed and joined the rest at the graduation square.

As the morning progressed, invited guests started to arrive. Several dignitaries, parents and guardians seemed to quietly talk about the previous announcement.

But the only person who was openly furious about the move was the then Minister for Education, Prof. George Saitoti. He approached me and stated that he had not been consulted about the transfer or change. He thought I had all along known but kept it a secret. It was not true. I told him that the announcement came as a surprise to me too. I, however, assured him that all would be well. In the meantime, my wife arrived and I acknowledged her presence as she was ushered to the pavilion by staff.

At about 9 am, my successor, Prof. Nick Wanjohi, arrived at the main administrative block. I had never met him before. He walked to me and introduced himself as the new Vice-Chancellor who was to take over from me. He actually thought that I would summarily hand over to him my graduation attire, the speech and all other instruments of authority immediately.

I was very diplomatic and handed him over to my registrar (AA) to usher him to the pavilion just like all other dignitaries. I knew protocol would not allow me to just surrender the whole graduation programme, including the inauguration of the second Chancellor of the University, to him since he had not officially taken over as VC. He was escorted there as the education minister, acting chairman of Council, Vice-Chancellors and myself waited in my office for the president to arrive.

The president and his entourage arrived and, as usual, we received him. He limped a bit as he walked and I talked to him, first wishing him a quick recovery and welcoming him to the great University. He was accompanied by the first lady, Lucy, whom I had met earlier and so it was easy for me to relate to her. We all sat in the room and tea and soft drinks were served. Again no reference was made to the immediate changes. I remained cool and extremely composed all along in my interactions with the Head of State of state. Maybe he had forgotten the announcement as everything seemed normal.

After a few minutes, when I noted some silence, and all had taken some drink, I announced that we all parade and form a procession. It was not a Chancellor's procession yet because he had not been robed as the new Chancellor. The rest of us were already dressed in the graduation gowns. I had told His Excellency the same the previous day and so he was aware of this procedure.

We all marched to the graduation square and took our seats. Being my tenth graduation ceremony, I was fully aware of the proceedings and protocols that such an event called for. Citing the relevant section of the University Act, I announced the 10th Congregation of JKUAT graduation ceremony. I further announced the Inauguration of the second Chancellor of JKUAT. The crowd sat pensively as I made the announcements.

After the national anthem by the Police Band, I read the relevant section which allows the Senate to award an honorary degree. I read the P resident citation and requested him to please stand up and step forward. The Acting Chairman of Council and myself robed the president and declared him JKUAT's Second Chancellor. The congregation appreciated him as the Police Band played fanfare music. He was my first graduand that time and we were ready to proceed with the rest of the formalities.

At the same period, the Senate had identified Prof. Risley Thomas Odhiambo (now deceased) for an honorary degree for the great scientific and innovative work that he had done. He was the founder of the International Centre for Insect Physiology and Ecology (ICIPE). Senate was very selective in its award of honorary degrees. We wanted to ensure that quality was the driving consideration in all our academic undertakings.

I was the first one to make the graduation speech before the Acting Chair of Council, Minister of Education and the new Chancellor. I confidently stood up, bowed to the newly-installed Chancellor and proceeded to the podium. I knew that this would be my last official speech in this university. The nation was perhaps watching my reaction keenly and monitoring what I would say as my parting words. They were in for a surprise. Normalcy prevailed!

I had changed my speech to highlight the many positive things that I had accomplished. I praised the international community for their support in various ways. I particularly singled out JICA and the Japanese government for the grants they gave the university which had assisted in technical, capital and human development.

I emphasized the university's financial gains because we now had substantial funds in our bank accounts. I mentioned by name the PhD students who were going to graduate that day as part of my great effort for availing scholarships to them. I also gave a brief narration of the successful story of placing JKUAT students. My milestones were many. Availability of jobs for my graduates was not an issue then.

We trained quality and not quantity. I gained immense courage and thrashed at the low salary packages for lecturers in public universities, and got applause on this score. I further deplored the unfortunate trend of Kenya's increasing poverty due to poor planning. Again I made no reference to the changes. I had known all along that public appointments were temporary and it was time for me to move on. This was perhaps one of my best speeches that I had ever delivered to a graduation congregation. The Chancellor spoke and made no reference to my job transfer. Later, my staff and students highly commended me for the wonderful and inspiring speech.

The graduation ceremony went on smoothly and all those scheduled to speak in the programme did so. I then called my Deputy Vice-Chancellor (academic affairs) to request the deans and directors to present the graduands for the award of diplomas and conferment of degrees. Within a period of two hours, we were through and I declared the congregation dissolved, again citing the relevant section of the University Act.

We left the podium and went for lunch. It was a bit difficult for the Chancellor to climb the stairs, but he slowly moved and made it to the dining room. The First lady, Mama Lucy, assisted him climb up the stairs. I sat next to the Chancellor and the vice-chair sat on the other side of the table. We shared a few issues as we enjoyed having the meal.

I recall His Excellency telling me of the good work that I had done in JKUAT, a statement he had made earlier at the graduation square. He now repeated it during lunch. I appreciated the compliment but still felt let down by the abrupt announcement of my job transfer.

The Chancellor made a lot of remarks soon after lunch. In fact he talked more here than during the ceremony at the pavilion. At some point, Mama Lucy, advised me to request him to wind up the speech. I could not respond. I thought it would be impolite. I therefore sat pensively and simply surveyed the guests as they all marvelled at the length and the impromptu content of the president's chat. I finally thanked the guests for gracing the occasion and requested them to leave at their own pleasure. I had at least hosted my second Chancellor for a few hours while I was still in JKUAT.

My guests, parents and students stayed around taking photos and sharing the happy moments. My wife and I mingled with staff and students and exchanged pleasantries until about 4 pm when I decided to know the exact genesis of the hurried job placement. I was told by a high-ranking officer from the State House that they wanted to create a space for their staunch supporters. I was then moved to a little-known, non-functional and inactive environmental authority. That was the public's perception. That was the only reason. I had no qualms and decided to comply.

I met my successor, Prof. Nick Wanjohi, and agreed to meet within the shortest time possible for the handing over. I asked my staff in various departments to write handing over notes for me. Meanwhile, I would take Prof. Nick Wanjohi around the campus to introduce him to the staff. Within three days, all reports were ready. I was particularly interested in the proper financial accounts documentation. The university had accumulated substantial amounts of money from various savings to use for capital and staff development.

I called the Council members to witness the handing over report to the incoming Vice-Chancellor. I also handed over all the university instruments: the Mace, the Logo, the Seal, the Act and Statutes. All these were done in the university boardroom with the Council witnessing.

I packed my few personal effects and had them delivered to my house. It was a smooth transition despite the sudden indecent announcement. I had no ill feelings whatsoever as I moved onto other national engagements. I knew that every hiccup had a positive image.

Reflections

The sudden announcement of change of guard was comparable to the first news I got when I was appointed as a deputy principal. It was a good surprise in December 1989. But the 25 March 2003 announcement was done with no regard for my feelings.

I considered the move to be primitive since it lacked human consideration. Human beings have esteem, plans and visions. Any sudden destabilization, therefore, leaves bitter memories even if the action was well-intended. Graduation ceremonies require careful planning and they do consume the Vice-Chancellor's time. Any disruption of the process does affect the entire system. Hence, any abrupt transfers or appointments of any kind should surely have a human face! The appointing authorities know all this too well.

I had my mentees in JKUAT whom I very much respected. The staff and the Council members whom I worked with liked my accomplishments. My belief in leaving a place better than you found it had now made sense. Here was a campus I had nurtured for several years as the chief executive. I had learnt a lot from the process. Certainly, I was very proud of the accomplishment.

I had two great institutions in one, the university and AICAD, both engineered by me with unwavering support from the Council and the Senate. In the process, I trained several persons who now manage almost all the current fully-fledged universities. I had earlier alluded to them. They are part of my legacy, and part of my pride considering that my services at JKUAT lasted a whole 13 years.

Informal Farewell

A day before I finally left JKUAT, and after handing over to my successor, I decided to bid farewell to my staff in the departments. I felt that it was inhumanly unfair to just leave the campus without quickly and promptly seeing them at their work stations. I remember visiting the farm workers, kitchen staff, security, water treatment personnel, the library, the hospital, the examination centre, halls of residence personnel, a few academic departments and finally AICAD office. This was my informal farewell visit.

I wished all of them well and promised that I would always cherish them. My parting words were simple, *"my sincere thanks to you all."* I was only leaving JKUAT for a more involving parastatal which dealt with all matters of the environment. They were welcome to visit my offices located off Mombasa Road in Nairobi South C. It was just that simple.

17

Leadership in the Corporate World

I did not waste time before reporting to my new offices as the second Director-General of the National Environment Management Authority (NEMA). Word had gone round that a professor from JKUAT would take over the roles of NEMA from Dr Korir Koech who had been the first director-general since NEMA's inception in 2000. I knew Dr Koech as a former lecturer at Kenyatta University during my tenure as a VC at JKUAT. I used to teach an MSc environmental course in his department where I supervised his two students, Samuel Ngugi and Dr Margaret Karembu. I also examined his courses then. That was my formal contact with Dr Koech.

I drove to the NEMA headquarters five days after my appointment. I actually passed the offices and had to ask a guard for direction. The area is located in a place where one has to drive on the Mombasa Highway for a reasonable distance before making a U-turn on a busy intersection.

My landmark was the Belle-vue Cinema premises and Kenya Bureau of Standards. I got to the headquarters unannounced. The former director-general was not in; he had also been informed about his departure. I introduced myself as the new DG and asked to see the former director-general.

The lady I found at the offices was very kind. She requested me to sit down for a cup of tea as she called him on his cell phone. As I sat there, some staff members passed by and never bothered to find out who I was. A young man who used to be my students came to me, greeted me and announced to the others around that I was the new director-general of NEMA. Word went round that the gentleman whose name was announced some four days earlier had arrived to take over NEMA.

The secretary called Dr Koech and we spoke on phone. He agreed to come over and see me at the office within an hour. I decided to wait around, read the papers and observe the premises. This was around 10 am. Dr Koech arrived and we had a discussion in the DG's office. His main concern was the reluctance of the parent ministry to release funds for the running of NEMA.

NEMA was an independent parastatal of the ministry and was to be treated under the new Environmental Management and Coordination Act (EMCA)1999. I wondered why the delay from 1999 to 2003. He gave me a rundown of the issues of environmental pollution and degradation. He lamented the slow pace at which Kenya was domesticating some of the crucial international protocols. He also told me about the lack of support staff in handling environmental matters. I listened carefully and only interjected whenever I thought it was absolutely necessary.

I requested that he should write comprehensive handing-over notes and spell out all the concerns that he thought I should tackle. He agreed to do so within three days. I had no problem, but hinted to him that I was ready to consult and work with him in any relevant area. I also assured him that I would be seeking his guidance in future. Meanwhile, I told him that I was now in charge and systems must continue. I was only waiting for his final handing over notes.

Dr Michael K. Koech brought a handing over report and a few copies of the EMCA 1999 Act. I perused through the report and kept it as my reference point once I assumed duty. He did not have a board.

The new NEMA Board of Management was gazetted the same week that I reported. The chairman was Prof. Canute Khamala who was a colleague from the University of Nairobi. Other members included Prof. Mohammed Mohamud, Prof. James Kahindi, Prof. Gemano Mwabu, Dr Dorcas Otieno, Dr Dominic Walubengo, and Mr Marsin.

The Minister for Environment and Natural Resources was Dr Newton Kulundu (deceased). We later on hired a legal officer, Prof. Francis Situma, also a colleague from the University of Nairobi. This was a pseudo-academic set-up.

I knew all the board members; and virtually all of them had worked under me during my tenure as a Vice-Chancellor of JKUAT. It was a kind of reunion. I was again their chief executive, the director-general. They were by law my employer. Some of them had been my external examiners and we had interacted at collegiate level.

My predecessor had not had a management board. He had therefore not been able to transact any business, financial or otherwise. I did not understand why he had not pushed for one. The government bureaucracy is so complex that if one does not push and follow through a request, one is forgotten completely. I suspect that my predecessor had just sat and had waited, without following up, for the board to be appointed. That was one reason why the functions of NEMA had been hampered.

Launching of the First NEMA Board of Management

Normally, boards, councils and trustees are launched by their relevant ministers once they are appointed. It was NEMA's turn to have its board launched once it was gazetted.

The chairman, the permanent secretary and I made plans to have the launch take place soon after the appointment. As the new chief executive, I had to make plans for the launch. This was not a problem.

My immediate concern was to understand my roles as the director-general. This was a new institution and I had never worked in a non-academic set-up previously. Before we were launched, I took the EMCA 1999 Act and read all of it. I highlighted the roles of each body specified in the Act. I picked on the most urgent requirements and prioritized them as urgent, and also noted the least-demanding ones. I also checked what other countries were doing in terms of environmental protection.

I proposed the launching date and the minister, Dr Newton Kulundu, was to be the guest of honour. He was to be accompanied by the permanent secretary, Mrs Rachel Arunga. Preparations were finalized and board members and staff assembled at the headquarters. Speeches were read, remarks were made, promises were made and the expectations of the authority were spelt out.

I was keenly listening to the speeches which were primarily directed at me, as the director-general. After the launch, the minister, board members, and the permanent secretary were paid their allowances and they left. We were now officially launched, given the mandate and powers to monitor, control, manage, and enhance Kenya's environmental issues.

Normally, the board would meet quarterly; its committees would meet once in a while on specific matters as spelt out in the parastatal's guidelines. The bulk of the promises were to be implemented by the DG and my staff. This is true even for universities and other parastatals. It behoved me to understand the laws and rules of the game.

The Minister for Environment, Mines and Natural Resources, Dr Newton Kulundu, the Permanent Secretary, Mrs Rachel Arunga, the Board Chairman, Prof. Canute Khamalla and two other board members came from the same ethnic community. The total number of appointed members was nine; and five of them came from one community.

This composition would later on have far-reaching implications on the general administration of the authority. Kenyan public appointments were so tribally-based that the vice was now haunting the society's fabrics. Boards and other senior appointments in society were, and still are, based on nepotism, on who one knew, whom one campaigned for during the general elections and on how much money one had "coughed out" towards the election of the appointing authority. One had to be loyal, familiar and related somehow. Seldom did credentials and merit matter those days despite their being clearly articulated in the requirements.

Nepotism, tribalism, loyalty and sycophancy therefore played crucial roles in the appointment of persons to key posts. Even the current appointments after the March 2013 general elections and despite the clear guidelines in the 2010 Kenyan

Constitution, professionalism and competencies were not adhered to. Also, people with financial power always get a free hand to secure all types of tenders. Such loopholes cannot enable the country to fully realize its developmental aspirations.

My Role as a Director-General

I took time familiarizing myself with the functions of NEMA. They were written in black and white. No serious work had been done previously and I sincerely did not have a reference point on which I would have based my entry. I had to start on a completely new slate.

First, I needed some parastatal autonomy from the parent ministry and had to transfer some of the staff to my authority. I needed a budget and clear terms and conditions of service for staff. I knew the roles of the board, the permanent secretary and the minister as far as parastatals were run. The head of civil service, Mr Francis Muthaura, had laid-down procedures on how the parastatals were supposed to be managed.

My first immediate headache was staffing. I had to engage staff from the parent ministry where they belonged and basically employ them. I prepared papers to justify the demand and staffing levels. The board met several times and directed me to come up with an equitable distribution of staff for NEMA and those who had to be left at the ministry headquarters.

The requirements for the director-general were so high that very few people would have qualified to fill the post. They included several years of experience, and having worked in a large institution. The person should also have had relevant academic achievements and experience in environmentally-related areas.

We had to think along these lines as we started to second staff from the parent ministry. These requirements were stipulated in the EMCA 1999 Act. It is one of the most comprehensive acts on environment and gave guidelines on sound management of Kenya's natural resources. There was one basic requirement which I always stressed to my students and whenever I make public address: *"Every Kenyan is entitled to a clean and healthy environment and we must all enhance it."* This is the basic cardinal rule as prescribed in the EMCA 1999 Act. Every other requirement on environment revolve around this statement. I again re-read the Act which created NEMA.

Another problem was to prepare and defend the budget to run the authority. Recruiting new staff to oversee the implementation of the law was also another issue. The most urgent problem was to publicize NEMA so that it would be visible within and outside Kenya. I hit the road running and occasionally stepped on people toes.

I quickly learnt the strategies that were required to see NEMA discharge its duties. I was expected to jump-start the authority. My brief from the president while I was still at JKUAT was that I should ensure environmental compliance for the country. I was aware of the loaded statement and hoped I would be accorded the necessary support.

The first board meeting made several requests for me to pursue. I took the urgent one of asking staff who were already seconded to NEMA to apply for formal absorption. All of them except one lady wrote for deployment and the board accepted them. I then signed the appointment letters after the board had met and approved the list. I never ever acted on anything until the board had endorsed and sanctioned the same. Even at JKUAT, my actions were dictated by the board or Council members.

The board approved the salary scale and other benefits which had been worked on by the parent ministry. It was slightly higher than the one that the staff in the ministry received. The board approved it. But later on the ministry disowned the figures, blaming it on the directorate of personnel. Staff had been earning the salary for over three years.

This caused a major misunderstanding between the authority and the accounting officer, the permanent secretary in the parent ministry. The blame was levelled on the board of management. I thought that was ridiculous because all monies were channelled through the ministry headquarters. The alleged accelerated payments were later adopted anyway.

The appointment of additional staff was processed by an approved consultancy, Manpower Consultants, whom the board approved and recorded. Later on, it was alleged that I was the owner of the recruiting company and used it for my personal gain. The allegations angered me so much that I got a lawyer to establish who owned it. I did not even know it until I was appointed to NEMA. I threatened to sue those behind the allegations but no one volunteered to own up.

We eventually hired competent fresh graduates who included one-third female graduates in total. The board, through my wise advice, had long complied with the balanced gender rule long before it was written in the Kenyan Constitution. I kept the philosophy of gender representation in all public appointments. Again, the board fully supported and endorsed my proposals.

The board interviewed and hired the staff; I was only giving letters of appointment and terms of service. I had done a similar management service before in JKUAT. I knew and was aware of the sensitivity attached to unfair appointments.

It was during my tenure in NEMA that we employed over 80 district environmental officers. This was one of the biggest, pronounced milestones in the promotion of NEMA's activities. We started to be recognized at the grassroots. As the authority was being popularized, I advertised for the appropriate logo and slogan just like I had done for JKUAT. We received excellent proposals and adopted one. We paid for the best competitor. "Our Health, our Environment" was the slogan we adopted.

The board was aware at all times of the events taking place in terms of publicity and documentation. Again, within a short period of time, I planned a seminar for both print and electronic media so that all matters on environment would be adequately advertised. I knew very well that without publicity, our efforts would pass unnoticed. Every requisition had to be approved prior to spending the money. I considered myself a prudent financial manager. I had left excess cash in my previous assignment as a Vice-Chancellor.

What was the reason for publicizing our activities? It is not possible to discuss all the problems that our country was experiencing as far as environmental pollution and degradation was concerned. I would like to highlight a few examples. The EMCA 1999 Act was not known by Kenyans. Not many people were aware that the law existed which gave them powers to sue or be sued by any citizen. They did not know their environmental entitlement. It was my duty as the DG to popularize the EMCA requirements.

Water pollution was witnessed in all our rivers, seas, oceans and lakes. We had to educate Kenyans on the relevant sections of the laws which governed protection of our water bodies. We gave guidelines regarding the rights of people as far as clean water provision was concerned. Later on, standards, regulations and guidelines were formulated to protect water sources.

Solid, liquid and gaseous pollution went on unchecked. Solid waste management still stands as a major challenge in all our urban areas. There is no single solution reached yet on how to control the solids in our towns and cities. This was an issue when I was a DG and it still is a problem which will hurt the economy of our country for a long time to come. Liquid waste management affects towns and cities due to burst pipes and poor connections. We had to make the public aware of the dangers associated with any type of pollution. We advised on health-associated pollutants and how the public could be protected.

The press was also persuaded to cover, where possible, any seminars, conferences, symposia and workshops related to the environment. I had a strong public relations office which was directly under me; and any news, whether positive or negative, had to be communicated at once to the media houses. District environmental officers had direct access to the local media stations. I also started the first magazine, NEMA News Magazine, which was published quarterly and covered the whole country. It was distributed in major offices free of charge. It was the equivalent of Agritech News of JKUAT which I had initiated during my tenure there as VC.

The authority also had a mandate of giving licenses for capital investments like: buildings, highways, airports and residential houses, among others. We had to give permission for all types of structures. Permission was only given once an environmental impact and social assessment (EISA) had been undertaken by competent registered individuals or firm and approved by NEMA.

Kenyans had to know that this requirement was necessary. We therefore had to license experts to carry out the assessment for proprietors at a fee. The press had to be aware of all these developments in order to educate the Kenyan citizens and prospective investors.

Protection of our forests was another important component in the conservation of our natural resources. It was imperative that NEMA highlighted the imminent dangers of deforestation and vegetation clearing. This was important especially on the hill-tops and slopes, water catchment areas and along the valleys. The need to repeat the same safety measures was necessary as part of public awareness. It was not that Kenyans were not aware of the dangers associated with deforestation. But impunity, greed and ignorance of the laws hampered any practice of conservation methods. There were even shameful cases of sacred places which had been grabbed by the wealthy for selfish, commercial interests.

The most disturbing environmental pollution was the polythene papers. NEMA was charged with the ban or control of the light gauge polythene bags which were usually scattered carelessly all over the urban areas and on highways.

The authority gazetted the recyclable bags of 30 microns. The gazette was later revoked because manufacturing companies feared loss of business and laying off of over 4,000 workers from their factories, as they was claimed. We have not managed to arrest the situation despite concerted efforts by environmental activists and lead agencies. The effect of polythene papers on our environment is devastating.

As an authority, we set up a public awareness and participation section to deal solely with educating the public, but few were receptive. I commend both public and private Universities which took up environmental challenges and quickly set up schools and departments to teach environmental courses. My constant interaction with institutions of higher learning eventually paid dividends. In fact, some Universities were very receptive of environmental programmes and I was personally involved in assisting them to draw up the curriculum. I regarded this development as the only way forward in environmental conservation.

Issues of coastal marine, coral reefs degradation and mangrove destruction were also given special attention. I realized the dangers of losing our pristine mangrove sand dunes and the green coral reefs. Hence, NEMA set up a powerful section to monitor the damage. All the coastal hotels and developments were required to carry out an environment audit and report to my office within six months' from the time I took over. I specifically made unannounced visits to some hotels and demanded to know whether they had sewerage treatment plants or they discharged their raw waste into the ocean. This was a tall order for several five-star hotels which were not prepared to comply with the EMCA 1999. Some of the hotels had some sewers but others did not.

We demanded immediate construction of sewerage treatment plants or we would go public and this would affect their businesses. The press was ready for any sensational news. Most of them complied and came up with excellent sewerage systems which discharged recycled water into the ocean.

They later appreciated our efforts and, again, NEMA gained positive publicity. We knew the damage it could have caused to our tourism industry if we went public. But we meant what we said and allowed for corrective measures. We could have not announced the rules, but just threatened to close the premises and/or take them to court. There were certain specific cases which I handled with the respective hotel managements. The environmental activists supported my efforts to the letter. In fact I had a stronger network with environmental groups than the press had. I used a carrot-and-stick technique to handle some delicate issues.

One area where we partially succeeded was the introduction of environmental classes and lessons at primary and secondary schools. I held a number of seminars with teachers and education officers to induce them to incorporate some lessons in their teaching curriculum. I had also requested the minister and permanent secretary to convince school heads to consider mainstreaming environmental issues in their curriculum. The Kenya Institute of Education accepted to have the matter considered in their normal review cycles. NEMA suggested basic topics which we thought were crucial for our country's conservation agenda. The topics were mainstreamed in the lesson plans.

But my happiness and satisfaction was when Universities heeded to my plea to have the programme introduced at BSc and MSc levels. Environmental courses are now the most popular and attractive programmes for students in many Kenyan Universities. During the publicity, I practically demonstrated the marketability of the courses by introducing several environmental officers we had employed and who were now enjoying doing their work. I planned to post environmental officers to divisions and locations to monitor the appalling land degradation.

The authority took up quarry inspection sites. There was an outcry that people and livestock drowned in open quarries. Developers never bothered to cover them; hence, the quarries posed environmental dangers. Cattle fell in open abandoned quarries; they were mosquito breeding grounds and emitted foul smell from the accumulated rain water.

The disused quarries were scattered all over the country with most of them located in and around Nairobi. I ordered that they should be secured by either having them fenced off or filled with construction soil. Each district, now county, had to identify those who excavated rock and hold them liable to prosecution.

Under normal circumstances, people who introduce or cause environmental pollution are held liable for their deeds. They therefore must make good of the damaged area. This is a common practice world over and is referred to as the *pollute pays principle*. It would have been applied to those who dug up the quarries. Unfortunately, they had already left the country or had wound up their businesses and it was difficult to locate them. We, however, made the public aware that those who excavate sand, soil, stones or any other mineral are responsible and had to refill the scarred site. We listed the affected areas and took statistics for each county.

NEMA was responsible for ensuring that any new major construction had to undergo environmental impact and social assessment (EISA). Under my watch, the regulations which were mandatory for carrying out the exercise were gazetted. At the same time, the environmental audit (EA) was gazetted. This was the first time to have the regulations, rules and guidelines which were meant for the public. It was a major intervention which had to be adhered to by the proprietors. It brought some order and a sense of conservation of our resources.

The law required that only registered experts had to carry out the EIA and the EA. They had to be registered by NEMA after undergoing a prescribed training course. I led the drafting of the courses to be taught and passed them over to the relevant training institutions.

To date, the EIA is mandatory before anybody develops any project. Environmental audit was supposed to be done on all existing firms, industries, institutions and any outfit in an area to ensure environmental compliance. NEMA-registered experts carried out the audits. I was lenient in this requirement because most institutions were put up several years ago. I only demanded compliance reports, at the same time corrective as measures on suspected environmental pollution and/or degradation were being taken.

Universities and training colleges were licensed to train experts and I was ready to provide the back-ups. Currently, we have over 5,000 full experts and a number of associates who carry out the assessments and audits. Audits are supposed to be carried out annually.

During these processes, I became a darling and an enemy of several people. But my belief in the long run was to leave a legacy of environmental compliance now and for the future generations. The biggest challenge was to deal with the politicians and well-to-do developers. They demanded licences immediately through intimidation. They used every method necessary to have me issue licences for project development. I remember receiving a threat letter signed, on the same page, by three ministers demanding to know why I had rejected the development of residential houses in Nairobi's posh Lavington area.

The same ministers, who had interest in the development had called me to their office, had a meeting with me and reprimanded me. I gently told them that the residents' association had refused any further development on the small plot, and a public hearing had been held to reject the same. I was following the law which they were party to when it was discussed and passed in Parliament. The matter stopped there and no further construction took place on the site.

I recalled my primary school days when I got a beating for reporting a teacher's wife who came late to class. But I never gave up, because for every decision I made, there were several people who supported me and only a few were aggrieved.

One major thing that bothered me was when senior officers from the ministry headquarters and others from within NEMA worked secretly to frustrate my efforts and undercut any decisions I made. I knew several cases whereby senior members of the board would incite proprietors to push for licences under dubious circumstances. I suspected corruption but always stood my ground. I was not going to sacrifice my integrity and country.

Another issue which I tackled as the DG was the electronic waste menace. I knew the country was going ICT fast and every company and institution wanted staff and management to be in line with the fast-growing digital-era nations. All sorts of obsolete equipment were being shipped into the country. Computers to schools were arriving in Kenya in numerous containers. Millions of them were donated from other countries.

A good number of them were old and had reached the end-of-use in their respectful countries. This equipment came in unchecked and were becoming functionless within the shortest period after installation. I must admit here that I had no control over their importation. The Kenya Bureau of Standards (KBS) was supposed to inspect and approve the functionality of the said equipment.

What I know is that we have millions of all types of e-waste in virtually every backyard and go downs. They are obsolete and have no use. They are an eyesore and we have no capacity to recycle or reuse them. The largest quantities of them were found in schools under the auspices of computers to schools as donations.

My role was to make the public aware of the imminent dangers caused by the components of e-waste chemicals and metals which would then affect our population, especially the children. E-wastes are major soil and water pollutants and should be handled carefully. I came up with a return-to-life of electron wastes. These included cell phone sets and batteries which usually become junk after a short period of use. The idea was considered brilliant, but currently there is no solution to this problem.

There are a few individuals and organizations which are collecting and dismantling e-waste, but the process leaves the components almost intact. Open empty boxes with exposed wires are seen in many go-downs or office spaces. The problem is far from being solved. The biggest victims of e-waste are children. as their safety becomes endangered.

I used to hold frequent meetings with my staff. I believed in direct communication with them and would leave the details to be expounded by the section heads. We used to have quarterly staff meetings and I would brief them on the latest environmental matters in Kenya and outside.

We wrote the first strategic plan for NEMA as a team and all staff participated. We used the strength, weakness, opportunities and threats (SWOT) analysis to come up with a strategic plan which would give us guidance in planning. Other considerations made during the process included political, economic, social, technological and legal implications (PESTEL). We considered that any environmental matter had some economic, political and legal implications. If our strategic plan was followed, Kenya would have been a world-class compliant country. The board was happy to have its first strategic plan launched.

We often had retreats and continuous seminars to help the staff bond and appreciate each other's work. We eventually felt that we owned the processes.

I inherited 232 members of staff from the parent ministry. Incidentally, this was about the number I had absorbed as staff when I was a principal of JKUCAT. I therefore knew the process of engaging them productively. The process in NEMA was easier than the JKUCAT one because there were no strict academic qualifications needed. Only one person opted to remain in the parent ministry. The remuneration package in NEMA was better than that of the ministry.

I travelled to many parts of Kenya and came across the pollution and degradation sites. The board employed 81 environmental district officers to oversee the districts. I used to contact them before going there and ask to visit the hot sports and hold public meetings there.

For example, we had to contend with oil spills from over-turned oil tankers on the highways. We had to mediate between disputes on quarry lands, public hearings on land sub-division in ranches, public hearings on complex structures in residential areas, numerous effluent discharges from factories and industries. There would also be some degree of insecurity depending on how I made certain judgements on issues. Many times we engaged the state administrators like the district or provincial commissioners and security personnel during public hearings. The situation would at times become volatile between the opposing sides of some of the projects.

I was the final decision-maker on controversial issues before the aggrieved party proceeded for further legal redress in the high court or in less arbitrator settings. The law was clear on the process one had to follow in case of unfair judgement.

I recall the case of a proponent who wanted to construct a five-star hotel in the famous Maasai Mara National Park. He had been cleared by environmental impact assessment to proceed and apply for the licence. I took a team to the site and found that there were several other hotels within a few kilometres apart. It was also going to be sited next to the sanctuary where elephants usually give birth.

I had to cancel the award of the licence despite numerous cries from interested parties. The hotel was not built and the foreign investor left. In any case, hotel bed occupancy analysis revealed that there was no need for additional hotels as annual occupancy was only 70 per cent for the entire year. I was justified in my rejection and the board supported me. I always kept the management fully informed of all the activities that I undertook. They were appreciative. Several other cases of large magnitude were presented to me and were amicably solved.

It should not be construed that I was out to reject any requests for development, far from that. I similarly endorsed numerous projects for development after going through the necessary procedures in documentation, assessment and approval. Only the complex cases which had not met the necessary requirements were rejected.

The law requires that NEMA conducts a whole country's environmental status and makes a comprehensive report on the same. The state of environment report is a detailed document which is compiled by lead agencies. The lead agencies, along with my staff, conducted surveys, visits, questionnaires and literature review to compile such reports. It covers the whole country and looks at all facets of environmental concerns. The lead agencies tackle specific problems in an area of their jurisdiction and document whether the environment is being enhanced and protected or polluted.

Each year's report tackled different themes. I remember that the first such report to come out covered all aspects of the environment. That was in 2004. The second one in 2005 dealt with waste management; subsequent ones covered biodiversity and climate change. These reports were the most comprehensive for Kenya and by law had to be tabled in parliament by the Minister of Environment.

It is in these reports that individual members of parliament would be aware of shortcomings in their respective constituencies. The reports were meant to monitor pollution and degradation in order to have those concerned address the issues. The report revealed and outlined the shortfalls and suggested corrective measures for each problem.

The state of environment reports were very significant milestones which I considered to be the best ever compiled. Nobody would claim that they did not know their areas as far as pollution, natural resources and poverty levels were concerned.

We developed instruments which gauged poverty levels in each constituency. We held public hearings during the data collection and ensured that whatever was captured was included in the reports. The stakeholders for each thematic group were eager to report problems which affected their well-being. The meetings were publicized and stakeholders assembled in halls to respond to our queries.

The reports covered water issues, solid waste management, refuse from hospitals and clinics, forests, natural resource, soil management, animals and crops in an area, processing industries, noise pollution, air pollution, electronic waste and any matters which would affect the environmental compliance of the nation.

The comprehensive annual reports are still being produced and cover specific concerns which NEMA considers urgent and need to be brought into focus for speedy action.

It was through these meetings that the role of NEMA became so popular and ordinary people got to know their right to a clean and healthy environment.

NEMA became a household name and my name was synonymous with NEMA. I took pride in advising Kenyans that we needed to take care of our country in all aspects of natural conservation. I warned the polluters and threatened them with court actions if they did not enhance the environment.

During my tenure as the DG, I accused the Nairobi City Council of neglecting solid garbage control and they instantly took action. I was instrumental in the closing down of the Dagoretti Abattoir, for being filthy and for discharging its wastes into the adjacent neighbourhood and the nearby river. I also took Kenyatta National Hospital to court for careless discharge of hospital wastes which was a danger to the community.

All these cases compelled the relevant organizations to respond to the charges. I involved the media in covering the processes and any action was an eye-opener to the rest of the country.

I used to get calls to address problems. I had several staff members who were assigned and responded to specific queries. It became routine for me to get at least five to ten requests per day and was able to address some and delegate others to my able deputies.

I had a team of excellent deputies who were very well versed on environmental issues. They included: Mr Maurice Mbegera; Prof Francis Situma, our legal officer; Mr. Benjamin Langwen; Mr. Reuben Sinange, a finance officer; and the public relations officer, Mr Titus Mungou. Mr Buigutt was an excellent editor for our magazines. They also had other officers below them who were active in accomplishing their assignments.

In rejecting to license more hotel structures in the famous Maasai Mara Wild Sanctuary, I explained that we already had enough human and vehicular traffic. The damage which would be caused by excess hotels would affect and threaten the sanctuary.

The eagerness to increase tourism destination was not matched by measures to preserve the famous ecosystem. The wildebeest migration, for example, attracts several visitors who cannot be controlled in pollution menace. I argued that both tourists and tour operators could be unmanaged and counter-productive as heavy traffic could take its toll on the flora and fauna of the famous sanctuary. I further argued that excess numbers would reduce the high-class tourist visitors who relish exclusivity and a pristine park, and who were ready to pay high premiums.

It was obvious that the amount of waste deposited in the park was destructive to the wildlife, their ecosystem and an eyesore to visitors. Worse still, the teeming humanity would interfere with the wildlife's routine lifestyle. We would expect stacks of wastes of all types in the park.

The problem would be exacerbated by the mushrooming of cheap campsites and restaurants owned by investors who would never pay any attention as far as pollution and degradation were concerned. This would be the case also at the watering points. At that time, I confirmed that despite enough hotel facility, there were not enough wardens and rangers to track the environmental violations by drivers, tourists and even local people selling their wares. There were guides and tourists who erected illegal campsites at non-designated areas and would drive too close to animals, destabilizing their routine behaviour. Indeed, the garbage at the park then was giving the Maasai Mara negative publicity in those countries whose residents frequented the park.

I chose to highlight the Maasai Mara Sanctuary because it is more endowed in habitat populations than any of the 22 parks and reserves. It is indeed famous for large populations of lions, rhino, cheetah, elephant, leopards, antelopes, giraffe, primates, and, of course, the wildebeest crossing. It is also the link to Serengeti Park in Tanzania which receives more visitors than any park conservancy in Kenya. If the rules were flouted, the encroachment on the ecosystem would be very devastating.

My appeal was that the Maasai Mara was unique. It had to be preserved. I knew for a fact that nowhere is nature's balance better portrayed than in the Great Migration for which the Maasai Mara is popular. The wildebeest migration into and out of the Mara is the planet's greatest scenario. It is now ranked as one of the seven world wonders. Every year close to 1.5 million wildebeest migrate from Tanzania to Kenya in search of the sweet green grass.

One cannot get satisfied of the scenery/view no matter how many times one visits the sanctuary! I have been there more than five times and I always long to go back. The Great Migration has no start and no end; it is an endless cycle. There is the need to appreciate and protect it.

Environmental pollution is a threat now. As the animals move towards the Mara from Serengeti in Tanzania, they face countless dangers from other animals, hunger, exhaustion, and harsh environmental conditions. It is the climax of Darwinian Theory where only the strong, able and lucky ones survive. The hyenas, crocodiles and lions feast on the animals, particularly the young calves which are quickly picked due to exhaustion and the physical strain of the long journey. It is estimated that over 500,000 to 800,000 are killed during one complete migration cycle!

The tour operators had to aim at a no-mass market approach to save the park. They had to go for quality tourists and not merely quantity. Human traffic is threatening the sanctuary. Pollution is on the increase. There should be on–the–spot-penalty for garbage littering, for example, as it is imposed in other countries.

There should also be incentives for anyone who takes care of garbage control, for instance. Some communities from certain countries are under fire for blatantly flouting our parks and littering garbage anyhow.

I wanted to institute rules and regulations for maintaining the standards and have them enforced. Laxity in enforcing the rules does contribute to the deterioration of the parks in the whole country. Lack of equitable access to benefit-sharing is the greatest hindrance in assuming ownership of such a lucrative resource.

Our natural resources seem to benefit external exploiters more than the local communities. This phenomenon alone discourages protection of the resources. Who benefits when an international crew shoots a wildebeest migration documentary? These are some of the basic questions which are not readily answered.

The exploitation of the resources progresses unabated and the communities within live in abject poverty. For example, certain plant species which have been proven to have medicinal properties are freely harvested and used in pharmaceutical industries without regard to the local communities where the plant is harvested. There is no sharing of any benefits which accrue from such a process. Those multinationals do not support any project which can assist in poverty reduction.

This is true for many products found in third world countries. Mineral mining also does not benefit the locals wherever it is carried out. Examples of such exploitations are many in Africa and many conflicts are resource-related. The well-to-do nations have, for years, contributed to the ever-increasing poverty in African countries by exploiting the natural resources for their selfish use. In Kenya, poaching by foreigners is a real menace.

I have deliberately delved into this particular subject because it was a concern that we could destroy the only world game tourist destination. Publicity did not mean excessive construction of hotels since we had an adequate number in place. Kenyans do not observe the basic laws which govern them. Tourists then follow suit and damage our very ecosystem and create negative publicity.

Yet the Maasai Mara still stands unique in Kenya as a world-famous wildlife sanctuary. It should be protected at all costs.

As I write this book, I have checked and found out that there are 225 temporary tented camps far beyond the number of 70 I knew. The moratorium was lifted and all sorts of camp sites were erected, creating uncontrollable vehicular and human traffic. With such practices, there is an imminent destruction of a world recognized game park. This is simply due to greed and non-conformity to the laws which govern game reserves.

The major problem in adhering to the laid-down regulations in the management of the park is sheer failure to impose the laws. Both the national and international communities do not care about the future generations. It is true that the number of wild game is decreasing steadily.

A typical example is the uncontrolled poaching of elephant tusks in Kenya. One of the big five is threatened with extinction due to high demand of tusks. The elephant tusk is made available by poachers who have an already existing market in other parts of the world. The incessant cries from many Kenyan go unheeded!

The vice continues despite the existing laws which ban poaching. International protocols have not deterred the menace either. The penalties which are imposed on poachers in Kenya are not deterrent enough and culprits usually go scott-free despite there being damning evidence. It will not be long before we lose all the elephants in Kenya.

The same thing applies to lions and other game. Pressure from human encroachment is a major threat to the survival of our game parks. I am saddened by this matter and efforts we tried to put in place to curb the heinous practice never yielded any good results.

This docket was overseen by the Kenya Wildlife Service and, in my view, they are unable to execute this task. They have been overpowered despite the good training they get. The external assistance has not been successful either. No tangible efforts seem to be undertaken as the nation loses precious natural resources. Such a loss will be irreversible.

This will be a terrible reflection on us if we sit by and watch a systematic loss of wild game in this modern age. The annals of history shall record and reflect that Kenya was once the cradle of the most abundant wild game in the world. Maybe there is something that can be done urgently. Time will tell.

Revitalization of National Environment Management Authority (NEMA)

After the announcement in the media that I should leave my post as Vice-Chancellor of JKUAT, I decided to make another jab on the activation of NEMA whose role was not pronounced. I got remote instructions from State House handlers and Mr Francis Muthaura that NEMA wanted a jumpstart. It was a moribund institution whose powers were not yet felt by Kenyans. I recalled my role in all the institutions that I had been assigned. The results were astounding.

I knew what role I played as a chairman to revitalize the Inter-university Council of East Africa; my leadership as a chairman to collect data for the establishment of new campuses to accommodate the 1990/91 double-in-take. I further reflected and revisited the technique I used to place my immediate University, JKUAT, onto the world map of performing Universities. I had enough living testimonies of successful projects. I saw a window of opportunity to make NEMA a great institution to tackle pollution and degradation in our country.

I decided to take a swipe at it. As alluded to earlier, I took the EMCA 1999 Act and studied it carefully. I picked out urgent areas which needed action and divided them in actionable blocks. The demand for environmental protection was important for the whole country, but other areas were more degraded than others and needed urgent action.

I also mapped out those which could create immediate and visible results for Kenyans. I knew the capability of the staff, their roles and expertise. I was also aware that the parent ministry was watching over my shoulders to size me up. I was aware that NEMA had already been branded a name that I had to turn into another JKUAT campus by swift actions. I was never deterred by negative publicity. My role was clear, to publicize NEMA and try to correct the environmental mess in all aspects.

The following achievements of NEMA are worth mentioning to demonstrate the role I played in my new assignment:

a) The establishment of formidable staff drawn from various universities at graduate level to man the districts. These were interviewed by the board and competitively hired. They were fully responsible for environmental complaints in their areas of jurisdiction.

b) The posting of staff who had attained their MSc or PhD degrees in natural science areas to be in charge of all the provinces of Kenya. Some staff were deployed from the headquarters and others were hired to man the provinces. Their remuneration packages were higher than those of their counterpart provincial commissioners. They had the required academic credentials and I felt constrained to pay them less. The board approved their packages. The visibility of NEMA and its roles were now prominent at the grassroots levels.

c) I held numerous training courses, seminars, workshops, conferences and symposia to popularize the role of NEMA. I wanted the public to be aware of the imminent dangers if we did not care to protect our environment, more importantly the natural resources. Again, the board was fully involved in conducting the target gatherings which brought together diverse stakeholders including lead agencies and the press.

d) I held a few workshops specific for the Kenyan electronic and print media personalities and trained them on how to write scientific articles on the environment. The money for the meetings was sourced from donors and the government.

e) I drafted a policy paper which was specific about the abolition of polythene paper. The idea was bought by many environmentalists including the Nobel Peace Winner, Prof. Wangari Maathai (deceased) who was the assistant minister then. She gave the policy paper full support and even proposed that we put a levy on each bag collected to deter excessive litter. I proposed a formula to the supermarkets for collecting and reusing the bags but the idea was killed by the manufacturers who wanted to continue producing them. The littering of polybags goes on unabated even today and untold environmental pollution keeps on escalating. Our water channels are choking as the previous productive pieces of wastelands are covered with polythene litter.

f) International linkages were boosted as I attended several United Nations conferences. Kenya's name was on the world map as far as environmental matters were concerned. I raised substantial amounts of money from donors by writing project proposals and encouraging staff to solicit for funding for various urgent needs. I even attracted international investors to bring business to Kenya.

g) My interactions at many international environmental meetings resulted into a number of conferences being held in Nairobi and this attracted foreign exchange earnings for the country.

h) I set various committees to tackle specific environmental tasks like marine and coastal pollution and degradation; wetlands committee to investigate the damage here; water towers committees to check on deforestation and land degradation; e-waste, compliance and enforcement committee; legal committee; finance and audit committee; solid and management committee.

i) I introduced a vigorous environmental and social impact assessment and audit section which was composed of professionals from the whole country.

j) I started giving licences for ESIA and EA experts at a fee to be able to assist NEMA in checking the mushrooming unapproved structures. The group was left on its own to regulate itself, but registration had to be done by NEMA after they had undergone basic training courses.

k) I drafted the curriculum for environmental and social impact assessment and audit and gave it to some institutions to use as a guideline for training. We had to source out this service as NEMA did not have enough staff to undertake the training courses. I gave the curriculum to the universities and private training colleges. This was part of the requirement before one was registered as an expert in environmental assessments.

l) I constituted a team of NEMA staff and lead agents to draft the state of environment reports which were a statutory requirement.

As has already been alluded to, the first report was produced in 2004, and thereafter annual ones have been compiled. They were supposed to be tabled in Parliament by the minister, but none of them did so.

By law, state of environment reports must be produced annually detailing all factors which affect the entire Kenyan environment. I personally chaired the meetings which drafted the reports as they tackled each sub-theme. The board was fully aware of the proceedings and we normally launched the reports in big Kenyan hotels at the coastal city of Mombasa.

I usually invited great personalities to launch the reports. The minister was invited and charged to table the 300-odd copies to parliament. Again, all the media houses were invited and made clips of the most damaging pollutants to present to the public. This was a major milestone for which I was proud. Launching of state of environment reports was the loudest and most popular technique that I used to issue alerts on the degree of pollution and degradation of our environment. All the funding to meet the expenses of committee members' allowances, publications, reproduction of copies, travel and accommodation were generously met by the United Nations Environment Programme (UNEP).

We used to write annual proposals to cover the costs which ran into millions of shillings (approximately US$600,000). This was a generous funding agreement which allowed me and my team to meet the requirement of producing the reports for all the period I was in NEMA. I really appreciated the support accorded by the UNEP.

    m) I usually held public hearings whenever there was a dispute in the construction of a project, sub-division of a ranch, diversion of a river or establishment of residential estates.

The law required that if there was any aggrieved party on any development, the two parties meet on the disputed site under the chairmanship of the district officer to debate on the pros and cons of the project in question. I used to announce publicly the intended hearing specifying the site, venue, date of hearing and time.

By law, I had to post the announcement in public places and in well-read, circulated dailies in Kenya. This had to be done 21 days before the date of the hearing. During such meetings, I had to arrange for security escort in case the meetings became volatile and the groups became unruly. I had a few such cases but I used some diplomacy to calm the groups down.

The final report and judgement was to be drafted by me, with a clear specific recommendation as to whether the project should proceed or not. The aggrieved party could then proceed to seek redress from the high court. I made many unpalatable and sometimes good decisions during and after the public hearings. I received many threats after delivering judgement. I usually held top management meetings before drafting the final letter. I kept the board informed of all the major decisions. I also copied the ministry headquarters for information.

Invariably, unsatisfied parties made their way to the minister's office for intervention. This was where the environmental mess started if, indeed, the minister overturned my ruling. This often happened but I could not be cowed or change my- well-searched decisions.

    n) I facilitated the acquisition of Global Environment Facility (GEF) for 77 programmes. The facility is based in Washington DC and gives funds to individuals and organizations to tackle specific environmental issues which the country considers important. NEMAs the – director-general was the focal point for Kenya and I had to vet the proposals for funding. Many NGOs, CBOs, and individuals benefited from this generous funding.

GEF is one arm of the World Bank's facilitation to assist small-to-medium environmental concerns. I was always to review and recommend the proposals which in turn assisted my work to be visible and practical.

o) World Environmental Days were dates set aside to make people aware of the environmental needs. UNEP set aside several dates for specific purposes to celebrate, or perform relevant duties in accordance with the theme and purpose. They were in the calendar of events and funds were set aside to meet the expenses. Successful meetings were held all over the country, with one major one attended by the minister to mark the ceremony. I do recall that we had about 20 similar commemorative days. Meetings were held across the country but the major ones used to be located in areas where I thought pollution was rampant.

p) Designated UNEP environmental calendar for creating awareness includes:
- World Day to combat Desertification;
- International Biodiversity Day;
- World Environment Day;
- International Day for Biological Diversity;
- Convention on the Conservation of Migratory Species of Wild Animals;
- Convention of International Trade in Endangered Species (CITES);
- Convention in Biological Diversity (CBD);
- World Water Day (WWD);
- World Wetlands Day;
- Earth Day;
- World Habitat Day;
- World Oceans Day;
- Clean up the World.

During these occasions, we invited dignitaries from all over the country to join us in celebrating the events. We mounted big road shows and drove through towns and cities to make an impact. There was no effective way to display the role of NEMA other than going public. Kenyans were made aware of the environmental dangers they caused. The objectives of the authority were being felt in all the corners of Kenya. I was confident that I had scored high on publicity and public awareness.

q) Mainstreaming the multilateral Environmental Agreements was a task I took up soon after I reported to NEMA. My role was to study and understand all the protocols, agreements, conventions and even statutes which Kenya had acceded to and domesticate them. This was not a mean task. I realized that hundreds of agreements had been signed by various government agents and needed to be implemented by several government ministries and agents. I selected those which were directly related to environment and mainstreamed them into our programmes. I assigned members of my management to take up specific agreements and advise on how best to implement the requirements. What was amazing was that nobody knew where the agreements were housed or who was responsible for their implementation.

The Attorney-General (AG) held some, while Ministry of Agriculture claimed others. My ministry headquarters also wanted to implement some. Wildlife Service housed others and Treasury kept a few. It was and still is confusing to know who is responsible for what.

Under normal circumstances, the authority should be the custodian of all the multilateral environmental agreements (MEAs). The documents deal with environmental matters and would have been discussed and ratified by Kenya. I just never understood why the confusion. We had a representative from the AGs Office in the board but, according to my judgement, the individual was wanting and never assisted in any way even in matters legal.

My legal officer, Prof. Francis Situma, was more versed in legal matters than the AGs' representative. Nevertheless protocols like Kyoto, Montreal were housed in NEMA and I domesticated them in our annual programmes. For example, climate change issues had my full attention and I had a large support staff headed by Ms Emily Massawa. This was true for a few other conventions to which Kenya was a signatory.

The confusion was not helping any agency but curtailing the implementation of the requirements. Under normal circumstances, all protocols and similar ones should be held by the attorney general's office with copies to relevant bodies for implementation. I was not sure who had what document and at what level of implementation it was.

r) Another milestone which I considered vital for the public was to come up with a quarterly publication. Just like I did in JKUAT where I introduced Agritech News, a magazine from the office of the Vice-Chancellor, I did the same at NEMA. We published the NEMA news magazine which covered environmental matters countrywide. It was one of the finest communication write-ups that I ever came up with. My public relations office headed by Mr Titus Mungou and Ruth Musembi formed an editorial board, of which I was the chairman and Mr K.S.A Buigutt the chief editor.

I knew the importance of timely communication. I knew that knowledge is power, and that our quarterly magazine would reach a reasonable number of readers. The NEMA magazine covered topical issues on environment and all the hotspots of Kenya as far as pollution was concerned. My parent ministry followed suit and started theirs despite the fact that we used to cover their events in the NEMA magazine.

The last quarter of December 2004, for example, highlighted both pictorially and by articles Prof .Wangari Maathai, the Nobel Peace Prize Laureate. The title of her article was "A Thunderous Homecoming for Nobel Peace Laureate". I was privileged to work with Prof. W. Maathai. I knew that documentation and publicity of my work was the only way to get to the Kenyan community.

    s) We trained judges, lawyers and advocates on environmental matters. Inadequate and poor judgement of environmental cases was one major contributor to environmental pollution. I solicited for funds from a donor and supplemented what we got from the Treasury and mounted countrywide training for judges, lawyers and magistrates. I tasked my legal services department to get funding from the Institute of Law and Environmental Governance to conduct several symposia. This initiative was an astounding success as we held meetings in Kenya's major towns, specifically to train the legal fraternity on the seriousness of the environment. I covered the EMCA 1999.

Act requirements and my staff delved in the penalties for offenders. The last symposium we held was in Mombasa for five days where the chief executive of the judiciary himself, Evan Gicheru, attended together with over fifty of his learned friends. The theme was, "Do not Delay Environmental Cases, says Chief Justice".

This was one of my happiest moments, to see cases expedited and appropriate fines apportioned to offenders. They were mesmerized when I took them through various concepts such as multilateral environmental agreements, suitable development, the precautionary principle, the polluter-pays principle and the application of the concepts. We further discussed the EMCA and its essential characteristics. I was emphatic when I discussed practical concerns related to health, life and the environment, hence the slogan, "Our Environment, our Health". I coined the slogan and it was approved by the board. It was brief and understandable.

The parting words at this important symposium were imploring the group to treat environmental cases with extreme urgency as they affected national health and life. Delayed decisions could jeopardize people's health and life, as well as cause irreversible damage in many cases. These were the words of Hon. Justice Evan Gicheru who stayed for the whole duration of the symposium. As usual, the media houses covered the proceedings extensively. Kenyans knew the importance of protecting our environment.

    t) I summoned organizations to submit their environmental audit (EA) reports. This requirement which had been ignored for years was to ask industries to declare their environmental compliance. I took this matter up with the urgency it required. The law, Sections 58 and 68 of EMCA, 1999 Act asks for the submission of the annual EA. The deadline is usually 31 December of each year. It was gratifying that after advising the affected groups, reports were submitted in thousands.

I gave a long list of categories of those who were supposed to comply and they came up with good reports on how their firms were complying with the law. I had a mammoth task to have the reports reviewed and the stakeholders informed of the outcome. I had to hire and outsource the reviewers. I also used several interns to review the reports using specific guidelines per category. We wrote to every respondent acknowledging receipt of their reports and advising accordingly. The requirements for environmental audit reports sounded awareness to those would-be polluters.

Reflections

Public awareness was my target and I accomplished that. I used past experience to ensure that Kenya and other countries knew of our concerns on pollution. The popularity of JKUAT nationally was translated here. I was cautioned many times that NEMA was not a second campus of JKUAT, but an intricately entwined parastatal to the parent ministry where the PS and the minister were the supreme decision-makers.

The continuous interference with all aspects of operations was nowhere closer to any I got at the university. In fact the chief executives of Kenyan public Universities have immense managerial, academic and decisive powers. My powers in NEMA were monitored and seen as retrogressive to the workers. Any decisions made by my officers and myself were suspect.

Prof. Wangari Maathai, the Nobel Peace Prize Laureate, and who was the assistant minister, worked with me very well. She would call me into her office and consult with me on actions to take on issues she considered needed consultations. I enjoyed working with her and we travelled together on several occasions on fact-finding missions. The ministers had their own personal agenda and politicized environmental matters.

The sad picture of the degradation and loss of biodiversity in Kenya is a worrying trend. It will be a sad day when our future generations will only read about the riches of Kenya's lost natural treasures. The past generations shall only be blamed for their naivety, ignorance and incompetence in the management and protection of the lost resources. The ever continuous increase of solid waste both on land and coastal areas is a matter which needs utmost attention. The occurrence of environmentally-related diseases in Kenya is invariably correlated with the pollution of our towns and cities. NEMA tried to educate the masses on the inherent dangers of pollution. The laws are provided but lack firm execution. My tenure in NEMA was very eventful.

I interacted with those who wanted to cut corners and have their demands met in the shortest time possible. Others were firm believers in the rule of law for our environment. The highest organs of the land were not committed to the conservation of our environment.

The annual, hyped tree-planting excises were showcases which never promoted or enhanced environmental conservation. The fact is that many of the seedlings which were planted annually were scorched in the sun and dried up for lack of care. The survival percentage was negligible countrywide. But this was considered the most important day of Kenya's calendar day of events. The sporadic clean-up days are never sustained either. Despite a designated World Environment Day, no serious follow-up is sustained to ensure compliance. The residents resume normal pollution practices unabated.

Water wars in Kenya are not new. A case in point is the wrangles of an Island in Lake Victoria where one community, which believes they own the island, is being asked to purchase residence permits. Senior government officers who are supposed to resolve the problem claim that one community owns the island and another one owns the waters that surround the island. These are trans-boundary conflicts which point out the crucial importance of water and its resources. The growing population demands water. The rivers, lakes and ocean are being polluted. Loss of marine life and continuous degradation of the coral reefs, mangrove swamps, and other aquatic ecosystems will affect future generations. The rising temperatures have caused a reduction in water flow.

When one talks of Maasai Mara, holiday memories come into mind. One thing I know clearly is that no matter how many times one visits the world-famous wildlife sanctuary, any subsequent trip offers cherished memories! The world views the conservancy as an inexhaustible pit of wild game. Those views have led to the construction of lodges and tented safari camps in every part of the sanctuary. This has caused the destruction of the fragile ecosystem. This is not necessary and I view the trend as destructive and selfish. The proprietors use terms like eco-friendly to simply convince the licensing authorities who may not have knowledge on the effect of human traffic in parks.

Until and unless the highest authority of the land takes a serious and continuous participation of conserving our environment, Kenya will stay a nation with its rich biodiversity diminishing. Our coastlines will no longer sustain the pressure and the terrestrial ecosystem will never hold the ever-increasing population.

International Linkages, Travels and Conferences

My Travelling Experiences

I enjoy travelling and interacting with different people from various cultures. I also enjoy visiting scenic places around the world. I recall one writer saying that the world is a big book and those who do not travel, read but the first page (paraphrased). Another great travel writer, Paul Thoreau, said that *"tourists do not know where they have been and travellers do not know where they are going"*. These are loaded statements. I would like to illustrate the importance of travel and the unique experiences specific to the traveller. Despite making contacts and adding new people in the list of collaborators, I had an opportunity to sit back and reflect on my job and enjoy interacting with new cultures.

I was able to learn that under our skin colour, native costumes, religion, different languages, tolerance of diversity, and eating habits, we all share core beliefs and it is only cultural differences that separate us; differences that should be celebrated and not feared. In my travels I relish them. I also learned that travel has shown me that we all love our homes, families and the country for what they are and not what they could be.

People notice this after cross interaction with many varied world cultures. It is not easy to sit in front of a television and judge others on the opposite side of the world, based on the unknown presenters and reporters. It is only the traveller who takes time to investigate such things for him or herself in order to dissipate ignorant impressions of others.

I was able to tell the good and the bad of the Nigerian, British, Comorian, Egyptian, Libyan, Japanese, American, South African, German or Zimbabwean culture. One understands the diversity of people's value system.

These conferences added value to me. I learnt a lot from my counterparts and the problems were very similar in the universities. I even encountered the ugly face of the apartheid regime in South Africa. During my many travels outside Kenya, I had an opportunity to undertake a consultancy service in the Comoros Islands. I used to bid for consultancies in my area of specialization and I won one which was advertised in the local dailies. Three of us won the consultancy but I was the only black, the others were Dutch nationals. We were to meet at Moroni, the capital city of the Comoros.

I was excited to travel to these remote Islands off the east coast of Africa. My colleagues were to join me there from Amsterdam. I was booked on Kenya Airways from Nairobi to Johannesburg and would then connect to South African Airways to Comoros. This was in 1983. I boarded the plane one early morning at Nairobi, alighted at Johannesburg for my connecting flight to Moroni, the Comoros capital city.

My morning flight from Nairobi was beautiful as I flew over the great Mt Kilimanjaro which was covered by snow (not there now), and over Lake Malawi to the industrial gold city of Johannesburg. It was a bright sunny day with no clouds. My four-hour flight was memorable. The Kenya Airways staff knew me as I was their frequent flier member and was booked in first class. I had no clue as to what I would encounter in South Africa.

As soon as I cleared my luggage and was ready to proceed to the transfer lounge, a white attendant told me that the flight to Comoros had already left and I was to wait for the next flight, 12 hours later. I was disoriented for a while and asked if he meant what he had said. He confirmed it.

The young man was polite and responsive. He requested me to follow him. I had been warned not to surrender my passport for stamping or else I would not travel to many countries thereafter. He gave me a piece of stamped paper for a visa. As we walked through the alleyways, I sensed something.

He was leading me with my luggage to an isolated waiting lounge. I never panicked but reminded him that I was travelling first class and had rights and privileges attached to it. That could not be so; the Johannesburg to Moroni leg was economy class. He showed me a new ticket.

I was taken to a large waiting lounge with two television sets. Both of them were broadcasting South African propaganda items. I found two black ladies also waiting. The young man told me to wait there for the next 12 hours, eat there and use all the facilities within until somebody came to give me further guidance, as to when my flight would take off.

History repeats itself; so goes the adage. I recalled my colonial experience at the Barclays Bank in 1970. Was I in for jailing or segregation? Perhaps the latter. It dawned on me that the lounge I was put in was for blacks only and we could not get out to even visit the duty-free shops! I had not planned a stop over here.

I sat pensively and started counting hours as repeats of South African propaganda blared from two TV sets. There were other people who waited there both blacks and whites but for a shorter duration. I was saddened by the turn of events. We were served lunch with white waiters who must have been told to watch over us. Later at 4 pm, tea was served, and dinner at 6 pm. It was indeed a prison-like set up.

I had not carried with me any books or local dailies which I had perused earlier in the morning. I convinced myself that all would be fine and I would accomplish my assignment. I did not have the capacity and means to call my dear wife, Esther, in Nairobi and tell her of my temporary predicament. I suffered quietly, but bravely.

The scheduled 10 pm flight was announced and I was relieved. My anxiety was over as I was ushered to the aircraft by another person. The Johannesburg airport, then the Jan Smuts Airport, has very powerful lights. We walked to the plane as if it was midday. One could pick a needle from the ground. I walked to the hangs of the plane and checked my luggage in as I carried my briefcase aboard.

I was the only black man in the whole plane. The rest of the passengers were all whites who stared at me as I settled down in my assigned seat. I travelled economy class. I was not bothered that I was the only black aboard. What bothered me were the constant stares!

This was not my first time in an all-white group. I had been in the US and UK but never noticed such behaviour. The crew appeared nervous and kept talking to themselves with quick eye glances at me. I also put on a stronger face and ensured direct eye contact with them.

My appearance in the aircraft caused confusion, agony, disbelief and anger to the passengers. The crew did not know what to do but kept on consulting one another as the aircraft was in motion taking off. I was not bothered an iota by this. I maintained my stony face and felt very relaxed. Just before take-off, an unusual solution was arrived at. To move me.

I was politely requested to move to another class, the first class. Due to the few occupants in that class, the crew decided to move all passengers who were less than twenty in number to join those in economy class so that I could be segregated to the first class. I obliged. All of first-class privileges and services were now accorded to me as the economy section steamed with white passengers. It was packed.

I was now locked up in a private comfortable cabin served with several attendants and I had an excellent four-hour night flight to the Comoros. In fact it turned out that the crew in my cabin became so free and friendly that we shared jokes and even chatted about cultures.

Indeed, I asked them why I was here alone. Of course I knew the answer but I wanted to be told the real story of apartheid. I had several glasses of South African wine and slept over the agony and psychological anguish. That was apartheid in practice and its ugly face in 1983. I developed a defensive disposition and totally fearless attitude for overcoming fear. I had to feel supreme confidence in myself and move on.

A bold act requires high degree of confidence and bravery. I knew that if I became overly sensitive, I would have been overwhelmed and would have lost my mind. My work would then have suffered. I was going to make substantial amount of money on a consultancy. I therefore needed a clear and focused mind.

I created a forward momentum and avoided verbal confrontations. I was tactful right from the waiting lounge to the Comoros Islands. It was crucial to know how to deal with a unique environment. I had to deal with the adversity and confront the unknown.

We landed in Moroni safely. The whites alighted first as I was still being entertained by the crew as we passed time. I knew well that it was a delay tactic and by 1 am, I was on my way to my hotel where my colleagues had been booked. They had travelled by Sabena Airlines from Brussels and, of course, with no hitches. I was happy to see many friendly Comorians of all races going about their daily business.

I slept soundly and reflected on the past 24 hours the following day, as we embarked on our project. I knew that it was the height of apartheid in South Africa. The struggle for independence was at its climax. The minor clash I had with one Mr Bird in Barclays Bank in the 1970s was nowhere comparable to the South African regime. I even reflected on my Animal Science lecturer, one Mr Raphael Mitchell, who wanted to give me a lower mark after scoring a high mark in his course. But still his action was easily sorted out after he realized that I would make a fuss about it.

We continued with our research work, hopping from one island to another in a military chopper. One of the four islands which constitute the beautiful Comoros was still under the French colonial government. The residents were divided; some wanted continued occupancy of the island while others were vehemently opposed to it. Grand Comoro (Ngazidja), Anjouan, Moheli and Mayotte make up the country.

Mayotte is regarded as not part of the country and retains ties with France. These islands constitute an archipelago of volcanic origin in the Indian Ocean. They had a population of approximately 500,000 people, with a total area of 2170sqkm. The information was important for us to advise the government about the planting densities of trees and where the project would be feasible. We finished in one week and I returned home via the same route I had travelled.

Surprisingly, the SAA flight this time was full of black and white travellers to Johannesburg. I was delighted that I would fly back in a more relaxed manner than previously. I realized that there were specific flights and times for both races to travel together.

When I told this story to my colleagues in Kenya, they were amazed at the. I recalled my having turned down a job offer to work with Monsanto Company in South Africa soon after my PhD graduation in 1978. My decision to decline the offer was justified and I would not have had peace in a country which was evidently racist.

Reflections

Extensive travel allows one to make new contacts and they keep chief executives well informed. They learn new techniques of managing their organizations. They open avenues for new ideas. I viewed all the linkages that I had initiated to have been beneficial to my university then. Whatever lessons I learned became additional treasure in the management of JKUAT. It was through these linkages that I was able to secure scholarships for staff development and later on produced persons of integrity to eventually take up leadership of several public and private universities in Kenya. Many of the mentees are now leaders and managers in their own right. When I reflect at my human resources contributions, I feel satisfied that my little role in practical involvement of staff training and appropriate mentoring bore fruits.

My encounter with apartheid was a lesson in itself. I did not know the extent of racial injustice and its magnitude. I learnt a lesson which I hope to narrate to generations to come. Every bad event has a lesson or two to learn. I later forged excellent relationships with several South African white academic colleagues with whom we developed meaningful programmes. The USHEPiA project which assisted my PhD staff training had its headquarters at the University of Cape Town. I later made several visits using South African Airways (SAA) and fondly remembered my unique trip to the Comoros Islands.

The trips and contacts I later made as director-general of NEMA were even more eventful than previous ones. These were group meetings with different agenda.

My tenure as the director-general of NEMA exposed me to local and international meetings which were unique in objectives and outcomes. To seriously address Kenya's pollution and degradation trends, I had to get to the hotspots and try to give solutions. Some of the areas which I visited were so degraded that any corrective measures would be costly.

I advised the authorities on the dangers of invasive weed species which were spreading all over the country. I advised on the physical removal of the water hyacinth weed from Lake Victoria when it had occupied only a small section of the lake at Hippo Point. But my advice was not taken seriously. I had been to Jinja Dam and seen the Uganda military scooping the entire weed from the lake to dry land for disposal. The operations were done under the close supervision of President Yoweri Museveni whom I found there giving directions. I thought it was an excellent move to deter further spread of the hyacinth weed on the waters.

The weed is now a menace on Kenyan waters, despite concerted efforts and exorbitant amounts of money spent to study alternative use for it. The weed can be used for other purposes, but its spread is so fast that it cannot be contained. My fear is that secondary vegetation is now invading the same area where the weed has inhabited. We have another weed, hippo grass, which is colonizing the area very fast. Lake Victoria is still threatened by the hyacinth weed despite the professional advice I had given on several occasions. In addition, the washing of vehicles along the beaches is contributing to the survival of the weed.

The other weed that has colonized our dry areas is the proposisjuliflora. Commonly known as Mathenge in Kenya. P. juliflora was introduced to Kenya as a greening plant in semi-arid regions of the country. It does beautifully there. My travels exposed me to the menace it had caused to pastoralists. Currently, thousands of hectares of land have been invaded by the weed. It is a threat in Ethiopia, Djibouti, Somalia, Southern Sudan, Eritrea and Kenya.

There are impressive projects which are being used to reduce it but, again, its rate of spread is too fast to be contained. There are uses such as charcoal burning, wood carving, and energy production among others. This is encouraging, but reduced forage where the plant is prevalent is an economic concern for pastoralists.

My visit to the heavily infested areas with *Proposisjuliflora* was disheartening to say the least. The negative effects far outweighed the positive ones. Another hotspot which I visited and advised the local authorities to negotiate for quick remedy was Wajir County in Kenya's arid and semi-arid region. Wajir town is a sprawling beautiful setting punctuated with short scattered acacia trees. It is a water-scarcity area but has a high water table (landscape). Very little agricultural activities take place here. The locals grow mainly vegetables and short-cycle crops.

The major environmental problem I witnessed here was lack of a sewerage system. The residents have no common waste treatment facilities. They rely on ferrying wastes to a distant common open site and dumping it there. When I visited the town, I tried to negotiate with the local authorities to write a proposal and seek donor support to construct a modern environmental sewerage treatment plant. I had thought of using the treatment to allow for fertilizer production through dehydration and processing.

Due to shallow water tables, pit latrines are not used. I never got to discuss the proposal because of other reasons. I hope somebody took it up to alleviate the miserable conditions of the town residents. Ferrying night soils in a bucket every so often was such an unsightly procedure that I thought something urgent needed to be done. A few houses, like where the district commissioner resided, had modern toilet facilities. I had to spend my two nights in the government house.

Environmental Hotspots

There were several environmental hotspots which I personally took on as the director-general of NEMA. It is not possible to enumerate all of them in this book. The calls and e-mail letters we received at the headquarters were responded to promptly. Action could be given in the same way or I could request my field officers to make follow-ups. The press could also pick areas which needed attention and call me directly. I was accessible and received calls on a 24/7 basis. I also attended to all my e-mails and got back to customers promptly. Responding to customers was my long-term habit which I still carry on up to this day.

I was instrumental in the setting up of the Nile Basin Initiative (NIB). This was a trans-boundary venture which was externally funded and housed in my NEMA offices. The main purpose was to set up management systems for the River Nile. The idea was to have all of the then 10 countries use this forum for solving the problems associated with the River Nile waters. The riparian countries included Tanzania, Rwanda, Burundi, Uganda, Kenya, Southern Sudan, Northern Sudan, Ethiopia, Djibouti and Egypt. Are the riparian states 10 or more?

My board appointed the first environmental officer to head the section. The lady we got did an excellent job in coordinating programmes for the riparian countries. To date, NIB is a strong trans-boundary set-up which tackles issues associated with the best uses and practices of the Nile waters. It rotates in various countries as needed.

Part of my responsibility as a DG NEMA involved international meetings, environmental governing Councils, conference of parties, meeting of parties, protocol negotiations, drafting of international agreements and many more; all fell under my office. They were in our calendar of events. Many staff and board members also had their opportunities to represent the country whenever need arose. I recall my board chairman attending many of these meetings and could come home with a different view on how we managed Kenya's environmental issues.

International Negotiations

The trips were either headed by the president, the minister for environment then, and the chief secretary at the time or me. I was in the lowest cadre of leading delegations, but I wrote all the technical papers for presentation. I actually led and guided the team from bottom up. In any case, technical papers and negotiations are usually conducted by technocrats who deliberate on issues and provide appropriate answers. I also had a competent paper drafting team headed by various section heads according to subjects. I, however, followed them up to ensure quality papers were produced.

Among the meetings I attended were those held in New York, Geneva, Tokyo, London, Buenos Aires, Dar-es-Salaam, Kampala, Paraguay, Bonn, Berlin, Bangkok, Hong Kong, Tripoli, Sirte, Mauritius, Arusha, Addis Ababa, Johannesburg, Paris, Prague, Del Monste, California and Montreal. Some of these meetings were headed by the Head of State of state which involved unbelievably large delegations. I never witnessed so many joyriders like those I saw at the UN General Assembly in New York City. Sometimes I wondered what necessitated some politicians to leave their work in Kenya and travel outside for weeks in the guise of joining a president in a ten-minute speech delivery.

A case in point was when the former president, Mwai Kibaki, had to personally attend a UN Governing Council in New York. I was officially invited to provide technical back-up on environmental matters, and I was number 25 on the official list of invitees. I was booked in the same hotel, the World of Astoria, an expensive classic hotel where Head of States of state stay. This was because I would be needed at any time to provide expertise before the president would deliver his ten-to-fifteen- minute speech. I was working closely with the Kenya's special ambassador to the UN, Prof. Judith Mbula Bahemuka, who was a colleague at the University of Nairobi and is now a Chancellor at the University of Eldoret. She was our representative in the Assembly and did everything possible to make our stay comfortable.

I was surprised when parliament back home raised the question of the numbers of delegates who travelled to New York! The taxpayers were also up in arms as to why 90 people accompanied the president. It was indeed true! Only 25 of us were officially cleared to accompany the president due to various security and technical assignments. When I got the full list, I honestly saw no reason why the other 45 joyriders had been present. This was when I knew that my nominal roll number was 25 and that my name was the last on the official list which had been cleared prior to our travel. The extra baggage, I think, had to pay from their ministries to avoid embarrassment.

During our stay in New York, my colleague, Prof. Bahemuka was so useful to us in providing information for Kenya to support certain aspects of the negotiations. She was so kind to us and invited the Kenyan delegation to her official residence in Rochester, on the outskirts of New York City. Among the many trips I made, Bahemuka and Amina Abdala who was then posted in Geneva, now a cabinet secretary, were the only individuals who had been so accommodating to the Kenyan delegation and who always invited us to their residences.

It did not matter how many times I went to these stations, the two would always extend Kenyan hospitality to us. Somebody might argue that it was their obligation to extend invitations whenever anybody showed up in their stations. That is not true. They are not obligated. They were there on duty and working as all of us were. I thought it was their character which I very much admired.

Cabinet ministers usually led delegations to represent Kenya. I had a chance to compare and internalize the capacity of our people in international negotiating meetings. We could have as many as 190 different countries to discuss an environmental issue like the Kyoto Protocol. Many a time, they could not sit in the plenary hall long enough to follow the proceedings. They could only show up at specific periods when they knew that their country was to deliver a statement.

This was also true to many African countries. The heads of delegations just disappeared for hours and days, leaving juniors to man the country's desks. My ministers, indeed, would not sit long enough to appreciate what other nations were contributing. Instead, the technocrats were put on their country tables taking notes, and would later brief the heads of delegations who could not even understand the jargon that had been used in the papers. There were times when they would miss completely and surface at the reception parties in the evenings.

This was not the case with other countries. China, for example, would oscillate in adopting positions; one time it would declare itself a developed country. It had to survive by taking advantage of positioning itself strategically. The staff from China were so many and their head of delegation used to stay the whole duration and intervene in almost all negotiations. Other countries had their heads of delegations sit throughout the proceedings.

I learnt one vital lesson which I always tell my family and students; as much as possible, sit through any meeting you attend. There is always something to learn. One needs to listen to others as one would like to be listened to. It is common respect and good human behaviour. I personally love to stay through meetings if, indeed, I attend them.

Such perseverance is also a good example to be emulated by the youth. There are university seminars and conferences where chief executives also disappear after opening meetings and workshops.

The Montreal Climate Change Conference of Parties Number Six (CoP6) was an episode to tell. I was requested to lead a one-man Kenyan delegation to Montreal in Canada. The reason was that my minister then, Stephen Kalonzo, would not attend as he was busy campaigning during the referendum on the new constitution.

I attended the conference and found other Kenyans there from various NGOs. I requested them to join me and form a visible Kenyan delegation led by me. I was the highest-ranking government officer there. The Kenyan High Commissioner was also present to give us logistic and protocol support. My main agenda here was to woo all the 190 countries to support us in holding CoP7 in Nairobi.

One of my officers, Mrs Emily Masawo, had given me the historical perspective as to why we needed to hold the conference in Kenya. We had requested for it several years earlier in a similar meeting. Two major deficiencies were noticed which would work against us.

One, the minister himself was not present to push for it through networking. Two, Kenya was viewed as unsafe as we were going to have elections for the referendum in 2005. It was now my duty to clear the air and plead to the parties. I did a lot of lobbying through the East African countries, African countries, Small Island States and the Far East.

We had to do the lobbying either during the breakfast meeting, dinner or during specific country or regional meetings. I learnt to mobilize heads of delegations for a purpose.

Meanwhile, Senegal very much wanted to host the conference. I recall that in one African countries meeting I had exchanged some bitter but friendly words with the Senegalese head of delegation when he was pushing for support to his country.

I had fully convinced Nigeria, our big brother, to support us. Their leader of delegation respected my request and vouched for Kenya. I had earlier in another meeting in Geneva supported them fully regarding hosting an African centre to manage hazardous wastes. They reciprocated and had high regard for me. I had also worked in Ibadan and remembered the good old days. I had respect for the Nigerian team as they always took pride in their tyranny of numbers to shut down small countries.

I had done enough lobbying and networking. When we got to decision-taking, I remember Mauritius and Thailand supporting Kenya to host the next conference of parties. I read the statement on behalf of the country. Canada supported us, and even Tanzania, through their minister of environment. We were actually endorsed to host the 7th Conference of Parties (CoP). I recalled one fact; courage is the ultimate career mover!

I returned home happy and confident only to be asked by my minister:

> "Who authorized your bid to have CoP7 come to Kenya?" Of course, I knew Treasury would cough out some initial money for the meeting. The net return far outweighed what Treasury would have spent. I wrote the report and handed it to the permanent secretary to brief the minister on the course of events toward the conference dates. I had to be in the committee of implementation.

I knew too well that they would not turn down the meeting. The country would benefit as the UNEP headquarters was located in Nairobi. We already had a meeting venue. We needed to beef up security, book hotels and clear litter from the streets. I envisaged that such an international conference would promote trade, business and tourism, and open up new avenues for country-to-country collaboration. That was eventually realized after the meeting was held.

I learnt some lessons from these practical experiences. First, some of the ministers who normally led negotiation teams contributed very little to the meetings. Many, especially from the African continent, did not actively participate in the proceedings; instead they went off to run their personal errands. Consequently, the countries lost in negotiations.

Secondly, attracting any meetings to one's country was a negotiated affair. It was a give-and-take game plan. It was horse-trading: "scratch my back, I scratch yours," as the saying goes. I learnt the technique and practice of public relations during any meetings I went to, whether local or foreign. What I detested most was to be involved in some shady discussions where matters had to be handled in a non-transparent manner or there was allusion to corruption.

If I supported a good idea for an organization or individual, I also expected similar treatment in return. It was a tit-for-tat in a good cause that was beneficial to many. I also learnt that attending, listening and finally participating in meetings gave me an advantage in any negotiations. I got fully informed of the matter at hand and could make informed decisions.

Every person has his price to pay for any decision made. Interpersonal relationship and genuine support of others far outweighs material gifts.

My previous positions as Vice-Chancellor, chairman of KARI, IUCEA, JAB, AICAD and KENET exposed me to enough intrigues that any negotiations, national or international, had a firm basis for me to refer to. Whenever I made a mistake, I quickly rectified. This act itself was a lesson for me. I was one of the eight African members who represented the continent in the United Nations Convention on Climate Change (UNFCCC) in Bonn, Germany, for three years, 2003 to 2006. We were part of those implementing the Kyoto Protocol. This was a powerful committee which vetted compliance soon after the ratification of the protocol. It was in these meetings that I learnt even more of the technique of tolerance in decision-making.

All the 40 of us from across the world had to agree unanimously on an issue before making a decision.

The decision had to be by consensus and not by voting. Every persuasion technique had to be employed to convince the opposing side to agree on an issue and avoid voting. I found this approach to be very useful in disciplinary matters where judgement was usually based on interviews and defence mechanisms. The chairman had to be a tolerant person who allowed the expression of divergent views.

I also learnt the fundamental problem which affects decisions, especially those that relate to the African continent was the number of experts who attend conferences. Meetings usually run well past midnight. The more difficult ones could run as late as 3 am, and even continue into the next day. The end results are exhausted negotiators. What other countries do is to come up with a large contingent of experts to attend to complicated matters in shifts. The USA, Japan, Germany, UK and China, for example, would come with enough negotiators to attend the meetings in shifts and brief each other appropriately. It was a relay for them, while a country like Kenya could afford one expert only who could easily retire early during negotiation time.

I do recall that I was unable to keep up with the talks we had in Canada on equitable access to benefit of natural resources. I retired to my hotel after long negotiations only to come in the morning to the plenary and receive shocking clauses in the document which were contrary to what had been agreed upon. Documents were doctored to suit certain groups of people or countries.

The truth is that certain crucial information could be included, excluded or altered in the late hours of the night and resolutions passed. It was also vital to have lawyers in such meetings. I recall a number of flawed decisions which would have been difficult to alter after they had gone through final drafting stage. One could look out of place if one participated in the final drafting and then start shooting questions in the plenary. You would be politely reminded that you were one of those who drafted the final communiqué and would therefore not be allowed to contribute.

These meetings were very enlightening. They opened my faculties and made me a tough negotiator and a competent manager. Indeed, character building was a continuum. I listened to one pastor in August 2013 during an SDA camp meeting in Nairobi. He made a very far-reaching statement which I thought could be useful to our youths.

He said that in his high school, their motto was, "Character comes before career". He was referring to the fact that an individual must be a person of integrity from the very beginning before he/she chooses a career. The pastor, Richard Brooks, was emphasizing the quality of a Christian in serving the nation. Integrity was a virtue that called for uprightness and Christians had to display it in their careers.

This was true in both local and international negotiations. One had to be absolutely clean in any dealings which could affect one's family or the country. The reality of life was that institutions and governments had to be headed by persons of the highest integrity. One will always be referred to in good or bad memories. The discipline and culture of a country was dictated by individuals who ruled and governed them. The youth would invariably emulate what their elders did.

The Gains

My travels for NEMA brought in funds for environmental concerns. My interactions with friendly nations and subsequent proposal writing assisted us to source funds for specific tasks. The foreign country ambassadors who resided in Kenya assisted in chosen environmental projects as long as we put that money in specified tasks.

Many ambassadors and high commissioners selected to fund sectors in agriculture, environment, water, afforestation, and urban problems, all of which are factors that affect the environment. It was through my interactions and Kenya's visibility in environmental issues that donors like the EU, DFID, JICA, IDRC, UNDP, UNEP, DANIDA and many others extended tremendous support to Kenya regarding environmental efforts.

We requested the British Environmental Agency to second some staff to train my staff on how to prosecute offenders. We were able to train prosecutors and vigilantes. The several Global Environment Facility (GEF) projects which were funded had my blessings. They complemented our efforts towards public awareness and participation. Funds for collecting information on the state of environment report were provided by the UNDP.

The fencing of Mount Kenya resources was also partially sourced by NEMA; so were those for the rehabilitation of the Mau Forest. Donors had confidence in the proper management of their funds and trusted my leadership just like JICA, which I had conducted projects with for over thirteen years, had entrusted its resources to me. There were times when the country was not favoured in borrowing money. But our projects went on uninterrupted and we were able to meet the expectations of the donors. We never had any audit queries in our financial management. By the time I left NEMA, there was substantial external funding for a number of projects.

Teamwork and timely communication on available opportunities are vital for successful implementation of programmes. I thought that one role as a leader was to shape the authority (NEMA). I worked around the clock to show that NEMA was operational. I raised it up and it became the envy of many. The authority's name had permeated in all corners of Kenya. Its roles were well understood and Kenyans started to discard the destructive habits which resulted into environmental pollution and degradation.

I rejected or approved several projects without favour or fear. I knew deep down my heart that for every decision I made, whether positive or negative, somebody somewhere would be affected. So what would be the solution? The law was there. It protected that somebody and me. I therefore made some decisions notwithstanding the consequences. My strongest leverage was that staff and the board were on my side on any decision I made. That was my shield.

Liberal Sharing of Information

My joy in NEMA was to share information freely with my staff and board. I was even more thrilled when I shared academic information with the scholars in Kenya and outside. Unlike other institutions where bureaucracy and protocol are followed to divulge non-sensitive information, NEMA believed in publicity and free sharing of information.

Some African countries had not embraced environmental concerns when I joined NEMA. Other than Uganda, Kenya's neighbour, several countries regarded pollution and degradation as a department within a larger ministry. However, the EMCA 1999 Act, gave the director-general authority to act.

I recall some countries sent delegations to my office to enquire how we were running NEMA. Delegations from Zimbabwe, Ethiopia, Somalia and Rwanda visited me. This is because we had made contacts in various meetings and Kenya was well noticed during the deliberations. Despite the bad environmental conditions in Kenya, I portrayed a positive image and took time to explain and freely highlight the good things that NEMA was doing. Some information was accurate I must admit. But some, like deforestation, was understated. I portrayed a positive image nevertheless.

As each of the countries' delegations visited NEMA, we discussed freely what our role was and how best our Act was functioning. On paper, the Act was, and still is, the most comprehensive piece of legislation. I was proud of it. After detailed explanation of what we did, I would happily furnish the delegations with copies of the EMCA 1999 Act. I would further stress the need for respective countries to enact a law which would safeguard the environment. I even advised them to involve universities to mainstream courses on the environment in their syllabi.

I emphasized the need to have a Head of State of state to be personally involved in enhancing the environment. This advice was taken seriously by Rwanda and it is one of the cleanest countries in Africa. My academic generosity was evidenced during these interactions. In any case, I knew that trans-boundary environmental problems were common.

What was good for Kenya could be great for Ethiopia. Furthermore, multilateral environmental agreements affected all African countries. I shared our pride and popularity. I also gave them my first report of the state of environment, 2004.

I do recall that while I was Vice-Chancellor at JKUAT, the Vice-Chancellor of Kigali University of Science and Technology, Prof Silas Lwakabamba, visited me. He wanted to see our academic programmes.

He told me that JKUAT's programmes were well established and known outside as the best in the region. He flattered me and we had a long conversation on the same. As a good friend and colleague, I freely shared the syllabi, explained to him the marketability of each course. I even pointed out any weaknesses about particular courses. He was most appreciative.

As if that was not enough, he requested me to give him a few staff in the technical subjects to go and start the programmes for six months as he organized himself. I cautiously accepted the request. Little did I know that I would lose some due to the higher offer he gave them. I quickly trained and hired others to replace those who had left JKUAT. I did not regret their departure as I considered this as normal brain circulation within the East African sister states.

I knew of the story when Prof. Silas Lwakabamba of Rwanda was bragging in Harvard that *"if anyone wanted quality trained lecturers, go to Michieka"*. After a while, I took my first-class graduates and trained them to take over. Although it took a little longer to get the full compliments of staff, we still maintained the quality I had jealously protected.

Now, Rwanda has staff who manage the environmental docket. In Rwanda, the Rwanda Environment Management Authority office is enjoying the best, enhanced and protected environment. The Head of State of state, President Paul Kagame is in charge, and clear instructions are posted and announced for the locals and visitors. It is a big crime to litter in Rwanda, for example. When I look back, I feel proud that whatever I shared contributed somehow to the well-being of other people. Most African countries have their laws on environmental agencies or authorities enacted by parliament. After all, the requirements are similar, although they differ in the implementation processes.

Tribulations at NEMA

No job is done without stress. No successful major undertaking is free of blame. Human nature is critical and usually fault-finding. My three-year tenure in NEMA (2003-2006) were marked by enhanced environmental awareness. Its articulation caused all those who did not know that there was an agency in charge of their environmental health to be curious. The NGOs were fully involved in environmental matters. Universities and schools started mounting several units in environmental courses; local governments doubled their efforts to control pollution.

I raised funds for research and for curbing pollution and land degradation. As a manager, I put together a working team of staff who knew their work and could defend with zeal any decisions we made.

The problems of NEMA arose merely because it had come of age and somebody who could be more accommodating to manipulation was sought. The available funds which I assisted to raise attracted the powers-that-be. Issues on environment were clear and anybody could talk about them.

Many proprietors were aggrieved when their projects were rejected due to various reasons. Those who were supposed to move environmental agenda forward were at times affected by the outcomes. They would then pick on the NEMA management and tarnish its name. No minister, for example, tabled the state of environment reports to parliament. We supplied information for the whole country and delivered the required 250 copies to the clerk of the National Assembly. We hoped that a minister in charge of environment would take the responsibility of tabling them in parliament.

Instead, the 250 copies gathered dust in the shelves of parliament. It was work done in futility, energies spent in agony! It was taxpayers' money and donor funds spent in vain. The country continued to be polluted and degraded.

Only very few National Environment Council (NEC) meetings were held. I recall attending only two during the entire period I was at NEMA. This was the ultimate policy body as far as decision-making was concerned.

It was, by law, a minister's supreme organ to endorse or reject deliberations which were brought to the NEC. The NEC would endorse or otherwise advise on controversial decisions I would have made.

I was left to deal with the board which also had no ultimate answers on technical matters. Again, the NEC was a dead outfit. Decisions we made at a lower-cadre level needed the NEC's blessings, failure of which, we were left at the mercies of the community. I nevertheless took the leadership risk and continued to implement decisions which were made by my technical staff bodies. The board was supportive but there came a time when nobody wanted to venture into decision-taking for fear of being sacked. Occasionally, there would be differences in decision-making, creating sharp divisions amongst board members and the technical staff. I usually laid facts as they were and requested for unbiased decisions.

The EMCA 1999 Act was clear on approval or rejection of projects. We used to have court litigations on decisions we made. The budget was not enough to undertake expensive cases once the authority was taken to court. We won many cases but lost a few. Luckily, we had competent lawyers who prepared well before the hearings.

I received several court summons. I delegated my legal officers, Prof. Francis Situma and his deputy Ms Ann Angwenyi, to handle them. Despite the good training we held for field officers, many projects which would normally undergo environmental impact assessment escaped our scrutiny, resulting into unnecessary litigations.

All the events which affected NEMA created public awareness. They were a blessing in disguise, and Kenyans learnt that licences for any project were a necessity.

They had to seek them before embarking on their implementation. I still pushed on the environmental agenda without relenting.

The assistant minister for environment, Prof. Wangari Maathai, was very supportive. On several occasions, the assistant minister would also confide in me the frustrations she faced even from the minister and other political operatives when it came to enforcing the environmental laws. She therefore appreciated the difficult terrain I was working in and the similar frustrations that I faced. She, however, encouraged me to follow the law despite resistance from well-to-do developers who wanted to undercut or circumvent legal approval procedures.

In fact most of the major decisions I made and were later challenged in a court of law were upheld. It was not my nature to reject progressive projects, but any that posed an environmental danger were turned down outright. It was through this hard stand that eventually saw NEMA control some project developments in fragile areas of the country.

Many Kenyans appreciated the work we did. Many NGOs and community-based groups formed coalitions to fight pollution. Several bodies came up as garbage collectors, conservationists and experts. I am not implying that there were no formalized groups to fight pollution and land degradation; there were individuals and even ministry sectors; but NEMAs stand strengthened and encouraged them. What they did not know was that there was a powerful Act, EMCA 1999, which carried immense powers to guarantee a safe and healthy environment for all.

Problems surfaced when laws were put in place and demanded compliance. The well-to-do developers termed our actions as an impediment to development. That was when trouble started.

18

Back to the Ivory Tower

My thirteen years in academic leadership at JKUAT came to an end when I was posted to NEMA as a Director-General. My three years in NEMA where I tried to cope with myriad environmental problems came to an end too. I had to revert to my original base with the vast knowledge and experience that I had gained from these postings.

This was not unusual. Any presidential appointment is not permanent. One could be relieved of one's duties as soon as one was appointed. My permanent professional domicile was the University of Nairobi. I did not sever my tenure there. I kept on renewing my employment as I was moving about the other jobs. It was not my choice that I became the deputy principal, principal and vice-chancellor of JKUAT. This was the choice of President Moi who thought I was doing a good job in nation-building.

It was not my choice either that I was appointed a director-general, NEMA. It was President Mwai Kibaki's choice. He thought I would do a good job in educating Kenyans about the value of environmental conservation and protection. It was not my choice either to be appointed the chairman of Kenyatta University Council. It was President Mwai Kibaki, and later overseen by President Uhuru Kenyatta. I thanked them for the appoinments.

As a responsible citizen, I never stopped conducting my research work and supervising postgraduate students who were registered with my parent University, the university of Nairobi. I kept on remitting my statutory deductions diligently and timely. I therefore kept my position as a professor of the university.

The Vice-Chancellors of the University during my job tour included: Professors Philip Mbithi, Francis Gichaga, Crispus Kiamba and George Magoha. I kept them fully briefed whenever I got a new appointment. They all understood the circumstances under which we worked. They themselves were under such arrangements.

What made my attachment to the University of Nairobi more stable was the fact that I continued teaching and supervising postgraduate students there. A good number of Master's and PhD degree graduates had been my students. We also published together. My linkage therefore was firm and active. The chairman of department appreciated my services, despite my absence from my station.

I never, at any time, solicited for any appointment to any of the positions referred to earlier. I think my actions and trust qualified me for appointment. All I knew was that I inherited new outfits which needed to be built on. I took risks and enjoyed moulding new outfits. Both JKUAT and NEMA were at their formative stages and they could be moulded in any way the moulder wanted.

I considered myself going through all academic and leadership cycles. The two are not the same, but complementary. I taught and trained students and led scholars. I also managed a department and a university. I led the academic staff, students and administrative cadre. I also headed one of the most demanding and thankless authorities, NEMA.

There, I led an initially young, non-performing organization to become a world-renowned environmental agency. I also guided the workers and provided leadership. This needed some wisdom and skills. It is for this reason that NEMA was referred to as an extended campus of JKUAT. The two were synonymous with my name. But I was not bothered.

Each institute, however, needed a different approach in its management. One was a university and the other, a strait-jacket government authority. It was my duty to adjust to these two different organizations.

In June 2006, I left NEMA for the University of Nairobi. I called my Vice-Chancellor, Prof. George Magoha, and made an appointment with him. I knew him earlier as a straightforward thinker and a no-nonsense administrator. He also knew me well as a former Vice-Chancellor and director-general of NEMA. He knew I was an accomplished scholar.

I told him that I had come to request him to allow me report back to my University of Nairobi. He was very receptive. We exchanged pleasantries before he responded to my request. He praised my exemplary work and told me that I was familiar with the job of a Vice-Chancellor.

Prof. Magoha is a down-to-earth individual. He is a serious worker and demands results immediately. He must get answers on the spot. I concur with these traits. Yet many of us seem have the bad culture of postponement of issues, requests and problems! As we had a cup of tea, the VC gave me a rundown of the work he needed to accomplish within his tenure in office. He told me of the self-sponsored students and the money the university was making. He referred to the many campuses I had started while in JKUAT and thought it was a good idea. He then answered my request.

> "Prof. Michieka," he started, "when a government requests you to assist in building institutions, you respond positively. When they use you well, they may not need you. Your usefulness wears out. You are no longer needed, despite the good work you may have done."

He commended my exemplary services to JKUAT, AICAD, KENET, IUCEA and NEMA which I assisted in setting up.

He was glad to have me back. He said that the government ceases to recognize you once you have accomplished its assignments. New blood takes over from where the old one leaves. He was categorical in the sense that if one was no longer considered useful by the borrower, one could always revert to the original employer. My records were clean with my original employer.

Prof. Magoha accepted me to return to my department and share my vast experience in teaching, research and outreach programmes. He called the chairman of my department, Prof. Florence Olubayo, and told her that I should follow the normal channels to be re-instated in the University of Nairobi system. I had to fill in staff movement documents to be re-instated as a teaching member of staff.

My brief interaction with the VC reminded me of the first discussion I had with the late Prof. J. M. Mungai in 1980. It was a friendly, collegiate meeting where we shared experiences. I found the two Vice-Chancellors comparable in their reception but different in action.

Prof. Magoha was a critic of a bad system and did not mince his words in condemning mediocrity. He hated non-performing systems. Prof. Mungai was a diplomat who would at times condone mistakes. Prof. Magoha acted on issues immediately. He was a performer and did not waste time. I found a big contrast between the two Vice-Chancellors with whom I worked so closely. The difference was that they worked under different government systems, Presidents Moi and Kibaki.

The University of Nairobi has the highest number of former Vice-Chancellors who have returned to service. They include professors: Raphael Munavu, Japheth Kiptoon, Shellemiah Keya, Francis Gichaga and Crispus Kiamba whom I had worked with previously. The university has a big heart, a big absorption capacity. It is very receptive to its staff who are appointed to various institutions as long as their records are clean. The statutes are liberal and accommodating. The workers understand that retention of manpower is key to a university's stability and visibility.

The former Vice-Chancellors teach, conduct research and supervise postgraduate students, just like any other lecturers. They also hold reputable positions in society as Chancellors of universities. They are always ready to take up any assignments given by their respective heads of department, the dean or college principal.

I find working with the young faculty members the most rewarding experience ever. I always attend meetings and give appropriate advice and guidance whenever needed. I enjoy the interactions with young students who are always ready to learn from senior professors.

Prof. Florence Olubayo was receptive and reminded me that I employed her in the department before I left for JKUAT as a deputy principal. It was a welcome return and I was happy to meet several young people that I had taught. They had become lecturers and were providing excellent services to the nation.

I could not have been to a better place than the University of Nairobi. A good number of workers whom I had assisted in one way or another were excited to see me back to lecture. We shared the past and reminded each other of the youthful days of the 1980s. I recall my excellent relationship with Professors Kimani Waithaka, Ole Mbatia, Margaret Wanyoike, Agnes Mwang'ombe (the College Principal) Willis Kosura-Oluoch, Steven Mbogoh, Samuel Mbugua, Susan Minae, Kareko-Gatere, some of whom were still teaching at greenery Upper Kabete Campus. We related very well.

They could occasionally refer to me as the former Vice-Chancellor, director-general and chairman, and would always ask for my intervention on certain issues when need arose. They would occasionally joke with me and refer to the bad terms of service which we could have changed while I was in the office. I would also respond to them that there was time and season for everything.

The most surprising thing that I found in our Universities was the large number of students admitted. The lecture theatres, laboratories and equipment were always fully occupied. Teaching and the conduct of practicals were never appropriate. The students' congestion did not allow for adequate lecture delivery. This was the case in all public Universities. The unprecedented number of students was certainly impacting negatively on the quality of education.

Over-enrolment of students in our public Universities has stretched the available resources to the limit. The teaching staff are overwhelmed with the crowded work places; residential halls are scarce; dining halls are overcrowded; the ablutions can no longer accommodate the large numbers. The living environment is therefore squalid. Courses that require practicals are no longer serviced. Both the teaching and support staff get overstretched to the furthest limit. The immediate overall effect is the reduced research by the lecturers and little concentration on students.

My transition back to my original career of teaching was smooth. I linked up with my former colleagues and joined a familiar group of academics. I must admit that there were several young and new faces including students that I had taught in the 1980s. Familiar names like Prof. John Kimenju, Dr George Chemining'wa, Dr Jesang Hutchinson, Dr Dorothy Kilalo were among the many lecturers whom I had taught and interacted with earlier. They were ready to receive me and I was glad to join them.

I knew of cases where former Vice-Chancellors would not freely re-join their departments because of past poor interpersonal relationships. This is a common occurrence in many African Universities. The reason is mainly due to poor past interactions where the incumbent would want to revenge on the past injustices. This was not the case when I returned to the University of Nairobi.

I resumed teaching in February 2006. I was assigned to teach courses in weed science and environment. I was competent in the subjects and used many live experiences during my lectures. I reviewed a number of articles to catch up with the latest discoveries especially in weed research.

The internet came in handy and I quickly learnt how to use the latest ICTs. I became a regular user of the library in addition to the internet. I enjoyed teaching the young people who were curious to know about my background. They had seen my pictures in the print media and also during some events. Some would boldly ask me why I opted to return to teaching instead of doing a more prestigious job either in the UNEP or FAO. I told them the importance of teaching them rather than be out there doing a desk job. Imparting knowledge to the youth had a better multiple effect than working for an organization.

I would practically tell them that a lecture to 100 students was more beneficial than lecturing to 10 people. I gave examples of best teaching practices and how that affects a community and eventually a country. I do also recall some colleagues asking me why I opted to return to Kabete Campus rather than go elsewhere for a more lucrative salary. I still insisted that the freedom of teaching and critical thinking was more satisfying than closing myself in an office alone, solving problems. Making huge salaries was not my priority as I was satisfied with my university work and the freedom therein.

I settled at the University of Nairobi quickly, continued educating my children and enjoyed local and international travels. Esther, my wife, was now working with the Higher Education Loans Board (HELB) in Kenya. She was happy that I was back to my original workplace which had fewer headaches than the previous ones.

A Stint in Politics

During the 2007 Kenyan general elections, I was persuaded by the local community from my home area to vie for a parliamentary seat. Nyaribari Masaba Constituency is my home area and I have a good development record there. I had been involved in various projects like promotion of education, support of poor children to school, church and school buildings, support of rural funeral arrangements and provision of reading materials in several needy cases.

It was in view of these projects that my community felt that I should vie for the seat. After long persuasion and having naïve understanding of the system of cheating applied then, I gave in. I submitted my papers and declared myself as one of the contestants of the parliamentary seat for Nyaribari Masaba. Twenty-two of us declared interest alongside the incumbent.

To make a long story short; all of us who tried the race were naïve. Many of the running mates were set up as spoilers. They were influenced to scuttle votes in favour of one individual. It was not worth going for. One spoiled ones standing in society.

The people one interacted with had no knowledge of any development agenda. They looked at the immediate gains and had no clue what development plans could bring about. As usual, bribery in Kenyan elections took the centre stage and voters would be bought for as little as Ksh.50 (five US cents) to vote for a non-progressive individual.

Politics in some African countries is a shame and it will take a long time for voters to be educated on their right to good governance and democracy. Tribalism and corruption would always lead the pack of vices in Kenya. Kenyan elections of 2007 resulted into massive rigging and brought about ugly tribal clashes. I was a victim of post-election violence in Molo South Constituency.

We owned a plot on which we had constructed three medium-size houses of three bedrooms each. They were all burnt down; trees felled and stolen; my chain link fence stolen; and the workers chased away. They missed death narrowly. Several personal items were consumed by fire. I never received any compensation as the meagre money that was supposed to be given to victims of post-election violence was distributed through corrupt administrators and never reached the genuine individuals.

I recall that my workers, several displaced people and they spent several nights in Cheptagum Secondary School because they had no place to sleep. Molo South is one of the coldest places in Kenya. Children caught colds, pneumonia and water-borne diseases because of poor hygiene. The truth of the matter is that the consequences of the violence are not fully understood by those who did not experience it.

I reverted to my profession and returned to the University of Nairobi where I resumed my teaching and research duties. I learnt another lesson of not indulging in an activity which was prone to dishonesty and cheating. I also learnt to appreciate that most people go for short-lived benefits. Long-term planning does not seem to be their priority.

"Easy-come-easy-go" philosophy is considered a manifestation of rural poverty. The little money that is dished out during the campaigns makes the people feel rich; they use it as quickly as possible and thereafter they need more. I concluded that for one to succeed in political campaigns, one needed substantial amounts of money. It could drain all the resources and leave one bankrupt and miserable.

Even after one succeeded, one had to keep the contacts alive by constantly "greasing" the electorate. They would abandon you if they were not catered for. This was certainly a backward way of surviving. I learnt the lesson fast and abandoned any future ambitious plans in politics.

Politics itself is not a dirty game as many claim. It is a developmental venture which many people use wrongly to enrich themselves. It becomes dirty only when the politician thinks he is above the constituents. It is a tactful game which can build or destroy individuals and even nations.

Indeed, in many developing nations, politics is seen as a licence for people to eat. It brainwashes many, and hampers development. Clanism, tribalism, nepotism and sycophancy take the centre stage. No country can reduce poverty and dependency if it does not condemn the said vices. Many countries seem to be content with political manoeuvres and can live with them. Few African countries are capable of phasing out election flaws.

19

Some Reflections on Leadership and Governance of Higher Education Institutions

This chapter examines the various traits that constitute good management and leadership. I have decided to write it using my experience over the many years of public involvement. I have been in four positions of human interaction and development: a lecturer, a college deputy principal and then principal, a VC and a director-general and a chairman. In all these positions, there were definite demands that were necessary in order to succeed. Each rank was unique and needed continuous management traits. I started engaging in leadership at an early age and continued to practise it without knowing. My primary and secondary school days and, finally, the university level, all gave me enough practical experience.

Leadership is not quite the same as management. They are different. Each one calls for different approaches as they both impact on human beings. I learnt the processes of being humane in both vocations. Indeed, they overlap and become complementary. Leadership in my view calls for proper discharge of responsibilities and well-balanced judgement of issues.

A leader would direct, influence, guide and be visionary. He/she follows through a certain route in order to achieve a goal. In the academia world a chairman of a department, Vice-Chancellor or rector is not necessarily a leader but an individual who can influence decisions brought up by staff. He/she has to guide, arbitrate and come to a consensus on issues that affect the institution.

The person is essentially a public relations leader who should have tremendous influential authority. For example, when Senates meet, they expect leadership and guidance. But many times members could come up with totally different agenda and put the chairman, who is normally the VC, into disarray. This is the time when leadership qualities are put to test. Tempers may rise; name-calling ensues but the Vice-Chancellor who is the chair must be able to control the meeting and give direction. Every Senate member considers him/herself an expert in a certain area and would always want to portray this character.

The chair, therefore, has to have the capacity, audacity, authority and vision to lead the otherwise heterogeneous scholars who are competent in their own areas of expertise. They call for respect but must be managed. In such circumstances, the chair may be intimidated, but must have the capacity to subdue the embarrassment. Leadership is manifested more to senior staff than perhaps to students. Students are more easily managed than led. Their reasoning capacity is highly influenced by circumstances and mob-psychology, on specific issues. They need managers.

A leader in a church congregation, a community, a union, a women group, a youth group, a political rally is always assertive and commandeering. The issues discussed in this scenario are generally predetermined. One does not get diverse views from the basic objects of the congregation. Meetings could be called to pass on messages or facts to the members.

Take the example of former Libyan president, Muammar Ghaddaffi, who preferred to be called a Leader of the People of Libya and led them for over 40 years is a good example to demonstrate the virtue of leadership. Despite his unprecedented/ excessive stay at the helm, Muammar gave the people of Libya certain confidence to have been allowed to stay in power that long. He led his people and provided them with basic facilities and fulfilled their needs.

I recall my last visit to Libya and in particular Sirte City, the birthplace of the late Muammar Ghaddaffi. I actually went round the city and noticed a big difference in the standard of living between the residents there and other African countries. They were provided with the basic needs like health services, water, good feeder roads, affordable food and electricity supply. I would be surprised to have an African country with these kinds of facilities provided to its ordinary citizens! They expected him to lead, provide the basics and give guidance. Hence, he was so revered. There are numerous examples of leadership which demonstrate the role that one plays to ensure a satisfied group of employees.

Good leadership, on the other hand, includes the following aspects:
- Visionary – frame the organization character and pursue it;
- Planning and generating potential solutions to the issues at hand;
- Deciding and making a commitment to a course of action;
- Explaining the rationale that led to this commitment and presenting the legitimate expectations;
- Executing the objectives to realization;
- Continuous evaluation of the progress with modifications as necessary;
- Integrity and accountability;
- Responsive public relations;
- Good sense of time management.

Management, on the other hand, is when one basically directs events. As a director-general of NEMA, for example, I had to manage the affairs of the environment and give orders. I had a duty to direct the environmental offenders to stop the vice. It was NEMA's duty as stipulated in the EMCA 1999 Act that we ensured all Kenyans were entitled to a clean and healthy environment. It was my responsibility to articulate and enforce the laws.

As a Vice-Chancellor, I also had a group of people to manage and to lead others. By and large, Vice-Chancellors or rectors manage and direct both staff and students. They may at times have to dictate.

For example, the kind of problems which Vice-Chancellors faced in the 1980s and 1990s did not allow much room for intelligent negotiations. The public universities were faced with numerous student strikes that did not allow for proper management. This circumstance called for either firm leadership or continuous university closures. This did not mean total authoritarian manoeuvres, but firm decision-making against the tides. As a Vice-Chancellor, therefore, I played the role of both a manager and leader through the university management board and Senate, and ensured continuous briefing of the Council.

Students appreciate facts, the truth, and fulfilled promises. I scored high on being firm if I knew I could not meet an obligation or promise. Luckily, I delivered virtually most of the promises that I had made and on time. This was being true to my staff and student community. I was known to fulfil promises and report back accordingly. However, my negative responses on many requests were equally respected, particularly when I knew I could not fulfil the demands. That was what I considered proper management.

The idea of managers fulfilling promises is rare because of perceived further consultations. What frustrates decision-making in many institutions is the notion that one has to make further consultations before one decides to honour the requests. Such delays leave room for speculations, rumours and uncertainty. The affected group becomes dissatisfied and disillusioned.

Leadership and management are complementary in their roles. There were situations in my career when I could make decisions for the benefit of the affected people. I encountered several situations that demanded action first and justification later. Sometimes, constructive reason on an issue would not be forthcoming and I would detect a danger looming due to delayed action. Management in many cases is orderly and negotiated. I never got stressed managing a team of people as long as we could communicate and exchange ideas.

Constructive debates yield productive results. This is the basic requirement of a good manager. A good manager can criticize staff and should also be positive and ready to receive and utilize constructive criticisms.

A combination of a good manager and a good leader is an imperative quality for running institutions in any country. There are several hidden qualities which make one a good manager or a good leader. A comparison between good leadership and good management has been researched by scholars who have published papers.

Leadership is service that seeks to meet the needs of a group of people by performing needed functions. Sometimes strong directive power of effective leadership is needed, such as when a group has lost its sense of direction or purpose. With another group, or at another time when the group is functioning well in its relationship and has its directions clear, non-directive styles of leadership are needed. Sometimes, the group needs to be encouraged and supported; at other times it may need to be oriented leadership serving the need of the group (Keating, 1982: 13)

Leadership, therefore, seeks to be more of service rather than dominate. It encourages others and mentors them. It does not exploit but respects them. It renders maximum service with a sense of unswerving and unceasing absorption of a belief. A leader brings out the best in people, makes them feel wanted and can be relied on. They accept criticisms and change positively according to circumstances. He /she must be a team leader, a team player, a coach and an inspirer. There are a number of leadership characteristics which are discussed by other scholars. The clusters of traits for leadership fall under the following categories:

- Capacity (intelligence, alertness, verbal facility, originality, judgement);
- Achievement (scholarship, knowledge, accomplishments);
- Responsibility (dependability, initiative, persistence, aggressiveness, self-confidence, desire to excel);
- Participation (activity, sociability, co-operation, adaptability, humour);
- Situation (mental ability, skills, needs and interests of followers, objectives to be achieved and tasks to be performed).

Certain general qualities in leadership such as courage, fortitude and conviction are a part of the list given above. Each individual in leadership may have all or some of the characteristics mentioned. In any case, I cannot decipher or identify which qualities I acquired. But from the role I played in the institutions I headed, I ranked well in many.

The two qualities, among many, that I consider vital for leadership are participation and ownership of processes. Workers enjoy identifying themselves with a venture or undertaking. They feel proud and could readily defend the new undertaking with dignity when and if it succeeds. Teamwork in any organization is important. Results are achieved and colleagues support the new ideas.

Bold stand and firmness also contribute to good leadership and management. Indecisive chief executives run into risks of dividing those they lead. I took firm positions in whatever issues I considered important. I could at times be torn between uncertainties, but I would consider views from both divides and steer the right decision to finality. This did not mean that I was correct in some matters, far from the truth.

I recall incidences where I rushed into making decisions but had to withdraw the verdict. This was not, however, a common occurrence. I always reflected over issues. The best decisions I took were the ones where I occasionally consulted with senior colleagues and then came up with a solution.

Many times, I could change, modify or withdraw my original decisions. These involved matters to do with students' affairs, disciplinary actions and staff problems. I thought that, as parents, we considered all aspects of human life and became more humane and sympathetic before arriving at a solution. It is difficult to lead and manage human beings. Each has his/her demands, hate and liking.

I am proud of the fact that during my thirteen years at JKUAT as a Vice-Chancellor, I did not expel any student. Not that they were all of the best behaviour. At times we had disturbing disciplinary cases. But I personally intervened in some of them just to advise. I counselled many students whenever they were involved in any mischief and made them sign agreements with me for good conduct and behaviour.

No student was expelled as long as they apologized in front of disciplinary committees and agreed to change for the better. I did not believe in student expulsions. My legacy rests in this humane approach to the youth. I, however, meted out harsh corrective punishments occasionally.

I learnt so many things as a leader. I went through adversity and rose up. I learnt that if I was so overly conscious of what people said and thought of me or how I performed my duties, I would never make progress. I also realized that if everyone agreed with me all the time, then I was not making progress or good decisions.

I had to elevate my thinking above an average manager and continually increase my exposure and visibility beyond the corridors of my working environment. I accomplished this by increasing my appetite for acquiring current information through any means necessary. It would have been through meetings, internet search, seminars, symposia, conferences or even staff bonding. Knowledge is power. This was the only key to confidence regardless of any circumstances. It was important that in spite of any adversity, I remained confident and planned strategically for the next move.

Leadership is a responsibility; and the buck stops at the chief executive's desk. In the end, it is important to note that knowledge graced with humility, education, and integrity is the only path that leads to intergeneration impact and the greatness that people vie for.

When I resumed my teaching role at the University of Nairobi in 2003, Kenya had seven public universities and five chartered private universities. There were also several university colleges and campuses. This was a great step towards providing education to all those who qualified and met the criteria for university admission. After a short period in 2013, the total number of public and private universities shot up to 66. These included 31 public universities, two university colleges and 33 privately-managed universities. This was massive growth of institutions in short time.

I was amazed at the speed of awarding the charters. What we have currently is the mushrooming of all types of university institutions purporting to provide quality education to unsuspecting students. This was the onset of the chaos. When reputable universities convert tertiary colleges into University campuses, the country negates chances for the middle-level colleges to train the much- needed middle-level cadre of technical staff. Areas like agriculture, engineering, architecture, health services, food processing, just to name a few, suffer in terms of technical personnel. This is the much-needed cadre of people to see Kenya jump- start its industrial agenda. I do not say graduates cannot propel the country's economy, but they need a large number of technical staff.

The idea of elevating middle-level colleges to universities is questionable. This was not the opportune time; it was a rushed decision. It would have been done systematically and selectively using a search team of experts. Perhaps a search committee like the one we used in 1994 should have been constituted to propose which colleges would be elevated to university status.

Such a select committee would have used the already existing technical inspection procedures and instruments to advise the relevant authority about the appropriate colleges that would be converted to full university status. In any case, the conversion of numerous middle-level colleges into universities was in itself retrospective. This was a sure way of slowing down industrialization and denying a certain cadre of school leavers a chance to progress. Universities could not absorb all the school leavers. The middle-level colleges were an excellent absorption entry for this particular group.

I used to chair a commission sub-committee on technical evaluation of private universities for accreditation. We developed a checklist which assisted us to qualify the universities. Among the questions we considered was the learning environment of the campus. To say the least, the current premises which house university campuses do not meet minimum requirements. Most of the students who are self-sponsored and virtually pay all the expenses find it tough to get good accommodation, water and power. Some lecture halls lack basic facilities and are poorly lit.

Currently, many universities have a section called Quality Assurance and Standards Departments. The sections are supposed to ensure that all the laid-down procedures for maintaining the standards as per the syllabi are adhered to. They also have a duty to prescribe rules and regulations to be followed for perfect course delivery. This is easily put on paper and committees are set up.

The actual practices leave a lot to be desired. Some universities may have no capacity to evaluate their peers in terms of content, materiel delivery, examination set-ups and even prescribed contact hours. The student numbers can be overwhelming and render such committees redundant. The government, on the other hand, has left the running and student enrolments to individual universities. It is basically detached from the vices that hurt the quality.

Other than the provision of funds to cater for the salaries and statutory deductions, the Ministry of Education has left the commission for university education to run the show solely. It is also not in a position to implement the numerous tasks it is charged with. The staffing is inadequate and incapable of judging the quality of education. Other than the government creating and changing tertiary campuses into university colleges, it has little, if any, role to play in ensuring that university statutes are followed.

Competition by individual universities for students as a source of income has led to the deteriorating quality of education. At the same time, this kind of education and poor student-lecturer interaction has led to incompetent graduates. The high number of students admitted into universities is disturbing. Critics claim that this has become a channel for robots, photocopiers, plagiarists and obscurantists. The students themselves are more concerned with getting grades using the shortest means possible. This has caused rampant plagiarism and hiring of "experts" to write papers and dissertations for others.

What is happening these days is totally the opposite of intellectualism and critical thinking and analyses of crucial national issues. A few people are engaged in nation-building through critical evaluations of the needs of the country. This is where the role of universities comes in.

The burdened lecturers who teach large numbers of students have no time to do scholarly work or research which are prerequisites. Lecturers keep moving from one campus to another and this is even more time-consuming and destructive to quality education. They do not get time to plan, mark, revise and even administer examinations. The introduction of certificate and diploma courses is a wrong move. I have never ever understood why senates stooped so low and approved such courses! A university has specific missions, to train degree students and conduct high-quality research for national development.

Surely, can a certificate or diploma youth come up with an advanced innovation while carrying out the several units they take? I am not saying they cannot; but they themselves have no time either. The middle-level colleges are best placed to absorb such students.

The Role of Boards and Councils

Generally speaking, Kenya has set up management system which uses committees to oversee functions of corporations, institutions and private companies. Boards and Councils are supposed to provide leadership and general guidance to their institutions. Government institutions are primarily parastatals, universities, colleges and schools. These are run by a group of people nominated by relevant authorities.

In many set-ups, members of the board are non-executive. Having been a chairman of KARI Board of Management, a Council member of National Council for Science and Technology (NACOSTI), a Commissioner for University Education and a board member of Kenya Marine and Fisheries Institute, I have had adequate experience to draw on in defining the parameters of good corporate governance. I have also been groomed by the JKUAT Council and finally NEMA Board of Management.

In all these engagements, no board chairman or member took the responsibilities of running the institutions in question. A board chairman who is executive in appointment is expressly indicated so. None of the one I cited above had executive powers. As a chairman also, I had non-executive powers. I chaired boards and I also had other persons chair mine. I worked in both worlds. I am using the board to include or infer the roles of Councils. I was also a chief executive.

Briefly, the role and responsibilities of a board/Council include:
- Effective leadership;
- Integrity in judgement;
- Monitoring and evaluation of progress;
- Approve strategic plans;
- Approval of budgets and accountable for financial statements;
- Compliance with relevant laws and codes for best practices;
- Employs staff and monitors their exits;
- Stewards of all property and resources;
- Accountability and disclosure;
- Approval of performance contract plans;
- Assist and work with the chief executive;
- Advise, protect and promote the chief executive and the institution;
- Assign the CEO duties as stipulated by law.

The board or Council meets quarterly as specified by law. The chief executive runs the institution on behalf of the board. He is the person in charge of the day-to-day running of the organization. The boards or Councils should not interfere with his roles unless requested to do so by a relevant authority of government. The boards or Councils transact business through their established sub-committees. These are also specified by law. In fact the board shields the chief executive whenever controversies arise.

The chairman is the spokesman of the Council or board. But he/she should not micromanage the organisation. Ideally, the chairman should possess the following attributes; courage, humility, integrity, diligence, conviction, optimism and discipline.

Suffice it to say that on all the boards and Councils I have chaired, I have kept the laws and regulations to the letter. I have had no interference with the CEO's work. I considered my relationship with members and the CEOs as perfect.

The other boards where I have had a chairman, the same mutual relationship prevailed. I had a perfect working relationship with the chairs and Council members, except one, NEMA – where confusion reigned sometimes. The NEMA chairman perhaps mistook the office as executive. He demanded an elaborate office with a fulltime secretary, a car and all other benefits as if for a fulltime officer.

Normally, there is an office for the chair but not as fulltime staff. The basic needs for a chair were provided, such as being picked and dropped for meetings or occasionally visiting the CEO for consultations.

This could be the beginning of problems which may culminate in ugly confrontations between the board members, the CEO and the parent ministry. The demands were not in line with the advisory roles as stipulated in the corporate governance Act. I had neither behaved so personally, nor had my previous chairmen.

Despite my making every effort to brief the board, particularly the chair, on their roles in NEMA, I still had a rough time convincing them that theirs was a bigger responsibility than clamour for the scarce office spaces and caused fear to the staff as they worked and hang around the offices. A case in point was to compare a university chairperson with those in parastatals. The current Education Act of 2010 is very clear on the type of chairperson that can head the Council. There is the chancellor who is ceremonial in nature and awards degrees and diplomas. He or she also inspects the university and may be called upon to advice.

The chairperson of Council therefore has more roles in the running of the university. The day-to-day functions are left to the chief executive. However, the Act is deficient in its stipulations of the qualifications of the Chancellor.

How can somebody who has not earned a PhD, for example, be empowered to award degrees and diplomas? What capacity does such a person have to advise on the quality of education? What practical research experience does such a person have to advise on the kind of research that future prospective candidates should undertake? Is that a role model to the upcoming young generation? Who are their academic peers?

I ask these questions because of the ever-deteriorating academic standards in many African universities. Chairpersons of Councils have some responsibilities to undertake. These responsibilities are not full-time in nature and do not deserve full-time personnel engagement – as the case was in NEMA.

Despite my concerted efforts to have a cohesive staff who worked as a team, the differences which were noticed between the board and management caused some visible division amongst staff. I had built up an excellent team to push the environmental agenda forward. Just like the university, many staff in NEMA saw the growing future prospects in terms of job professionalism and upward mobility. We bonded well with predetermined notions of creating environmental awareness for Kenyans.

My staff's salaries were approved by the board, having gone through the normal board committee. They were paid higher than their counterparts in the parent ministry and this was viewed as unfair. For example, provincial, now county, environmental officers were paid better than their counterparts. One difference was that I demanded a second degree for one to be employed in the province, which their counterparts did not have.

Unfortunately, appointments in Kenya's plum positions at the time were not based on merit but sycophancy, tribalism, nepotism and favours. NEMA had clearly laid down appointment criteria for various posts. The board was supportive on these demands. Board members comprised eminent scholars: three professors and the rest were PhD graduates. The authority was proudly nicknamed a JKUAT extended campus.

This did not bother me as I had scored well in running the university for thirteen years. In fact the respect and promotion of NEMA's mandate was directly linked to my success in managing JKUAT. I had already left a legacy in JKUAT. My immediate permanent secretary was not a graduate and hence differences would occur in numerous undertakings. We worked very well despite the slow pace and unbelievable bureaucracy in government operations/mechanisms. Approval of documents, for example, would take ages to be effected. This was the biggest hindrance in implementation of programmes. I honestly did not understand why a small ministry like environment, as it was considered, would take weeks if not months to respond!!

My working desk in NEMA was always clear. I never left pending work. Just as in JKUAT, I never postponed any paper work for the following day. I did not carry any office work home either. It was not in my nature to delay decisions. I took action no matter how unpopular it would be. I made a few mistakes and corrected them along the line, but I surely made decisions. If an action needed consultations, I just did that and had my minutes written clearly in case of litigation.

Delays in decision-making may result in unrest, strikes, mistrust, despondency, corruption, all of which are retrogressive. Institutions in developed nations have a system of expediting their decision-making on time. This avoids loss of opportunity costs in various aspects of the economy.

Clearing of my desk was one habit that I developed early even when I was at the bank. I needed my freedom after work. I therefore disengaged my office matters from those of my family. That was what I was known for; and I was unhappy with colleagues who delayed to take action. Such tendency hampers processes and kills the morale of workers. A few exceptions exist, of course. Reading a student's thesis or reviewing academic papers may demand to be done at home. I would be flexible in this regard.

Notwithstanding all these shortcomings, our mandate to create environmental awareness was felt. The basic questions I always asked myself was: Why do some changes fail while others succeed? How do we manage change among a new group of staff who do not regard new ideas as useful? How can one lead change successfully both professionally and in personal concerns?

Resistance to change borrows a lot from past successes or failures. My past record was good and NEMA employees were positive in realizing new ideas. I had engaged my employees in the change process. They owned and participated in environmental activities. I recall staff being very enthusiastic during major environmental days when we mounted road shows all over the country to educate the public. The road shows were embraced by the workers and NEMA largely succeeded in the promotion of the environmental agenda for and in Kenya.

I always stressed eight characteristics of a good worker. I said they included:

> courage, humility, integrity, loyalty, diligence, conviction, optimism and discipline. If my staff possessed these characteristics, I knew that we would meet our challenges and obligations without fail. I used to emphasize to them the importance of a good public image and name. These were the same messages I used to deliver while I was the Vice-Chancellor of JKUAT.

www.ingramcontent.com/pod-product-compliance
Lightning Source LLC
Chambersburg PA
CBHW070809300426
44111CB00014B/2459